Handbook of C–H Transformations

Volume 1

*Edited by
Gerald Dyker*

Further Titles of Interest

R. Mahrwald (Ed.)

Modern Aldol Reactions

2004, ISBN 3-527-30714-1

A. de Meijere, F. Diederich (Eds.)

Metal-Catalyzed Cross-Coupling Reactions
2 Vols.

2004, ISBN 3-527-30518-1

M. Beller, C. Bolm (Eds.)

Transition Metals for Organic Synthesis

2004, ISBN 3-527-30613-7

K. C. Nicolaou, S. A. Snyder (Eds.)

Classics in Total Synthesis II

2003, ISBN 3-527-30685-4

M. M. Green, H. A. Wittcroft

Organic Chemistry Principles and Industrial Practice

2003, ISBN 3-527-30289-1

C. Reichardt

Solvent and Solvent Effects in Organic Chemistry

2003, ISBn 3-527-30618-8

Handbook of C–H Transformations

Applications in Organic Synthesis
Volume 1

Edited by
Gerald Dyker

WILEY-VCH Verlag GmbH & Co. KGaA

Editor

Prof. Gerald Dyker
Department of Chemistry
Bochum University
44780 Bochum
Germany

■ All books published by Wiley-VCH are carefully produced. Nevertheless, authors, editors, and publisher do not warrant the information contained in these books, including this book, to be free of errors. Readers are advised to keep in mind that statements, data, illustrations, procedural details or other items may inadvertently be inaccurate.

Library of Congress Card No.: applied for
British Library Cataloguing-in-Publication Data
A catalogue record for this book is available from the British Library.

Bibliographic information published by Die Deutsche Bibliothek
Die Deutsche Bibliothek lists this publication in the Deutsche Nationalbibliografie; detailed bibliographic data is available in the Internet at <http://dnb.ddb.de>.

© 2005 WILEY-VCH Verlag GmbH & Co. KGaA, Weinheim

All rights reserved (including those of translation into other languages). No part of this book may be reproduced in any form – nor transmitted or translated into machine language without written permission from the publishers. Registered names, trademarks, etc. used in this book, even when not specifically marked as such, are not to be considered unprotected by law.

Printed in the Federal Republic of Germany.
Printed on acid-free paper.

Typesetting Kühn & Weyh, Satz und Medien, Freiburg
Printing betz-druck GmbH, Darmstadt
Bookbinding J. Schäffer GmbH, Grünstadt

ISBN-13: 978-3-527-31074-6
ISBN-10: 3-527-31074-6

Preface

The direct transformation of C-H bonds is a fundamental task in organic synthesis, regularly facing reactivity and selectivity problems but simultaneously promising substantial benefits. The intention of this handbook, written by renowned authors who have contributed substantially to this research area, is to present, very concisely within its 66 sections, the whole range of modern methods for C-H-transformation.

Most of the sections follow a general concept and are therefore divided into five parts which cover the most important features of the reaction in focus. "Introduction and Fundamental Examples" gives general information about the reaction, especially the scientific background and related reactions. This part also includes reactions which might be important to understanding although not necessarily of preparative value. "Mechanism" presents current mechanistic considerations, eventually including critical remarks. "Scope and Limitations" concentrates on examples which lead to interesting structures, usually with yields in excess of 50%. "Experimental" presents instructive, comprehensible examples, including work-up procedures. Information about appropriate methods for monitoring the reaction (TLC data or diagnostic NMR spectroscopy) are also given. If a special catalyst is needed, the procedure for its synthesis is also included. "References and Notes", of course, leads to significant publications where further details are available.

You may notice that this preface is as concise as the contents of this handbook. Nevertheless, as editor I should not forget to thank all authors and the team from Wiley-VCH, who made this project possible. The transformation of C-H bonds is certainly one of the most important fields of research in preparative organic chemistry; let us hope this handbook will further motivate research, simultaneously accelerating the change from new developments to established synthetic tools.

Gerald Dyker   Bochum, April 2005

Contents

Volume 1

Preface V

List of Contributors XVII

I   **General** *1*

1   **What is C–H Bond Activation?** *3*
         Bengü Sezen and Dalibor Sames

1.1   Introduction *3*
1.2   Activation or "Activation" *3*
1.3   The Origin and Historical Context of the "Organometallic Definition" *4*
1.4   What Do We Do With Two Definitions? *6*
1.5   Conclusions *9*

2   **C–H Transformation in Industrial Processes** *11*
         Leslaw Mleczko, Sigurd Buchholz, Christian Münnich

2.1   Introduction *11*
2.2   Alkane Activation *11*
2.3   C–H Transformation at Olefins *17*
2.4   Basic Chemicals from Aromatic Hydrocarbons *19*
2.5   Fine Chemicals *22*
2.5.1 Fine Chemicals by Organometallic Catalysis *23*
2.5.2 Metal-free Synthesis of Fine Chemicals *24*

II	**C–H Transformation at sp-Hybridized Carbon Atoms** 29

1	**C–H Transformation at Terminal Alkynes** 31

1.1　Recent Developments in Enantioselective Addition of Terminal Alkynes to Aldehydes 31
Tobias Ritter and Erick M. Carreira
1.1.1　Introduction 31
1.1.2　Background 32
1.1.3　Enantioselective Addition of Terminal Alkenes to Aldehydes 33
1.1.4　Applications 37
1.1.5　Conclusion 42
　　　Experimental 42
1.2　The Sonogashira Coupling Reaction 45
Herbert Plenio and Anupama Datta
1.2.1　Introduction and Fundamental Examples 45
1.2.2　Mechanism 46
1.2.3　Scope and Limitations 48
1.3　Glaser Homocoupling and the Cadiot–Chodkiewicz Heterocoupling Reaction 53
Peter Siemsen and Beatrice Felber
1.3.1　Introduction and Fundamental Examples 53
1.3.2　Mechanism 56
1.3.2.1　Oxidative Homocoupling 56
1.3.2.2　Nonoxidative Heterocoupling 57
1.3.3　Scope and Limitations 58
1.3.3.1　Oxidative Homocouplings of Tetraethynylethene Derivatives 58
1.3.3.2　Nonoxidative Heterocoupling of Terminal Alkynes with Haloalkynes: Cadiot–Chodkiewicz Reaction 60
1.4　Dimerization of Terminal Alkynes 62
Emilio Bustelo and Pierre H. Dixneuf
1.4.1　Introduction and fundamental examples 62
1.4.1.1　Simple Dimerization of Alkynes 62
1.4.1.2　Dimerization of Alkynes and Propargyl Alcohols into Functional Dienes or Cyclobutenes 66
1.5　*anti*-Markovnikov Addition to Terminal Alkynes via Ruthenium Vinylidene Intermediates 72
Christian Bruneau
1.5.1　Introduction 72
1.5.2　Application to the Synthesis of Vinylcarbamates 73
1.5.3　Application to the Synthesis of Enol Esters 73
1.5.4　Application to the Isomerization of Propargylic Alcohols 75
1.5.5　Application to the Synthesis of Vinylic Ethers 76
1.5.6　Application to the Synthesis of Unsaturated Ketones 76
1.5.7　Application to the Synthesis of Cyclic Enol Ethers and Lactones 77

| 1.5.8 | Application to the Synthesis of Aldehydes 78 |
| 1.5.9 | Scope and Limitations 78 |

2 Asymmetric Hydrocyanation of Alkenes 87
Jos Wilting and Dieter Vogt

2.1	Introduction 87
2.1.1	Cyclic (Di)enes 88
2.1.2	Vinylarenes 88
2.2	Mechanism 89
2.3	Scope and Limitations 92

III C–H Transformation at sp²-hybridized Carbon Atoms 97

1 C–H Transformation at Arenes 99

1.1	Direct Oxidation of Arenes to Phenols and Quinones 99
	Vsevolod V. Rostovtsev
1.1.1	Introduction 99
1.1.2	Radical Hydroxylations 99
1.1.3	Electrophilic Hydroxylations 102
1.1.4	Nucleophilic Hydroxylations 104
1.1.5	Direct Synthesis of Quinones from Arenes 105
1.2	Metalation of Arenes 106
1.2.1	Directed ortho and Remote Metalation (DoM and DreM) 106
	Victor Snieckus and T. Macklin
1.2.1.1	Introduction and Fundamental Concepts 106
1.2.1.2	Mechanism 110
1.2.1.3	Scope and Limitations 112
1.2.1.4	DoM Methodology for Substituted Aromatics 113
1.2.1.5	DoM in Total Synthesis 115
1.2.2	Electrophilic Metalation of Arenes 119
	Vladimir V. Grushin
1.2.2.1	Introduction 119
1.2.2.2	Mercuration 119
1.2.2.3	Thallation 121
1.2.2.4	Plumbylation (Plumbation) 122
1.2.2.5	Stannylation 124
1.2.3	Iridium-Catalyzed Borylation of Arenes 126
	Tatsuo Ishiyama and Norio Miyaura
1.2.3.1	Introduction and Fundamental Examples 126
1.2.3.2	Mechanism 128
1.2.3.3	Scope and Limitations 129
1.2.4	Transition-metal Catalyzed Silylation of Arenes 131
	Fumitoshi Kakiuchi

1.2.4.1 Introduction and Fundamentals *131*
1.2.4.2 Mechanism *133*
1.2.4.3 Scope and Limitations *133*
1.3 Alkylation and Vinylation of Arenes *137*
1.3.1 Friedel–Crafts-type Reactions *137*
1.3.1.1 Comparison of Classical and Fancy Catalysts in Friedel–Crafts-type Reactions *137*
Gerald Dyker
1.3.1.2 Lanthanoid Triflates in Catalytic Amounts for Friedel–Crafts-type Reactions *142*
Shu Kobayashi
1.3.1.3 Enantioselective Friedel–Crafts Type Alkylation Reactions *150*
Marco Bandini, Alfonso Melloni, and Fabio Piccinelli
1.3.1.4 Gold-catalyzed Hydroarylation of Alkynes *157*
Manfred T. Reetz and Knut Sommer
1.3.2 Alkylation and Vinylation via Intermediary Transition Metal σ-Complexes of Arenes *166*
1.3.2.1 Ruthenium-catalyzed ortho-Activation of Carbonyl-substituted Arenes *166*
Fumitoshi Kakiuchi and Shinji Murai
1.3.2.2 Ruthenium-Catalyzed alpha-Activation of Heteroarenes *175*
Naoto Chatani
1.3.2.3 Ruthenium(II)- and Iridium(III)-catalyzed Addition of Aromatic C–H Bonds to Olefins *180*
T. Brent Gunnoe and Roy A. Periana
1.3.2.4 Catalytic Functionalization of N-Heterocycles via their Rhodium–Carbene Complexes *187*
Sean H. Wiedemann, Jonathan A. Ellman, and Robert G. Bergman
1.3.2.5 Fujiwara Reaction: Palladium-catalyzed Hydroarylations of Alkynes and Alkenes *194*
Yuzo Fujiwara and Tsugio Kitamura
1.3.2.6 Palladium-catalyzed Oxidative Vinylation *203*
Piet W. N. M. van Leeuwen and Johannes G. de Vries
1.3.3 Minisci Radical Alkylation and Acylation *212*
Ombretta Porta and Francesco Minisci
1.3.3.1 Introduction *212*
1.3.3.2 Mechanism *213*
1.3.3.3 Scope, Limitations and Fundamental Examples *214*
1.4 Aryl–Aryl Coupling Reactions *223*
1.4.1 Intermolecular Arylation Reactions *223*
1.4.1.1 Intermolecular Arylation Reactions of Phenols and Aromatic Carbonyl Compounds *223*
Masahiro Miura and Tetsuya Satoh

1.4.1.2 Palladium-Catalyzed Arylation of Heteroarenes 229
Masahiro Miura and Tetsuya Satoh
1.4.1.3 Palladium-Catalyzed Arylation of Cyclopentadienyl Compounds 235
Gerald Dyker
1.4.2 Palladium-catalyzed Arylation Reactions via Palladacycles 238
1.4.2.1 Intramolecular Biaryl Bond Formation – Exemplified by the Synthesis of Carbazoles 238
Robin B. Bedford, Michael Betham, and Catherine S. J. Cazin
1.4.2.2 Carbopalladation–Cyclopalladation Sequences 245
Marta Catellani and Elena Motti
1.4.3 Oxidative Arylation Reactions 251
Siegfried R. Waldvogel and Daniela Mirk
1.4.3.1 Introduction and Fundamental Examples 251
1.4.3.2 Mechanism 254
1.4.3.3 Scope and Limitations 256

2   C–H Transformation at Alkenes 277

2.1 The Heck Reaction 277
Lukas Gooßen and Käthe Baumann
2.1.1 Introduction and Fundamental Examples 277
2.1.2 Mechanism 278
2.1.3 Scope and Limitations 280
2.1.3.1 Substrates 280
2.1.3.2 Heck Reactions of Aryl Bromides and Iodides 280
2.1.3.3 Domino Reactions Involving Carbometallation Steps 281
2.1.3.4 Enantioselective Heck Reactions 282
2.1.3.5 Heck Reactions of Aryl Chlorides 283
2.1.3.6 Heck Reactions of Diazonium Salts 284
2.1.3.7 Heck Reactions of Carboxylic Acid Derivatives 284
2.1.3.8 Miscellaneous Substrates 285
2.1.3.9 Industrial Applications 286
2.2 Wacker Oxidation 287
Lukas Hintermann
2.2.1 Introduction and Fundamental Examples 287
2.2.2 Mechanism 289
2.2.3 Scope and Limitations 291
2.2.3.1 Reactions Initiated by the Addition of Water to Terminal Alkenes 291
2.2.3.2 Reactions Initiated by Addition of Water to Internal Alkenes 294
2.2.3.3 Reactions Initiated by the Addition of Alcohols or Carboxylic Acids to Alkenes 296

3	C–H Transformation at Aldehydes and Imines *303*
3.1	Inter- and Intramolecular Hydroacylation *303*
	Chul-Ho Jun and Young Jun Park
3.1.1	Introduction and Fundamental Examples *303*
3.1.2	Mechanism *306*
3.1.3	Scope and Limitations *309*
3.2	Cyclization of Aldehydes and Imines via Organopalladium Intermediates *309*
	Xiaoxia Zhang and Richard C. Larock
3.2.1	Introduction *309*
3.2.2	Mechanism *310*
3.2.3	Scope and Limitations *312*

Volume 2

IV	C–H Transformation at sp^3-hybridized Carbon Atoms *317*
1	C–H Transformation at Functionalized Alkanes *319*
1.1	C–H Transformation in the Position α to Polar Functional Groups *319*
1.1.1	Transition Metal-catalyzed C–H Activation of Pronucleophiles by the α-Heteroatom Effect *319*
	Shun-Ichi Murahashi
1.1.1.1	Introduction *319*
1.1.1.2	The C–H Activation of Tertiary Amines *320*
1.1.1.3	The C–H Activation of Nitriles *320*
1.1.1.4	Aldol Type Reactions and Knoevenagel Reactions of Nitriles *321*
1.1.1.5	Addition of Nitriles to Carbon–Carbon (Michael Addition) and Carbon–Nitrogen Multiple Bonds *321*
1.1.1.6	Catalytic Thorpe–Ziegler reaction (Addition of Nitriles to Nitriles) *323*
1.1.1.7	The C–H Activation of Carbonyl Compounds *324*
1.1.1.8	The C–H Activation of Isonitriles *325*
1.1.1.9	Acid and Base Ambiphilic Catalysts for One-pot Synthesis of Glutalimides *326*
1.1.1.10	Application to Combinatorial Chemistry *326*
1.1.2	Palladium-Catalyzed Addition of Nitriles to C–C Multiple Bonds *328*
	Yoshinori Yamamoto and Gan B. Bajracharya
1.1.2.1	Introduction and Fundamental Examples *328*
1.1.2.2	Mechanism *330*
1.1.2.3	Scope and Limitations *332*
1.1.3	Asymmetric Catalytic C–C Coupling in the Position α to Carbonyl Groups *339*
1.1.3.1	Direct Catalytic Aldol Reactions *339*
	Claudio Nicolau and Mikel Oiarbide

1.1.3.2	Michael Addition Reaction 347
	Yoshitaka Hamashima and Mikiko Sodeoka
1.1.3.3	Direct Catalytic Asymmetric Mannich Reactions 359
	Armando Córdova
1.1.4	Oxidative Free-Radical Cyclizations and Additions with Mono and β-Dicarbonyl Compounds 371
	Barry B. Snider
1.1.4.1	Introduction and Fundamental Examples 371
1.1.4.2	Mechanism 373
1.1.4.3	Scope and Limitations 374
1.1.4.4	Common Side Reactions 376
1.1.5	Radical α-Functionalization of Ethers 377
	Takehiko Yoshimitsu
1.1.5.1	Introduction and Fundamental Examples 377
1.1.5.2	Mechanism 379
1.1.5.3	Scope and Limitations 380
1.1.6	Aerobic Oxidation of Alcohols 385
	Francesco Minisci and Ombretta Porta
1.1.6.1	Introduction 385
1.1.6.2	Mechanism 385
1.1.6.3	Scope, Limitations and Fundamental Examples 387
1.1.7	Kinetic Resolution by Enantioselective Aerobic Oxidation of Alcohols 393
	Brian M. Stoltz and David C. Ebner
1.1.7.1	Introduction and Fundamental Examples 393
1.1.7.2	Mechanism 395
1.1.7.3	Scope and Limitations 397
1.2	C–H Transformation in the Allylic and Benzylic Positions 402
1.2.1	C–H Transformation at Allylic Positions with the LICKOR Superbase 402
	A. Ganesan
1.2.1.1	Introduction and Fundamental Examples 402
1.2.1.2	Mechanism 403
1.2.1.3	Scope and Limitations 405
1.2.2	Heterogeneous C–H Transformation with Solid Superbases 409
	Stefan Kaskel
1.2.2.1	Introduction and Fundamental Examples 409
1.2.2.2	Mechanism 411
1.2.2.3	Scope and Limitations 414
1.2.3	Sequences of Hydro- or Carbometalation and Subsequent β-Hydrogen Elimination 416
1.2.3.1	Borate Isomerizations 416
	Jesús A. Varela
1.2.3.2	Heck-Type Reactions with a Migrating Double Bond 427
	Gerald Dyker

1.2.3.3 Enantioselective Olefin Isomerizations *430*
         Andrea Christiansen and Armin Börner
1.2.3.4 Palladium-catalyzed Deuteration *438*
         Seijiro Matsubara
1.2.4   Copper- and Palladium-catalyzed Allylic Acyloxylations *445*
         Jean-Cédric Frison, Julien Legros, and Carsten Bolm
1.2.4.1 Introduction *445*
1.2.4.2 Copper-catalyzed Allylic Acyloxylation *446*
1.2.4.3 Palladium-catalyzed Allylic Acyloxylation *450*
1.2.5   Transition Metal-catalyzed En-yne Cyclization *454*
         Minsheng He, Aiwen Lei, and Xumu Zhang
1.2.5.1 Introduction and Fundamental Examples *454*
1.2.5.2 Mechanism *460*
1.2.5.3 Applications and Limitations *462*
1.3   C–H Transformation at Functionalized Alkanes via Palladacycles *465*
         Gerald Dyker
1.3.1   Introduction and Fundamental Examples *465*
1.3.2   Mechanism *467*
1.3.3   Scope and Limitations *468*
1.4   CH Transformation at Functionalized Alkanes via Cyclometalated Complexes *470*
         Bengü Sezen and Dalibor Sames
1.4.1   Introduction and Fundamental Examples *470*
1.4.2   Mechanism *471*
1.4.3   Scope and Limitations *473*

2 C–H Transformation at Unfunctionalized Alkanes *497*

2.1   **C–O Bond Formation by Oxidation** *497*
2.1.1   Gif Reactions *497*
         Pericles Stavropoulos, Remle Çelenligil-Çetin, Salma Kiani, Amy Tapper, Devender Pinnapareddy, and Patrina Paraskevopoulou
2.1.1.1 Introduction and Fundamental Examples *497*
2.1.1.2 Mechanism *499*
2.1.1.3 Scope and Limitations *505*
2.1.2   Oxidation of Unactivated Alkanes by Dioxiranes *507*
         Waldemar Adam and Cong-Gui Zhao
2.1.2.1 Introduction and Fundamental Examples *507*
2.1.2.2 Mechanism *510*
2.1.2.3 Scope and Limitations *511*
2.1.3   Selective Enzymatic Hydroxylations *516*
         Bruno Bühler and Andreas Schmid
2.1.3.1 Introduction *516*
2.1.3.2 Mechanisms of Oxygenase Catalysis *518*
2.1.3.3 Applications of Oxygenase Catalysis in Organic Syntheses *524*

2.1.3.4	General Conclusion and Outlook	528
2.1.4	Transition Metal-catalyzed Oxidation of Alkanes	529

Gaurav Bhalla, Oleg Mironov, CJ Jones, William J. Tenn III, Satoshi Nakamura and Roy A. Periana

2.1.4.1	Introduction	529
2.1.4.2	C–H activation and Functionalization by Pt(II)	531
2.1.4.3	Electrophilic C–H activation by Hg	533
2.1.4.4	Electrophilic C–H Activation by Au	536
2.1.4.5	Oxidative Carbonylation of C–H bonds by Pd(II)	538
2.1.4.6	Summary	541
2.2	Radical Halogenations of Alkanes	542

Peter R. Schreiner and Andrey A. Fokin

2.2.1	Introduction and Fundamental Examples	542
2.2.2	Mechanisms	544
2.2.3	Scope and Limitations	546
2.3	Preparative SET C–H Transformations of Alkanes	548

Andrey A. Fokin and Peter R. Schreiner

2.3.1	Introduction and Fundamental Examples	548
2.3.2	Mechanisms	550
2.3.3	Scope and Limitations	553
2.4	Photochemical Processes	554
2.4.1	The Mercat Process	554

Robert H. Crabtree

2.4.1.1	Introduction and Fundamental Examples	554
2.4.1.2	Industrial Applications	560
2.4.1.3	Conclusion	560
2.4.2	Rhodium-catalyzed C–H Bond Transformation Under Irradiation	561

Toshiyasu Sakakura

2.4.2.1	Introduction and Fundamental Examples	561
2.4.2.2	Mechanism	562
2.4.2.3	Scope and Limitations	564
2.4.3	Stereoselective Photocyclization of Ketones (Norrish–Yang Reaction)	569

Pablo Wessig

2.4.3.1	Introduction and Fundamental Examples	569
2.4.3.2	Mechanism	570
2.4.3.3	Scope and Limitations	575
2.4.4	The Barton Reaction	579

Hiroshi Suginome

2.4.4.1	Introduction and Fundamental Examples	579
2.4.4.2	Mechanism	583
2.4.4.3	Scope and Limitations	585
2.5	Heterogeneous Catalysts for the C–H Transformation of Unfunctionalized Alkanes	589

Robert Schlögl

2.6   Transition-metal Catalyzed Carboxylation of Alkanes 599
        Yuzo Fujiwara and Tsugio Kitamura
2.6.1   Introduction and Fundamental Examples 599
2.6.2   Mechanism 600
2.6.3   Scope and Limitations 601
2.7   Photochemical and Thermal Borylation of Alkanes 605
        John F. Hartwig and Joshua D Lawrence
2.7.1   Introduction and Fundamental Examples 605
2.7.1.1 Borylation of Alkanes 605
2.7.1.2 Thermodynamics of Alkane and Arene Borylation 606
2.7.2   Mechanism 606
2.7.2.1 Photochemical Borylation of C–H bonds 606
2.7.2.2 Thermal Borylation of C–H bonds 607
2.7.3   Scope and Limitations 608
2.7.3.1 Photochemical Borylation of Methyl C–H Bonds 608
2.7.3.2 Thermal Borylation of Methyl C–H Bonds 609
2.7.3.3 Selectivity Between Methyl C–H Bonds 611
2.7.3.4 Borylation with the Substrate as the Limiting Reagent 612
2.7.3.5 Borylation of Polyolefins 614
2.8   Preparation of Olefins by Transition Metal-catalyzed Dehydrogenation 616
        Alan S. Goldman and Rajshekhar Ghosh
2.8.1   Introduction and Fundamental Examples 616
2.8.2   Substrates Other than Simple Alkanes 620
2.9   Rhodium-catalyzed Enantioselective Carbene Addition 622
        Huw M.L. Davies
2.9.1   Introduction and Fundamental Examples 622
2.9.2   Mechanism 623
2.9.3   Scope and Limitations 625
2.9.3.1 Intramolecular Reactions 625
2.9.3.2 Intermolecular Reactions 627

Index 653

List of Contributors

Editor

Gerald Dyker
Fakultät für Chemie/
AG Organische Chemie
Ruhr-Universität Bochum
Universitätsstr. 150
44780 Bochum
Germany

Authors

Part I

Sigurd Buchholz
Bayer Technology Services GmbH
Process Technology-RPT
Geb. E 41
51368 Leverkusen
Germany

Leslaw Mleczko
Bayer Technology Services GmbH
Process Technology-RPT
Geb. E 41
51368 Leverkusen
Germany

Christian Munnich
Bayer Technology Services GmbH
Process Technology-RPT
Geb. E 41
51368 Leverkusen
Germany

Dalibor Sames
Department of Chemistry
Columbia University
3000 Broadway, MC 3101
New York, NY 10027
USA

Bengü Sezen
Department of Chemistry
Columbia University
3000 Broadway, MC 3101
USA

Part II

Christian Bruneau
UMR 6509 : CNRS – Université de Rennes
Organometalliques et Catalyse
Campus de Beaulieu, Bât. 10C
Avenue du Général Leclerc
35042 Rennes Cedex
France

List of Contributors

Emilio Bustelo Gutiérrez
UMR 6509 : CNRS – Univeristé de Rennes
Laboratoire de Chimie de Coordination et Catalyse
Campus de Beaulieu, Bât 10C
Avenue du General Leclerc
35042 Rennes Cedex
France

Erick M. Carreira
Laboratorium für Organische Chemie
ETH-Hoenggerberg HCI H 335
8093 Zürich
Switzerland

Anupama Datta
Institut für Anorganische Chemie
Petersenstr. 18
64287 Darmstadt
Germany

Pierre H. Dixneuf
UMR 6509 : CNRS – Univeristé de Rennes
Laboratoire de Chimie de Coordination et Catalyse
Campus de Beaulieu, Bât 10C
Avenue du General Leclerc
35042 Rennes Cedex
France

Beatrice Felber
Schillerstraße 9a
85386 Eching
Germany

Herbert Plenio
Institut für Anorganische Chemie
Petersenstr. 18
64287 Darmstadt
Germany

Tobias Ritter
Laboratorium für Organische Chemie
ETH-Hoenggerberg HCI H 335
8093 Zürich
Switzerland

Peter Siemsen
Schillerstraße 9a
85386 Eching
Germany

Dieter Vogt
Laboratory of Homogeneous Catalysis
Eindhoven University of Technique
STW3.29, P. O. Box 513
5600 MB Eindhoven
The Netherlands

Jos Wilting
Lab. of Homogeneous Catalysis
Eindhoven University of Technique
STW3.29, P. O. Box 513
5600 MB Eindhoven
The Netherlands

Part III

Marco Bandini
Dipartimento di Chimica G. Ciamician
Università di Bologna
Via Selmi 2
40126 Bologna
Italy

Käthe Baumann
Bayer HealthCare AG
Chemical Development –
Process Research
Business Group Pharma
42096 Wuppertal
Germany

Robin B. Bedford
School of Chemistry
University of Exeter
Stocker Road
EX4 4QD Exeter
UK

Robert G. Bergman
Department of Chemistry
University of California, Berkeley
Berkeley, CA 94720-1460
USA

Michael Betham
School of Chemistry
University of Exeter
Stocker Road
EX4 4QD Exeter
UK

Marta Catellani
Dipartimento di Chimica Organica e Industriale
Parco Area delle Scienze 17/A
43100 Parma
Italy

Catherine S. J. Cazin
School of Chemistry
University of Exeter
Stocker Road
EX4 4QD Exeter
UK

Naoto Chatani
Department of Applied Chemistry
Faculty of Engineering
Osaka University,
Suita
Osaka 565-0871
Japan

Jonathan A. Ellman
Department of Chemistry
University of California, Berkeley
Berkeley, CA 94720-1460
USA

Yuzo Fujiwara
Department of Chemistry
Graduate School of Engineering
Kyushu University
Hakozaki
Fukuoka 812-8581
Japan

Lukas J. Gooßen
Max-Planck-Institut
für Kohlenforschung
Kaiser-Wilhelm-Platz 1
45470 Mülheim
Germany

Vladimir V. Grushin
DuPont de Nemours & Co., Inc.
Central Research and Development
Experimental Station, E328/306
Wilmington, DE 19880-0328
USA

T. Brent Gunnoe
Department of Chemistry
North Carolina State University
Raleigh, NC 27695-8204
USA

Lucas Hintermann
Institut für Organische Chemie
der RWTH
Prof.-Pirlet-Str. 1
52074 Aachen
Germany

Tatsuo Ishiyama
Division of Molecular Chemistry
Graduate School of Engineering
Hokkaido University
060-8628 Sapporo
Japan

Chul-Ho Jun
Department of Chemistry
Yonsei University
Seoul 120-749
Korea

Fumitoshi Kakiuchi
Department of Applied Chemistry
Faculty of Engineering
Osaka University
Suita
Osaka 565-0871
Japan

Tsugio Kitamura
Department of Chemistry
Graduate School of Engineering
Kyushu University
Hakozaki
Fukuoka 812-8581
Japan

Shu Kobayashi
Graduate School of Pharmaceutical
Sciences,
University of Tokyo
Hongo, Bunkyo-ku
113-0033 Tokyo
Japan

Richard C. Larock
Department of Chemistry
Iowa State University
Ames, Iowa 50011
USA

Piet W. N. M. van Leeuwen
DSM Pharma Chemicals
Advanced Synthesis, Catalysis &
Development
PO Box 18
6160 MD Geleen
The Netherlands

Alfonso Melloni
Dipartimento di Chimica G. Ciamician
Università di Bologna
Via Selmi 2
40126 Bologna
Italy

Francesco Minisci
Dipto. di Chimica del Politecnico
via Mancinelli 7
20131 Milano
Italy

Daniela Mirk
Organisch-Chemisches Institut
Universität Münster
Corrensstr. 40
48149 Münster
Germany

Masahiro Miura
Dept. of Applied Chemistry
Osaka University
2-1 Yamada-oka
565-0871 Osaka
Japan

Norio Miyaura
Division of Molecular Chemistry
Graduate School of Engineering
Hokkaido University
060-8628 Sapporo
Japan

Elena Motti
Dipartimento di Chimica Organica e Industriale
Parco Area delle Scienze 17/A
43100 Parma
Italy

Shinji Murai
Department of Applied Chemistry
Faculty of Engineering
Osaka University
Suita
Osaka 565-0871
Japan

Young Jun Park
Department of Chemistry
Yonsei University
Seoul 120-749
Korea

Roy A. Periana
Department of Chemistry
North Carolina State University
Raleigh, NC 27695-8204
USA

Fabio Piccinelli
Dipartimento di Chimica G. Ciamician
Università di Bologna
Via Selmi 2
40126 Bologna
Italy

Ombretta Porta
Dipto. di Chimica del Politecnico
via Mancinelli 7
20131 Milano
Italy

Manfred T. Reetz
Max-Planck-Institut für Kohlenforschung
Kaiser-Wilhelm-Platz 1
45470 Mülheim an der Ruhr
Germany

Vsevolod V. Rostovtsev
Research Chemist
DuPont Central Research and Development
Experimental Station
P.O. Box 80328
Wilmington
DE 19880-0328
USA

Tetsuya Satoh
Dept. of Applied Chemistry
Osaka University
2-1 Yamada-oka
565-0871 Osaka
Japan

Victor Snieckus
Department of Chemistry
Queen's University
K7L 3N6 Kingston
Canada

Knut Sommer
Max-Planck-Institut für Kohlenforschung
Kaiser-Wilhelm-Platz 1
45470 Mülheim an der Ruhr
Germany

Johannes G. de Vries
DSM Pharma Chemicals
Advanced Synthesis, Catalysis & Development
PO Box 18
6160 MD Geleen
The Netherlands

List of Contributors

Siegfried R. Waldvogel
Organisch-Chemisches Institut
Universität Münster
Corrensstr. 40
48149 Münster
Germany

Sean H. Wiedemann
Department of Chemistry
University of California, Berkeley
Berkeley, CA 94720-1460
USA

Xiaoxia Zhang
Department of Chemistry
Iowa State University
Ames, Iowa 50011
USA

Part IV

Waldemar Adam
221 Guajataca Street
Villas de la Playa
Vega Baja, Puerto Rico 00693
USA

Gan B. Bajracharya
Department of Chemistry
Graduate School of Science
Tohoku University
Sendai, 980-8578
Japan

Gaurav Bhalla
Loker Hyocarbon Research Institute
Department of Chemistry
University of Southern California
Los Angeles, CA 90089-1661
USA

Carsten Bolm
Institut für Organische Chemie
RWTH Aachen
Professor-Pirlet-Str. 1
52056 Aachen
Germany

Armin Börner
Institut für Organische
Katalyseforschung
Universität Rostock e.V.
Albert-Einstein-Str. 29a
18059 Rostock
Germany

Bruno Bühler
Department of Biochemical and
Chemical Engineering
University of Dortmund
Emil-Figge-Strasse 66
44227 Dortmund
Germany

Jesus Angel Varela Carrete
Departamento de Quimica Organica
Facultade de Quimica
Universidade de Santiago de
Compostela
15782 Santiago de Compostela
Spain

Remle Çelenligil-Çetin
University of Missouri
Department of Chemistry
315A Schrenk Hall
Rolla, MO 65409
USA

Andrea Christiansen
Institut für Organische
Katalyseforschung
Universität Rostock e.V.
Albert-Einstein-Str. 29a
18059 Rostock
Germany

Armando Cordova
Departement of Organic Chemistry
Stockholm University
Arrhenius Laboratory
Arrhenius gatan
106 91 Stockholm
Sweden

Robert H. Crabtree
Department of Chemistry
Yale University
225 Prospect Street
New Haven, CT 06201-8107
USA

Huw M. L. Davies
Department of Chemistry
University at Buffalo
The State University of New York
Buffalo, NY 14260-3000
USA

David C. Ebner
Division of Chemistry and Chemical Engineering, M/C 164-30
California Institute of Technology
1200 East California Boulevard
Pasadena, CA 91125
USA

Aney A. Fokin
Institut für Organische Chemie
Justus-Liebig-Universität
Heinrich-Buff-Ring 58
35392 Giessen
Germany

Jean-Céic Frison
Institut für Organische Chemie
RWTH Aachen
Professor-Pirlet-Str. 1
52056 Aachen
Germany

A. Ganesan
Department of Chemistry
University of Southampton
Highfield
SO17 1BJ Southampton
UK

Rajshekhar Ghosh
Rutgers University
Chemistry Department
610 Taylor Road
Piscataway, NJ 08854-8087
USA

Alan S. Goldman
Rutgers University
Chemistry Department
610 Taylor Road
Piscataway, NJ 08854-8087
USA

Yoshitaka Hamashima
Institute of Multidisciplinary Research
For Advanced Materials (IMRAM)
Tohoku University
Katahira
Miyagi 980-8577
Japan

John F. Hartwig
Department of Chemistry
Yale University
225 Prospect Street
New Haven CT 06520-8107
USA

Minsheng He
152 Davey Lab, C28
Department of Chemistry
Penn State University
University Park, PA 16802
USA

C. J. Jones
Loker Hyocarbon Research Institute
Department of Chemistry
University of Southern California
Los Angeles, CA 90089-1661
USA

Stefan Kaskel
Max-Planck-Institut für
Kohlenforschung
Kaiser-Wilhelm-Platz 1
45470 Mülheim an der Ruhr
Germany

Salma Kiani
University of Missouri
Department of Chemistry
315A Schrenk Hall
Rolla, MO 65409
USA

Joshua D. Lawrence
Dept. of Chemistry
Yale University
P.O. Box 20 81 07
06520-8107 New Haven
USA

Julien Legros
Institut für Organische Chemie
RWTH Aachen
Professor-Pirlet-Str. 1
52056 Aachen
Germany

Aiwen Lei
152 Davey Lab, C28
Department of Chemistry
Penn State University
University Park, PA 16802
USA

Seijiro Matsubara
Kyoto University
Graduate School of Engineering
Department of Material Chemistry
Kyoutodaigaku-Katsura
Nishikyo, Kyoto 615-8510
Japan

Francesco Minisci
Dipto. di Chimica del Politecnico
via Mancinelli 7
20131 Milano
Italy

Oleg Mironov
Loker Hyocarbon Research Institute
Department of Chemistry
University of Southern California
Los Angeles, CA 90089-1661
USA

Shun-Ichi Murahashi
Department of Applied Chemistry
Okayama University of Science
Ridai-cho 1-1, Okayama 700-0005
Japan

Satoshi Nakamura
Loker Hyocarbon Research Institute
Department of Chemistry
University of Southern California
Los Angeles, CA 90089-1661
USA

Claudio Palomo Nicolau
Departamento de Química Orgánica I
Facultad de Químicas. Universidad del
País Vasco
20018 San Sebastián
Spain

Mikel Oiarbide
Departamento de Química Orgánica I
Facultad de Químicas. Universidad del
País Vasco
20018 San Sebastián
Spain

Patrina Paraskevopoulou
University of Missouri
Department of Chemistry
315A Schrenk Hall
Rolla, MO 65409
USA

Roy A. Periana
Loker Hyocarbon Research Institute
Department of Chemistry
University of Southern California
Los Angeles, CA 90089-1661
USA

Devender Pinnapareddy
University of Missouri
Department of Chemistry
315A Schrenk Hall
Rolla, MO 65409
USA

Ombretta Porta
Dipto. di Chimica del Politecnico
via Mancinelli 7
20131 Milano
Italy

Toshiyasu Sakakura
National Institute of Advanced
Industrial Science & Technology (AIST)
1-1-1 Higashi, Central 5
Tsukuba 305-8565
Japan

Dalibor Sames
Department of Chemistry
Columbia University
3000 Broadway, MC 3101
New York, NY 10027
USA

Robert Schlögl
Fritz-Haber-Institut der Max-Planck-
Gesellschaft
Faradayweg 4-6
14195 Berlin
Germany

Andreas Schmid
Department of Biochemical and
Chemical Engineering
University of Dortmund
Emil-Figge-Strasse 66
44227 Dortmund
Germany

Peter R. Schreiner
Institut für Organische Chemie
Justus-Liebig-Universität
Heinrich-Buff-Ring 58
35392 Giessen
Germany

Barry Snider
Department of Chemistry MS 015
Brandeis University
415 South Street
Waltham, MA 02454-9110
USA

Mikiko Sodeoka
Institute of Multidisciplinary Research
for Advanced Materials (IMRAM)
Tohoku University
Katahira
Miyagi 980-8577
Japan

Pericles Stavropoulos
University of Missouri
Department of Chemistry
315A Schrenk Hall
Rolla, MO 65409
USA

Brian M. Stoltz
Division of Chemistry and Chemical Engineering, M/C 164-30
California Institute of Technology
1200 East California Boulevard
Pasadena, CA 91125
USA

Hiroshi Suginome
Organic Synthesis Division
Hokkaido University
Kita-ku, 060 Sapporo
Japan

Amy Tapper
University of Missouri
Department of Chemistry
315A Schrenk Hall
Rolla, MO 65409
USA

William J. Tenn III
Loker Hyocarbon Research Institute
Department of Chemistry
University of Southern California
Los Angeles, CA 90089-1661
USA

Jesús A. Varela
Dept. de Chimica Orgánica
Univ. des Santiago de Compostela
15782 Santiago de Compostela
Spain

Pablo Wessig
Institut fuer Chemie
Humboldt-Universität zu Berlin
Brook-Taylor-Str. 2
12489 Berlin
Germany

Yoshinori Yamamoto
Department of Chemistry
Graduate School of Science
Tohoku University
Sendai, 980-8578
Japan

Takehiko Yoshimitsu
Meiji Pharmaceutical University
2-522-1, Noshio, Kiyose
Tokyo 204-8588
Japan

Xumu Zhang
152 Davey Lab, C28
Department of Chemistry
Penn State University
University Park, PA 16802
USA

Cong-Gui Zhao
221 Guajataca Street
Villas de la Playa
Vega Baja, Puerto Rico 00693
USA

I
General

1
What is C–H Bond Activation?

Bengü Sezen and Dalibor Sames

1.1
Introduction

The possibility of direct introduction of a new functionality (or a new C–C bond) via direct C–H bond transformation is a highly attractive strategy in covalent synthesis, owing to the ubiquitous nature of C–H bonds in organic substances. The range of substrates is virtually unlimited, including hydrocarbons (lower alkanes, arenes, and polyarenes), complex organic compounds of small molecular weight, and synthetic and biological polymers. Consequently, selective C–H bond functionalization has long stood as a highly desirable goal. The introduction of transition metals to the repertoire of reagents unlocked entirely new opportunities in this area. As such, novel reactions have been discovered and the term "C–H bond activation" has been coined and used to describe certain C–H cleaving processes, initially in the context of saturated hydrocarbons. With time this term has become popular, if not fashionable, and its frequent and liberal usage has led to some uncertainty about its definition and meaning. Complex organic substrates contain a plethora of C–H bonds of different acidity and reactivity, and consequently many mechanistic modes exist for an overall C–H functionalization process (e.g. radical, electrophilic substitution, deprotonation, metal insertion).

Naturally, the question of which processes can be described as "C–H bond activation" arose. After numerous discussions with colleagues in the broad chemical community, we felt compelled to provide some thoughts on this topic, including a historical perspective.

1.2
Activation or "Activation"

In lay language, "activation" means making an object or a person active. A number of fields of science and engineering have adopted this term to describe various processes and phenomena (e.g. regeneration of inorganic catalyst, transformation of inactive enzyme to an active form, excitation by heating or irradiation) [1]. In the context of chemical reactions, "activation of a substrate" or "activation of a bond" refers to, in a most general sense, any process or phenomenon by which the reactivity of a substrate or a bond is increased. Thus, this represents a rather open term and as such is used by the chemical community in many different ways and contexts; for instance, activation of bonds by substituents (cf. activated C–H bonds in malonate esters) or activation of bonds by formation of a discrete intermediate between the substrate and a reagent (cf. alkene activation by Lewis acids). Although distinction between "activation" and "reaction" can in principle be made, "bond activation" is frequently equated with bond cleavage. For instance, activation of strong bonds (C–H, C–C, C–F) is often understood as cleavage of these bonds with transition metal reagents. Similarly, "nitrogen activation" describes a variety of processes for reduction of N_2 to hydrazine or ammonia.

In this light, we can appreciate the wide spectrum of interpretations and uses of this terminology. To bring some clarity to our discussion, we first need to make a clear distinction between "activation" and "reaction". In harmony with the general understanding of the term "activation", "*bond activation*" should refer to any chemical process which increases the reactivity of a bond in question ("general definition") [2]. On the other hand, bond-cleaving processes should be labeled by a separate term, for instance "*bond transformation*". We should emphasize that both of these terms, used in this general sense, cast no limits on the actual activation or reaction mechanism.

Nevertheless, in addition to this general understanding of the term, "bond activation" has acquired specific meaning in various subdisciplines. Most notably, "C–H bond activation" is frequently used as an organometallic term to describe certain metal-mediated processes ("organometallic definition"). Before we address this inconsistency, let us first elucidate the origin and historical context of "C–H bond activation" as an organometallic term.

1.3
The Origin and Historical Context of the "Organometallic Definition"

One of the early uses of the "C–H bond activation" term appeared in the chemical literature in 1936 to describe the H–D exchange in methane catalyzed by a heterogeneous Ni^0 catalyst (Scheme 1) [3]. Although no definition of the term was provided, this work implied that a new mode of chemical reactivity was operative at the metal surface, enabling cleavage of alkane C–H bonds. With some insight, an analogy between this new process and cleavage of a hydrogen molecule on hydrogenation surfaces was proposed.

A few decades later in 1968, Halpern formulated the need for new approaches to the activation of C–H bonds with a particular focus on saturated hydrocarbons. C–H bond activation, equated with "dissociation of carbon–hydrogen bonds by metal complexes", was identified as one of the most important challenges in catalysis [4]. Perhaps the most influential discovery in this area was made in the late 1960s by Hodges and Garnett, who demonstrated that a *homogeneous* aqueous solution of platinum(II) salts catalyzed deuteration of arenes and alkanes [5]. Subsequently, Shilov extended this work by using mixtures of platinum(II) and platinum(IV) salts to achieve hydroxylation and chlorination of alkanes, including methane (Scheme 2) [6]. This work inspired numerous mechanistic studies which established an alkylplatinum species as a reasonable intermediate. Most notably, unusual chemo-selectivity was observed, because rate constants for oxidation of an unactivated methyl group were occasionally greater than those for the oxidation of an alcohol (Scheme 2) [7]. Clearly, a new reactivity mode, other than radical or ionic substitution, had been discovered and the term "activation of saturated hydrocarbons" was used.

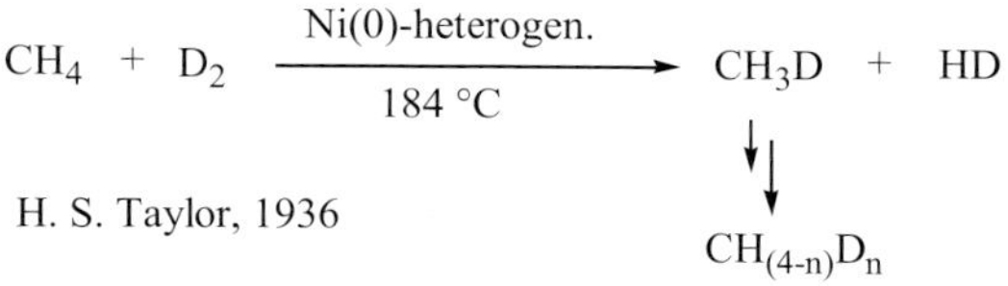

$$CH_4 + D_2 \xrightarrow[184\ °C]{Ni(0)\text{-heterogen.}} CH_3D + HD$$
$$\downarrow$$
$$CH_{(4-n)}D_n$$

H. S. Taylor, 1936

Scheme 1

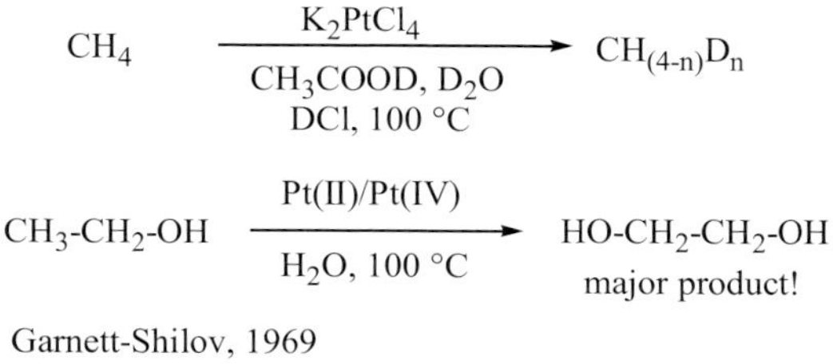

$$CH_4 \xrightarrow[\substack{CH_3COOD,\ D_2O \\ DCl,\ 100\ °C}]{K_2PtCl_4} CH_{(4-n)}D_n$$

$$CH_3\text{-}CH_2\text{-}OH \xrightarrow[H_2O,\ 100\ °C]{Pt(II)/Pt(IV)} HO\text{-}CH_2\text{-}CH_2\text{-}OH$$
major product!

Garnett-Shilov, 1969

Scheme 2

Transition metal complexes have also unlocked new mechanistic possibilities for cleaving arene C–H bonds. Hydrogen–deuterium exchange at the benzene nucleus, catalyzed by homogeneous metal hydride complexes, was demonstrated by Parshall (Scheme 3) [8]. Interestingly, it was observed that electron-deficient arenes underwent the labeling reaction at faster rates. These results (reaction rates and regioselectivity) were inconsistent with electrophilic substitution; rather, the metal complexes had nucleophile-like properties which pointed to a new mechanism. The intermediacy of arene–metal hydride species, similar to those observed earlier by Chatt and Davidson [9], was proposed (Scheme 3). By analogy with the reaction of alkanes, these new processes were described as "C–H bond activation", to distinguish them from electrophilic metalation and electrophilic substitution reactions.

Thus, the historical context reveals that the term "C–H bond activation" was introduced with a clear purpose to distinguish metal-mediated C–H cleavage from traditional radical and ionic substitution, and as such was essentially a mechanistic term [8]. As a result we may formulate the "organometallic definition": *the term "C–H bond activation" refers to the formation of a complex wherein the C–H bond*

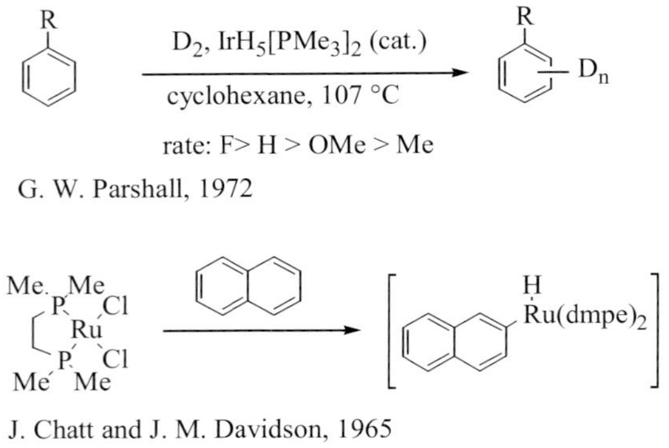

R—⟨C6H5⟩ + D₂, IrH₅[PMe₃]₂ (cat.), cyclohexane, 107 °C → R—⟨C6H4⟩-Dn

rate: F > H > OMe > Me

G. W. Parshall, 1972

[Me₂P(Cl)Ru(PMe₂)Cl] + naphthalene → [naphthyl-Ru(dmpe)₂-H]

J. Chatt and J. M. Davidson, 1965

Scheme 3

interacts directly with the metal reagent or catalyst. These complexes often afford a C–M intermediate in the absence of free radical or ionic intermediates.

We believe this definition captures the essence of numerous proposals in the organometallic literature [10]. Indeed, the term "C–H bond activation" is used routinely to differentiate, for instance, between oxidative addition pathways and deprotonation [11].

1.4
What Do We Do With Two Definitions?

Equation (1) depicts an early example of an intermolecular addition of an alkane C–H bond to a low valent transition metal complex [12]. Mechanistic investigations provided strong evidence that these reactions occur via concerted oxidative addition wherein the metal "activates" the C–H bond directly by formation of the dative bond, followed by formation of an alkylmetal hydride as the product (Box 1). Considering the overall low reactivity of alkanes, transition metals were able to "make the C–H bonds more reactive" or "activate" them via a new process. Many in the modern organometallic community equated "C–H bond activation" with the concerted oxidative addition mechanism [10b,c].

Strictly speaking, however, in addition to the concerted pathway, oxidative addition can also proceed via radical or ionic mechanisms [13]. Although these alternatives are less likely for alkanes (cf. Eq. 1) they must be considered with substrates containing reactive C–H bonds. For example, proton transfer is a readily available process for acidic C–H bonds (Box 1). Insertion of low valent transition metals has been reported in substrates including alkynes, ketones, and nitriles. As an example, the synthesis of iron hydride complex **5** was accomplished by treating a terminal alkyne with Fe(dmpe)₂, generated in situ (Eq. 2). This reaction, assumed to proceed via concerted oxidative addition, stands in stark contrast to deprotonation by a strong base. The label "C–H bond activation" was used to make this distinction and we may argue that it serves well as a qualitative mechanistic term.

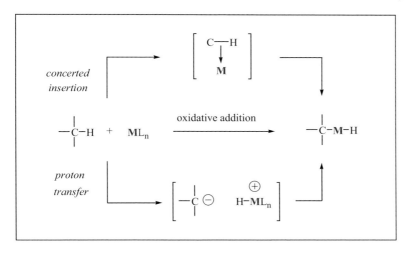

Box 1. Oxidative addition of C–H bonds.

Difficulties arise, however, when the organometallic definition is to be applied in a rigorous mechanistic sense. This point is illustrated by comparing the reactions of the iron complex in Eq. (2) with an alkyne or with HCN [14]. Although a metal hydride is the product in both reactions, a significantly faster rate was observed with HCN. This observation suggests that addition of HCN proceeds via proton transfer. Which of these processes can be described as "C–H bond activation"? According to the organometallic definition proton transfer as an ionic process would be disqualified. What, however, if oxidative addition proceeds via proton transfer, followed by very fast ion recombination? What if a new experiment suggests that the iron metal interacts with the alkyne triple bond before a proton-transfer step? These questions are often contentious and debated issues and experimental measurements from two different laboratories may favor different mechanistic proposals. This case illustrates the types of problematic issue that arise when attempting to define "C–H bond activation" as a rigorous mechanistic term.

Furthermore, the inconsistency between the restrictive organometallic definition and the general understanding of "bond activation" will pose further problems. Let us discuss this issue in the context of a concrete example – alkyne cupration. It is thought that copper complexation and base-assisted deprotonation work in concert ultimately forming the alkynyl cuprate (Eq. 3). Thus, the proposed cupration mechanism may be viewed as a variation of the deprotonation mechanism (π-acid/base-promoted deprotonation) [15]. Experimental evidence shows that Cu^I salts increase the acidity of the terminal alkyne C–H bond by coordination to the π-bond [16]. Hence, it is clear that copper metal activates the alkyne C–H bond; following the organometallic definition, however, would lead to an absurd linguistic situation; i.e. copper activates the alkyne C–H bond but it is not "C–H bond activation".

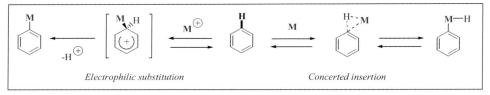

Box 2. Arene metalation. Electrophilic versus concerted insertion.

Another instructive scenario may be found when considering the metalation of arenes. There are two distinct mechanisms for the metalation of aromatic C–H bonds – electrophilic substitution and concerted oxidative addition (Box 2). The classical arene mercuration, known for more than a century, serves to illustrate the electrophilic pathway whereas the metal hydride-catalyzed deuterium labeling of arenes document the concerted oxidative addition mechanism [8, 17]. These two processes differ both in kinetic behavior and regioselectivity and thus we may appreciate the need to differentiate these two types of process. However, the choice of "C–H bond activation" to designate only one, the oxidative addition pathway, creates a similar linguistic paradox. Indeed, it is hard to argue that the C–H bond in the cationic σ-complex is not activated.

These examples clearly illustrate that "bond activation", whether it refers to C–H bonds or other bonds, is a poor choice for designation of certain reaction types and mechanisms.

1.5
Conclusions

The analysis of the origin and usage of the term "C–H bond activation" revealed dichotomy between the organometallic definition of this term and the general understanding of the word "activation". As discussed in this essay, "C–H bond activation" is frequently used in the organometallic sense and indeed serves well as a qualitative mechanistic term. Although distinction between the organometallic "C–H bond activation" and general "activation" could be made (and is made intuitively by many), this is clearly degenerate and inconsistent terminology. We have, furthermore, shown that it is difficult to find a rigorous mechanistic basis for defining a "C–H bond activation" class of processes according to the organometallic definition discussed herein.

Consequently, we were faced with the task of formulating a widely acceptable and consistent definition of "bond activation". Our research, discussions, and analyses led to a conclusion that "bond activation" should refer to a process of increasing the reactivity of a bond in question and as such encompasses an entire spectrum of possible mechanisms. Also, we argue that "activation" is not equivalent to "reaction" or, in other words, that "activation" of a bond is not the same as cleavage of a bond. For the latter process we proposed the general term "bond transformation". It should be emphasized that both "bond activation" and "bond transformation" are general terms and, therefore, information about the reaction and mechanism category should be specified by additional descriptors (cf. C–H bond arylation via electrophilic metalation, C–H bond metalation via concerted metal insertion).

Acknowledgment

We acknowledge Professor Gerald Dyker for providing critical contribution to this manuscript. We also thank many colleagues in the broad chemical community for their stimulating questions and discussions on this topic. Dr J. B. Schwarz is acknowledged for editorial assistance. This work was supported in part by the National Science Foundation.

References and Notes

1. (a) R. Grant, C. Grant, *Grant and Hackh's Chemical Dictionary*, 5th edn, McGraw–Hill, New York, **1987**; (b) A. D. McNaught, A. Wilkinson, *Compendium of Chemical Terminology, IUPAC Recommendations*, 2nd edn, TSC, Cambridge, UK, **1997**; (c) *International Encyclopedia of Chemical Science*, Van Nostrand, Princeton, **1964**.
2. This view has previously been suggested in print. G. Dyker, J. Heiermann, M. Miura, *Adv. Synth. Catal.* **2003**, *345*, 1127–1132. We also thank Professor G. Dyker for sharing his thoughts on this subject.
3. K. Morikawa, W. S. Benedict, H. S. Taylor, *J. Am. Chem. Soc.* **1936**, *58*, 1445–1449. According to our search, the first time the phrase "activation of C–H bond" was used dates to **1929**. Incidentally, authors referred to activation of arene C–H bonds by the substituents. F. Swarts, *Recl. Trav. Chim. Pays. Bas.* **1929**, *48*, 1025–1028.
4. J. Halpern, *Discuss. Faraday Soc.* **1968**, *46*, 1–20.
5. R. J. Hodges, J. L. Garnett *J. Catal.* **1969**, *13*, 83–98.
6. N. F. Gol'dshleger, M. B. Tyabin, A. E. Shilov, A. A. Shteinman, *Russ. J. Phys. Chem.* **1969**, 1222–1223.
7. (a) J. A. Labinger, A. M. Herring, D. K. Lyon, G. A. Luinstra, J. E. Bercaw, *Organometallics*, **1993**, *12*, 895–905; (b) A. Sen, M. A. Benvenuto, M. Lin, A. C. Hutson, N. Basickes, *J. Am. Chem. Soc.* **1994**, *116*, 998–1003.
8. G. W. Parshall, *Acc. Chem. Res.* **1975**, *8*, 113–117.
9. J. Chatt, J. M. Davidson, *J. Chem. Soc.* **1965**, 843–855.
10. (a) M. L. H. Green, D. O'Hare, *Pure Appl. Chem.* **1985**, *57*, 1897–1910; (b) R. H. Crabtree, *J. Chem. Soc., Dalton Trans.* **2001**, 2437–2450; (c) W. D. Jones, *Acc. Chem. Res.* **2003**, *36*, 140–146.
11. (a) E. J. Hennessy, S. L. Buchwald, *J. Am. Chem. Soc.* **2003**, *125*, 12084–12085; (b) C.-H. Park, V. Ryabova, I. V. Seregin, A. W. Sromek, V. Gevorgyan, *Org. Lett.* **2004**, *6*, 1159–1162.
12. (a) A. H. Janowicz, R. G. Bergman, *J. Am. Chem. Soc.* **1982**, *104*, 352–354. For other related processes: (b) W. D. Jones, F. J. Feher, *J. Am. Chem. Soc.* **1984**, *106*, 1650–1663.
13. (a) J. Halpern, *Acc. Chem. Res.* **1970**, *3*, 386–392; (b) C. Amatore, F. Pfluger, *Organometallics*, **1990**, *9*, 2276–2282.
14. S. D. Ittel, C. A. Tolman, A. D. English, J. P. Jesson, *J. Am. Chem. Soc.* **1978**, *100*, 7577–7585.
15. For related zincation of alkynes: D. E. Frantz, R. Fässler, C. S. Tomooka, E. M. Carreira, *Acc. Chem. Res.* **2000**, *33*, 373–381.
16. J. G. Hefner, P. M. Zizelman, L. D. Durfee, G. S. Lewandos, *J. Organomet. Chem.* **1984**, *260*, 369–380.
17. Similarly, recent computational studies suggested that the iridium-catalyzed borylation of arenes also proceeded via the insertion mechanism. H. Tamura, H. Yamazaki, H. Sato, S. Sakaki, *J. Am. Chem. Soc.* **2003**, *125*, 16114–16126 and references therein.

2
C–H Transformation in Industrial Processes
Leslaw Mleczko, Sigurd Buchholz, Christian Münnich

2.1
Introduction

The general aim of C–H transformation is to introduce groups with a higher complexity to hydrocarbon structures. Industrial processes therefore usually involve transformation of C–H groups starting from simple molecules. The reactions employed are selective oxidation, substitution (radical, electrophilic), nitration, ammoxidation, and sulfonation. The functionalized molecules are then further converted to more valuable products and intermediates by different reaction pathways. The latter often comprise further steps of C–H-activation.

This chapter can only give a brief summary of the many reactions used for C–H transformation in commercial chemical processes. Further detailed information about the reactions can be found in the general literature about industrial chemistry [1–3]. A general survey of C–H transformation of alkanes, olefins, and aromatics will be given in Sections 2.2 to 2.4. In Section 2.5 the synthesis of fine chemicals will be considered separately. The production of basic chemicals, raw materials, and intermediates described in the first three sections differs significantly from the production of fine chemicals and highly functionalized products such as pharmaceuticals. The difference originates not only because of the quantity produced and reactor size but also because of the significantly different chemistry. It is therefore the aim of this chapter to summarize the main characteristics of the different reactions, i.e. to illustrate the similarities and differences and current and future needs for process development.

2.2
Alkane Activation

An overview of the main reactions and processes is given in Table 1. Functionalization of lighter hydrocarbons to basic chemicals is performed by thermal activation, oxidation, sulfoxidation, ammoxidation, and chlorination. Reactions are carried out either in the gas phase or under milder conditions in the liquid phase.

Handbook of C–H Transformations. Gerald Dyker (Ed.)
Copyright © 2005 WILEY-VCH Verlag GmbH & Co. KGaA, Weinheim
ISBN: 3-527-31074-6

Thermal activation of light hydrocarbons, e.g. CH_4, is used for the production of acetylene at high temperatures and extremely short residence times (Table 1, entry 1). Because of the thermodynamics and kinetics of the reaction the production of acetylene by conversion of light hydrocarbons, preferably C_1, is conducted at temperatures above 1000 °C and with extremely low residence times of less than 10 ms. The heat for the reaction is supplied by burning a part of the feedstock in a burning zone followed by a rapid quenching of the product mixture with oil or water (partial combustion processes). Various processes operating according to this basic principle are known [4, 5].

Perhaps the most important industrial reaction is the production of synthesis gas, a gas mixture containing CO and H_2 in different proportions (Table 1, entry 2). Natural gas-based processes have successfully replaced coal-based routes to synthesis gas. The desired, more valuable product generated is hydrogen. Most of the hydrogen produced (>70%) is used in the production of ammonia and methanol and in refinery processes. The reforming reaction occurs over a heterogeneous catalyst in a fixed-bed tubular reactor. Synthesis gas is usually generated by steam reforming, first described in patents in 1912 [6], and the main carbon

Table 1. Industrial processes for C–H transformation of alkanes.

	Educt	Condition/reactant	Product
1	C_1	$T > 1000$ °C	Acetylene [4]
2	C_1 and higher	H_2O, O_2	Synthesis gas [6]
3	C_1	O_2/NH_3	Hydrogen cyanide [7, 8]
4	C_{2+}	$T = 750–900$ °C, cat.	Olefins, diolefins, aromatics [10]
5	C_4	Catalyst/O_2	Maleic anhydride [11]
6	C_1	Cl_2	Chlorinated hydrocarbons [13, 14]
7	$i\text{-}C_4$	O_2	tert-Butylhydroperoxide [15]
8	Linear/cycloalkane	H_2O (H_3BO_3)	Secondary alcohols [16, 17]
9	Linear/cycloalkane	O_2	Alcohols, ketones, carboxylic acids [19]
10	n-Alkanes	Cl_2, SO_2	Alkanesulfonychlorides [18]
11	n-Alkanes	O_2/SO_2	Alkanesulfonic acids [18]
12	Cyclohexane	NOCl	Cyclohexanol, cyclohexanone (oxime), caprolactam [19, 20]
13	Paraffins	HNO_3	Nitro compounds
14	C_1	S	Carbon disulfide
15	n-Alkanes	O_2, nutrients microorganism	Proteins

source nowadays is natural gas. Alternatively synthesis gas can be produced by partial oxidation. Partial oxidation (POX) is employed in the new generation of gas-to-liquid processes. Finally, autothermal reforming, i.e. a combination of partial oxidation and steam reforming must be mentioned. All these processes are performed at high temperatures (>900 °C) and high pressures (up to 15 MPa).

The direct conversion of methane with oxygen and ammonia results in the formation of hydrogen cyanide (Table 1, entry 3); the dehydration of formamide is of minor importance only. The reaction is highly endothermic and the various processes differ mainly in the method of energy supply to the reaction system. Different processes have been developed for this short-contact-time reaction, with the most important being the Andrussow process with platinum-based gauze as catalyst and a reaction temperature higher than 1000 °C [7], the Degussa BMA process carried out in tube bundles with a thin layer of platinum catalyst [8], and the Shawinigan process in a fluidized bed in the absence of a metal catalyst at $T = 1500$ °C [9].

The thermal and catalytic conversion of different hydrocarbon fractions, often with hydrotreating and other reaction steps, is characterized by a broad variety of feeds and products (Table 1, entry 4). New processes starting from natural gas are currently under development; these are mainly based on the conversion of methane into synthesis gas, further into methanol, and finally into higher hydrocarbons. These processes are mainly employed in the petrochemical industry and will not be described in detail here. Several new processes are under development and the formation of BTX aromatics from C_3/C_4 hydrocarbons employing modified zeolite catalysts is a promising example [10].

The oxidation of butane (or butylene or mixtures thereof) to maleic anhydride is a successful example of the replacement of a feedstock (in this case benzene) by a more economical one (Table 1, entry 5). Process conditions are similar to the conventional process starting from aromatics or butylene. Catalysts are based on vanadium and phosphorus oxides [11]. The reaction can be performed in multitubular fixed bed or in fluidized bed reactors. To achieve high selectivity the conversion is limited to <20% in the fixed bed reactor and the concentration of C_4 is limited to values below the explosion limit of approx. 2 mol% in the feed of fixed bed reactors. The fluidized-bed reactor can be operated above the explosion limits but the selectivity is lower than for a fixed bed process. The synthesis of maleic anhydride is also an example of the intensive process development that has occurred in recent decades. In the 1990s DuPont developed and introduced a so called cataloreactant concept on a technical scale. In this process hydrocarbons are oxidized by a catalyst in a high oxidation state and the catalyst is reduced in this first reaction step. In a second reaction step the catalyst is reoxidized separately. DuPont's circulating reactor-regenerator principle thus limits total oxidation of feed and products by the absence of gas phase oxygen in the reaction step of hydrocarbon oxidation [12].

The direct chlorination of methane is carried out as a radical reaction in the gas phase and the highly exothermal reaction produces a mixture of chlorinated methanes (Table 1, entry 6). Higher chlorination is achieved by recycling lower chlorinated products into the process [13, 14]. If dichloromethane is the desired product, a large excess of methane must be used, as dichloromethane is more rapidly chlorinated than methane. Several process modifications (bubble columns, fluidized bed, tubular reactor, and photochemical initiation) have been developed to overcome the obstacles of high exothermicity, run-away-problems, low product selectivity, and corrosion as a result of formation of hydrochloric acid.

As is apparent from the examples described above, process development of gas-phase reactions involving alkanes is limited by several obstacles:
- corrosion and low stability of reactor materials
- high process temperatures and temperature control
- high pressure
- explosion limits
- limited stability of desired products, decrease in selectivity because of consecutive reactions
- economic necessity of large-scale production
- catalyst deactivation

These obstacles are also crucial for conversion of olefins and aromatics, as will be discussed in Sections 2.3 and 2.4. It is the task of reaction engineering experts to develop reactors and designs for large-scale synthesis while controlling the kinetics of reactions. The kinetics of gas-phase processes are often in the range of milliseconds (HCN synthesis) to seconds (maleic anhydride synthesis). The high reaction temperatures, in particular, require careful selection of reactor materials that are stable and do not act as catalysts or initiators of gas-phase reactions. Processes operated at high pressures make the choice of reactor material even more difficult. Energy supply and, more often, removal of heat from the reactor system is difficult, especially in large scale reactors used for highly exothermic reactions. This has led to a number of solutions, for example fluidized bed reactors, catalytic wall reactors, short-contact-time reactors and quenched reactors. Uncontrolled release of heat in the reactor always favors consecutive reactions reducing the overall selectivity. It can, furthermore, even lead to a run-away and destruction of the reactor.

A problem that must be carefully considered is the occurrence of side- and consecutive reactions. This is especially important for alkane activation, because severe reaction conditions are necessary to activate the C–H bonds. When reactions are fast, as in the HCN and acetylene syntheses, rapid quenching of the reaction products is possible. Another way of affecting selectivity is to limit the partial pressure of reactants, thus also reducing the partial pressure of the desired product. In this way in the maleic anhydride synthesis conversion is limited by diluting the gas and limiting the amount of oxygen available for the reaction.

When employing higher alkanes as a feedstock, process economics are often determined more by selectivity than by conversion per single pass. Most of these reactions are therefore conducted in the liquid phase, which enables easier control of reaction rates and also of side and consecutive reactions by adjustment of the temperature. However, activating alkane C–H bonds always requires reactions

2.2 Alkane Activation

conditions which also favor these unselective reactions. If the rates of production of the desired product and by-products are similar, influencing product selectivity by process design becomes very difficult. As the examples summarized in Table 1 illustrate, few processes are used industrially for C–H transformation of alkanes.

Inserting oxygen into the C–H bond of an alkane initially leads to hydroperoxides. When this reaction is performed with atmospheric oxygen it is also called autooxidation. It usually leads to a multitude of products, because of further spontaneous reactions, so this reaction is of limited synthetic use. An exception is oxidation of *iso*butane with oxygen, which leads to 70 % yield of *tert*-butyl hydroperoxide at a conversion of 80 % (Table 1, entry 7). Hydrogen bromide is used, among other compounds, as an initiator [15]. *tert*-Butyl hydroperoxide is used as an oxidant in propylene oxide production by the Halcon process. In the formation of phenol by the cumene process cumene is oxidized into the corresponding hydroperoxide in a similar way.

The Bashkirov oxidation (liquid-phase oxidation of *n*-alkanes or cycloalkanes in the presence of boric acid and hydrolysis) yields the corresponding secondary alcohols [16, 17]. The reaction is used industrially for oxidation of C_{10} to C_{18} *n*-alkanes, providing raw materials for detergents and for oxidation of cyclododecane to cyclododecanol as an intermediate for the production of Nylon 12 (Table 1, entry 8). The process is not of much commercial importance in the western world, however. Oxidation in the absence of boric acids usually leads to mixtures of alcohols, ketones, and carboxylic acids (Table 1, entry 9).

Of the large number of possible ways of synthesizing alkanesulfonates, only sulfochlorination of alkanes (conversion with sulfur dioxide and chlorine to form alkane sulfonyl chlorides and their saponification with sodium hydroxide; Table 1, entry 10) and sulfoxidation (reaction with sulfur dioxide and oxygen and neutralization of the sulfonic acids; Table 1, entry 11) are of industrial importance [18].

The catalytic oxidation of cyclohexane is performed in the liquid phase with air as reactant and in the presence of a catalyst. The resulting product is a mixture of alcohol and ketone (Table 1, entry 12) [19]. To limit formation of side-products (adipic, glutaric, and succinic acids) conversion is limited to 10–12 %. In a process developed by Toray a gas mixture containing HCl and nitrosyl chloride is reacted with cyclohexane, with initiation by light, forming the oxime directly (Table 1, entry 12). The corrosiveness of the nitrosyl chloride causes massive problems, however [20]. The nitration of alkanes (Table 1, entry 13) became important in a liquid-phase reaction producing nitrocyclohexane which was further catalytically hydrated forming the oxime.

Another example of the use of natural gas is the reaction of methane with sulfur over silica gel as catalyst at temperatures around 650 °C forming carbon disulfide (Table 1, entry 14). CS_2 is still used for production of CCl_4.

Alkane activation can also be achieved under very mild reaction conditions. This is demonstrated by the use of microorganisms for transformation of *n*-alkanes into proteins by oxidation (Table 1, entry 15).

The activation of alkanes is becoming increasingly relevant because of the expected shortage of crude oil. With the demand for energy and chemicals rising, there is, especially in Asia, a tendency to go back to the use of coal as a raw material. Furthermore, biomass as a renewable resource starts to become interesting for ensuring future resources of different carbon fractions. New processes which are close to commercialization are summarized in Table 2. The direct use of natural gas is of particular interest. Direct conversion to methanol (Table 2, entry 16) would facilitate transport of natural resources from the respective deposits to large-scale production sites. Direct synthesis would also make the step of synthesis-gas production (Table 1, entry 2) no longer necessary in routes leading to higher hydrocarbons, for example the methanol-to-olefins (MTO) process [21]. Direct synthesis of formaldehyde, as one of the most reactive intermediates in industrial chemistry is also an aim of process development. Another way to produce higher hydrocarbons from natural gas is the transformation of C–H bonds in methane into C–C bonds by (oxidative) coupling of methane (Table 2, entry 17). Extensive research work has been conducted on this process [22]. It seems, however, there is an internal barrier that limits yields to values below 25%, which would make the coupling of methane economically feasible.

Table 2. New processes for alkanes under development and close to commercialization.

	Starting material	Reactant/condition	Product
16	C_1	O_2	CH_3OH (CH_2O) [21]
17	C_1	O_2	C_{2+} [22]
18	C_x	Metal catalyst	Carbon nanotubes/carbon nanofibers/buckyballs [27-29]
19	C_3	O_2	Acrolein, acrylic acid [23, 24]
20	C_3	O_2/NH_3	Acrylonitrile [25, 26]
21	C_3/C_4	Catalyst	BTX aromatics [10]

Since the early 1990s several methods have been used to transform the C–H bonds of methane and higher hydrocarbons directly into highly ordered C–C bonds in a nano-scaled product; such work has attracted much attention in recent years in attempts to manufacture carbon nanotubes, nanofibers, and buckyballs [27–29]. The production of nanoscaled carbon material is already well established – annual production of carbon black is approx. 8×10^6 metric tons per year. In contrast with carbon black, nanotubes and fibers have highly ordered structures. These are of special interest because of their unique properties, for example high mechanical strength, excellent electrical and thermal conductivity, and very high ratio of length (up to tens of microns) to diameter (usually 3–150 nm). Furthermore, the possibility of functionalizing the carbon atoms on the outer layer of nanotubes, thus building complex structures, makes many new applications possible. Processes are based on the decomposition of hydrocarbons, e.g. CH_4, at high temperatures (600–1200 °C), on metal catalysts based on, e.g., iron or nickel [27–29]. Process economics and yields must still be improved to make carbon nanotubes in a way that can compete with established products in the mass markets.

Further examples of attempts to replace olefins by alkanes as a starting materials, as in the maleic anhydride process, are the development of processes for selective oxidation and ammoxidation. Examples are processes for acrolein, acrylic acid (Table 2, entry 19) and acrylonitrile (Table 2, entry 20) using propane as a feedstock [30]. As discussed at the beginning of this section, the difficulty of activating short alkanes causes selectivity problems in these oxidation reactions. Intermediates formed in the, usually, consecutive reaction mechanisms are less stable than propane against further oxidation, and total oxidation and cracking occur as side reactions limiting the selectivity compared with the first vanadia-based catalysts [31]. The development of a complex mixed-oxide catalyst brought the activity and selectivity of the catalyst into a range in which the process may become economically feasible and it has been announced that pilot and demonstration plants are to be brought into operation, partly working with a mixed feed of olefin and alkane [32].

Dehydrocyclodimerization of liquefied petrol gas (propane and butane) can be performed to yield BTX-aromatics (Table 2, entry 21). Modified ZSM-5 based catalysts are used, for example in the UOP cyclar process [33], and the process will become attractive for extraction of aromatics from natural gas fields containing further C_3–C_4 fractions.

2.3
C–H Transformation at Olefins

In many olefinic reactions activation of the C=C double bond occurs, although in many reactions at least one C–H bond is transformed. Established processes are summarized in Table 3. Examples of liquid-phase reactions are the synthesis of ethers, especially methyl *tert*-butyl ether by reacting olefins (*iso*butene) and alcohol (methanol) in the liquid phase at slightly elevated temperature and pressure (Table 3, entry 22). Different processes developed differ only slightly in feed composition and design, which is optimized for heat removal [2].

In the direct oxidation of ethylene to yield acetaldehyde (Table 3, entry 23), aqueous solutions of a catalyst and slightly elevated temperatures (100–150 °C) are used [34]. For process development the corrosion caused by the $CuCl_2$ and $PdCl_2$ solutions used as catalysts was crucial.

In the industrial production of acetic acid the main production routes are based on the carbonylation of methanol, a process which was first developed in the 1940s. Although the process remains cheap, the availability and pricing of raw materials have forced the development of new processes based on the direct oxidation of ethylene to form acetic acid (Table 3, entry 24). Complex multimetal oxide catalysts are suitable [35, 36], and routes employing ethane have also been investigated [37]. Although C_2 is an inexpensive raw material and high selectivity has been achieved [23], it is unlikely that direct oxidation, based on ethane, will compete with carbonylation of methanol in the near future. The oxygen concentration must be limited for safety reasons and, therefore, conversion per single pass is limited.

Table 3. C–H transformation processes of olefins.

	Starting material	Reactant/condition	Product
22	C_x	ROH	MTBE, ether [2]
23	C_2	$O_2/PdCl_2$	Acetaldehyde [34]
24	C_2	O_2	Acetic acid [35–37]
25	C_3	O_2	Acrolein, acrylic acid [38–40]
26	C_3	O_2/NH_3	Acrylonitrile [42, 43]
27	C_4	O_2	Maleic anhydride
28	$C=CH_2$	Cl_2, dehydrochlorination	$C=CHCl$, e.g. chloroprene [44, 45]

The direct heterogeneously catalyzed oxidation of propylene yielding acrolein was commercialized in the 1960s (Table 3, entry 25). Complex metal oxide catalysts [38–40] were developed, and oxidation with air is performed in tubular reactors. Propylene is diluted with air and steam. The multitubular reactors are cooled by molten salt baths to remove the heat of reaction [41]. The ammoxidation of propylene is also a well established process (Table 3, entry 26). In the so called Sohio process high conversion and selectivity are achieved by controlling the temperature in fluidized bed reactors with internal cooling [42, 43]. In his process higher conversions are possible because of the greater stability of acrylonitrile compared with acrolein against total oxidation. Ammonia and propylene are fed separately into the reactor to avoid explosive mixtures. Other processes known employ cooled fixed bed multitubular reactors.

As described above, butylene is employed as part of the feedstock in maleic anhydride synthesis (Table 3, entry 27). Process conditions are similar but, because of the pricing of feedstocks, butane oxidation is mainly employed.

Chloroprene is of high industrial importance for manufacture of synthetic rubbers. For a long time the synthesis was based on acetylene. More recent processes are based on butadiene as a feedstock, which is substantially cheaper [29]. The initial step is a gas-phase free-radical chlorination at 250 °C and temperature control is ensured by use of excess butadiene (molar ratio of Cl_2 to butadiene: 1:5 to 1:50) [44]. To limit side reactions, short contact time reactors operating at higher temperatures and residence times below one second are also known [45]. Good mixing of reactants is crucial, to prevent formation of soot. The excess butadiene is recycled into the chlorination reactor. The chlorination yields a mixture of the 1,2- and 1,4-adducts. The latter can be isomerized to the 1,2-adduct by use of catalytic amounts of copper or iron salts. In a second step dehydrochlorination is performed in a dilute alkaline solution. Addition of phase-transfer catalysts increases the reaction rate.

2.4
Basic Chemicals from Aromatic Hydrocarbons

The transformation of C–H bonds in aromatic hydrocarbons has numerous applications, yielding a broad variety of chemicals. The variety of reactions does not enable description of process characteristics and chemistry in detail here. As already stated in the introduction, more details are readily available in the general literature [1–3].

Table 4. Basic functionalization of aromatic hydrocarbons.

	Starting material	Reactant	Product
29	Benzene	Olefin/Catalysts	Ethyl-, isopropylbenzene
30	Benzene	Propene/O_2	Phenol/acetone
31	Benzene	Cl_2, catalyst	Chlorinated benzenes
32	Toluene	Cl_2	Chlorinated toluenes
33	p-Xylene	O_2/Co/Mn	Terephthalic acid
34	o-Xylene/naphthalene	O_2/VPO	Phthalic acid
35	Benzene	HNO_3/H_2SO_4, H_2	Nitrobenzene, aniline
36	Toluene	HNO_3, H_2, $COCl_2$	Nitrotoluene, TDI
37	Toluene, xylene	Cl_2	Side-chain chlorinated products
38	Benzene	SO_3/H_2SO_4	Benzenesulfonic acid, tensides
39	Phenol	CO_2	Salicylic acid
40	Toluene	O_2	Benzoic acid
41	Nitrotoluene	O_2	Nitrobenzoic acid

The alkylation of benzene is the most important reaction in the further use of this compound (Table 4, entry 29) [1]. The reaction is performed in the liquid phase with Friedel–Crafts catalysts or in the gas phase with acid catalysts such as H_3PO_4, aluminum silicates, or BF_3. More than 50 % of benzene consumption goes into the production of ethylbenzene, which is further converted to styrene by direct catalytic dehydrogenation. It is important to note that in the dehydrogenation process, carried out at high temperatures, once again the partial pressure of ethylbenzene is reduced by co-feeding water vapor to limit side reactions. The alkylation of benzene with propene (Table 4, entry 30) yields cumene, which is almost exclusively used in the Hock process for the manufacture of phenol and acetone, a process commercialized in the 1950s. Higher alkylbenzenes are starting materials for the corresponding sulfonates, which are converted further into surfactants and detergents.

Chlorination of benzene was industrially established by liquid phase reaction with $FeCl_3$ catalysts. In the oxychlorination process (Raschig–Hooker process) HCl/Air mixtures are used at a reaction temperature of 240 °C and atmospheric pressure. Fixed bed catalysts ($CuCl_2/FeCl_3/Al_2O_3$) are used and conversion must be limited to below 20 % to suppress formation of higher chlorinated products. In the Raschig–Hooker process hydrolysis occurs catalytically yielding HCl, avoiding the production of NaCl, and limiting the amount of HCl needed. Disadvantages of the gas-phase process are the high investment costs for corrosion-resistant plants, large energy costs, and the need for frequent regeneration of the hydrolysis catalyst, because of coke formation. Chlorination of toluene (Table 4, entry 32) is important in the production of chlorotoluene, which is a basis for further production of cresols.

The oxidation of *p*-xylene to terephthalic acid (Table 4, entry 33) would only result in the formation of *p*-toluic acid. To convert the second group in high yield and selectivity, the oxidation must be influenced in one of three ways:
- conversion of the carboxyl group into an ester followed by oxidation of the second methyl group;
- addition of a co-catalyst (often a bromine compound); or
- co-oxidation in a process in which an auxiliary substance capable of supplying peroxides (e.g. acetaldehyde, paraldehyde, methyl ethyl ketone) is simultaneously oxidized

These dimethyl terephthalate and terephthalic acid processes illustrate that it can be necessary to modify the reaction chemistry and mild (liquid phase) conditions to obtain the product in the desired yield and purity.

In the oxidation of naphthalene or *o*-xylene forming phthalic anhydride simple oxidation is sufficient (Table 4, entry 34). As in all heterogeneously catalyzed oxidation reactions with high exothermicity, oxidation of naphthalene or *o*-xylene over vanadium oxide catalysts is carried out in multitubular or fluidized bed reactors, enabling removal of the heat of reaction. It is also necessary to cool the reaction mixture from the reaction temperature (360–420 °C) to temperatures below the dew point of phthalic anhydride to avoid consecutive reactions. This task is difficult in large-scale operation and led, e.g., to the development of a two-stage fluidized-bed process. In the fluidized bed process the reaction is operated within flammability limits, the catalyst particles acting as an effective dispersal medium preventing an explosion. *o*-Xylene oxidation is also performed in the liquid phase using Co, Mn, or Mo-based catalysts. This technique is of minor importance for larger scale production (a phthalic anhydride world scale plant operates at 50 000 metric tons/year in one reactor).

Nitration of benzene (Table 4, entry 35) is usually conducted using a mixture of nitric acid and sulfuric acid (nitrating acid). Sulfuric acid promotes the formation of the nitronium ion, prevents dissociation of nitric acid and enhances solubility between phases. The process is still performed batch-wise at temperatures of 50–60 °C. Vigorous stirring of the two-phase reaction mixture is necessary for both mass and heat exchange and residence times are in the range of several hours. Continuous nitration plants consist of cascades of vessels. The purified nitrobenzene is reacted with hydrogen using a heterogeneous catalyst to form aniline. Fixed beds (Bayer) and fluidized beds (BASF) are used as reactors in this exothermic hydrogenation. In an analogous process toluene is nitrated to dinitrotoluene (Table 4, entry 36). Hydrogenation is carried out catalytically with copper, palladium, or nickel catalysts in solvents like methanol. In contrast with aniline, gas-phase hydrogenation is not possible.

Side-chain chlorinated alkyl aromatics based on toluene and xylene play an important role as chemical intermediates. They are used in the manufacture of a variety of chemical products including plastics, pharmaceuticals, flavors, pesticides, catalysts, and more. Monochlorination of the sidechain (Table 4, entry 37) is rather difficult.

The sulfoxidation of benzene (Table 4, entry 38) yields benzenesulfonic acids and the respective derivatives. The electrophilic aromatic substitution reaction gives high yields and aqueous sulfuric acid or oleum is used for the sulfonation reaction, which is performed in cascades of reactor vessels.

The carboxylation of phenols is a well established process for synthesis of salicylic acid according to the Kolbe–Schmitt method (Table 4, entry 39). The exothermic reaction is carried out at slightly elevated temperatures around 150 °C and pressures of approximately 5 bar. Batch processes are still mainly used. The main task is to exclude water from the reaction mixture, because this would release the alkali metal hydroxide from the phenoxide salt.

The liquid-phase oxidation of toluene with molecular oxygen is another example of a well established process (Table 4, entry 40). A cobalt catalyst is used in the process and the reaction proceeds via a free-radical chain mechanism. Heat of reaction is removed by external circulation of the reactor content and both bubble columns or stirred tanks are employed. It is important to note that air distribution is critical to prevent the danger of a runaway. Another example of direct oxidation is the commercial production of nitrobenzoic acid by oxidation of 4-nitrotoluene with oxygen (Table 4, entry 41).

Apart from development of new catalysts with enhanced activity, few processes with innovative chemistry are currently developed for C–H transformation in aromatics. Though new processes using cheaper raw materials or reducing the number of reaction steps may seem attractive at first glance, efforts for process development including research, scale-up, pilot plant, and timeline must be considered. Especially for large scale-synthesis of bulk chemicals process economics

Table 5. New processes for C–H transformation in aromatic hydrocarbons.

	Starting material	Reactant/condition	Product
44	Benzene	N_2O	Phenol [46, 47]
45	Cresols	O_2	Hydroxybenzaldehydes [49]

must be taken into account. Potential new processes therefore have to show a significant potential in reduction of production costs in the early stage of development to go into further scale-up.

An interesting new process is the conversion of benzene to phenol with N_2O developed by the Boreskov Institute for Catalysis and Solutia [46, 47]. Important for the process is the in-situ generation of active oxygen on the catalyst – so called α-oxygen – minimizing the content of free oxygen in the gas phase. This elegant approach generates an active oxygen species which is inserted into the C–H bond. Thus highly selective oxygen is generated on the catalyst. The amount of free oxygen in the gas phase is on a very low level and attack of the aromatic ring and total oxidation is minimized. The catalyst must be regenerated periodically, because of deactivation after coke-formation. Though the yield and selectivity of the process are high and only one step is required for conversion of benzene, it is not yet commercialized. One reason is the need to construct a world-scale size plant to be economically competitive. Furthermore N_2O has to be generated in sufficient amounts for the reaction, e.g. by oxidation of ammonia. The ability to insert this oxygen was also demonstrated for C–H bonds in methane, forming methanol [48], but the overall space–time yield was too low for large-scale synthesis.

Another example of minimizing the reaction steps necessary for synthesis is the direct conversion of cresols into hydroxybenzaldehydes, replacing the Reimer–Tiemann reaction in which phenol is reacted with chloroform in the presence of KOH. Hydroxybenzaldehydes are used in pharmaceuticals, perfumes, and colors. In the direct oxidation cresols are reacted with oxygen in the presence of Cu/Co/C catalysts forming the salicylaldehyde [49].

2.5
Fine Chemicals

In the synthesis of fine chemicals the same types of reaction often employed in the formation of basic chemicals (see Sections 2.5.1 and 2.5.2) are used. Furthermore, the same or similar starting materials may be used. There are usually two routes to obtain fine chemicals from C–H transformation that differ with regard to the basic chemicals. One way is to start with a simple substrate e.g. methane and perform a complex functionalization in one step. Most research in the field of fine chemistry is focused in this field. Another way is to start with a more complex starting material and carry out a C–H transformation step with very high selectivity to reduce the amount of wasted starting material and increase the economic value of the synthesis.

Furthermore, in the synthesis of fine chemicals typical process technology considerations, for example space-time-yield, are less important than for bulk chemicals. Because of the relatively small production outputs batch reactors are the most common apparatus in which to perform the synthesis. New chances in the field of fine chemistry may be offered by micro-reaction technology. Microstructured systems can be used to improve heat transfer which may be critical for highly exothermic reactions and, furthermore, they may be also useful in reactions where fast mixing of components is recommended.

The C–H transformation in the synthesis of fine chemicals can be separated in reactions employing catalysis by organometallic compounds and metal-free synthesis. Organometallic catalyzed functionalization is usually performed in liquid or gas/liquid reactions whereas the metal-free synthesis of fine chemicals often occurs as a gas-phase reaction.

2.5.1
Fine Chemicals by Organometallic Catalysis

Functionalization of alkanes to produce fine chemicals uses many different reaction types that cannot be listed completely within this introduction. An overview of typical reactions, for example halogenation, dehydration, borylation, oxidation, and hydroxyalkylation, is given in Table 6.

Table 6. Organometallic catalyzed functionalization of alkanes.

	Starting material	Reactant/conditions	Product
46	Alkane	$Cl_2/[Pt(Cl)_4]^{2-}$	Chloroalkanes [50]
47	Cyclooctane/vinyl-*tert*-butane	$ReH_7(PR_3)_2$	Cyclooctene/ethyl-*tert*-butane [51]
48	Methylcyclohexane	$IrH_2(O_2CCF_3)(PR_3)_2$, UV	Methylenecyclohexane [52]
49	Alkane	$R_2B-BR'_2$, $CpRh(C_2H_4)_2$, UV	Boroalkane [53]
50	Cyclohexane	CH_3OH, Hg, UV	Hydroxymethylenecyclohexane [54]

The platinum-catalyzed reaction of alkanes with chlorine leads to alkyl chlorides and alcohols (Table 6, entry 46) with modest rates and conversions [50]. Cyclooctane can be easily dehydrogenated (Table 6, entry 47) in the presence of a stabilized vinylalkane by use of the neutral rhenium compound $ReH_7(PR_3)_2$ [51]. By employing an iridium-based catalyst, the photochemical dehydrogenation of methylcyclohexane to methylenecyclohexane is performed at room temperature

(Table 6, entry 48) whereas alkene isomerization is very slow [52]. Photochemistry is also used in the rhodium-catalyzed alkylborylation of alkanes in which boroalkanes (Table 4, entry 49) are obtained [53]. An example of a gas-phase reaction is the photochemical mercury-catalyzed hydroxymethylation of cyclohexane to hydroxymethylenecyclohexane (Table 6, entry 50). This reaction is an example of the abstraction of a hydrogen atom from an alkane by an excited metal atom [54].

The activation and transformation of C–H bonds in alkanes by homogeneous transition metal catalysts should to be a topic of further research. No processes are close to commercialization on a technical scale. So far the organometallic approach by Shilov is one of the most promising ways of producing a practical system [55].

2.5.2
Metal-free Synthesis of Fine Chemicals

An overview of some metal-free reactions, for example oxidation, amination, and halogenation, is given in Table 7.

Table 7. Metal free reactions.

	Starting material	Reactants/conditions	Product
51	2,3-Dimethylbutane	O_2, UV	Acetone, 2,3-dimethylbutanehydroperoxide [56]
52	Methane	O_2/NO_2	Formaldehyde, CO, nitromethane [57]
53	Cyclohexane	O_2/NO_2	Cyclohexanal, nitrocyclohexane [58]
54	Adamantane	$NHPI/O_2$	Adamantanol [59]
55	Adamantane	H_2O_2/O_2	Adamantanol/adamantanal [60]
56	Cyclohexane	O_2/Zeolite NaY, UV	Cyclohexanehydroperoxide/cyclohexanal [61]
57	Cyclohexane	$NHPI/NO_2/O_2$	Nitrocyclohexanes [62]
58	Ethane	NH_3/Hg, UV	Imines [63]
59	Cyclohexane	HCI_3/NaOH/phase transfer catalyst	Iodocyclohexane [64]
60	2-Chloroacetic ester	$HCOOH/F_2/N_2$	2,2-Chlorofluoroacetic ester [65]
61	Methane	$NH_3/O_2/N_2$	HCN [66]

The gas phase oxidation of 2,3-dimethylbutane (Table 7, entry 51) leads to number of products even under mild conditions [56]. The selectivity of alkane oxidations, e.g. methane oxidation (Table 7, entry 52) increases substantially in the presence of nitrogen oxides but are still below 10% [57]. For many alkanes, e.g. cyclohexane, the gas-phase oxidation (Table 7, entry 53) is unselective because of fragmentation reactions [58]. The selectivity of alkane oxidations can be improved by using oxygen and NHPI (N-hydroxyphthalimide; Table 7, entry 54), often in the presence of a cobalt catalyst under mild conditions [67]. By using NHPI, which forms PINO (phthalimide-N-oxyl) radicals in situ, adamantane can be oxidized to adamantanol with 52% yield [59]. Classical reagents, for example H_2O_2/O_2, lead to a mixture of adamantanol and adamantanal (Table 7, entry 55) [60]. Another approach to selective photochemical oxidation of cyclic alkanes is the oxidation of cyclohexane (Table 7, entry 56) using zeolite NaY and leading to conversion of 40% [61]. Furthermore, NHPI may be used in the nitration of alkanes. Functionalizations with the $NHPI/NO_2/O_2$ system are very effective in catalytic selective nitration of alkanes (Table 7, entry 57) under mild conditions [62]. Direct photochemical-induced imination of short-chain alkanes in the gas phase has been developed using alkane/ammonia mixtures in the presence of mercury vapor (Table 7, entry 58), leading to a mixture of methylenimine and higher imines [63]. Selective halogenation of alkanes may be achieved by using more complex halogen compounds (Table 7, entry 59). The reaction of cyclohexane with iodoform (HCI_3) in the presence of a base and a phase-transfer catalyst leads to the formation of iodocyclohexane [64]. Fluorination (Table 7, entry 60) is usually the most difficult halogenation reaction, because high exothermicity often makes the reaction uncontrollable in conventional reactors [68]. The reaction of ethyl 2-chloroacetoacetate in a microreactor resulted in conversion of 59% and a yield of 74%, these values are significantly higher than for conventional reactors [65].

This example shows another trend in the field of fine chemicals – the use of micro technology, especially for reactions that must be performed under extreme conditions because of thermodynamic limitations. Another example of reactions under extreme conditions carried out using micro reaction technology is the Andrussov synthesis. This is the reaction of methane in the presence of ammonia and oxygen/nitrogen to form HCN that is performed at temperatures above 1000 °C (see also Section 2.2) [66]. In addition to the examples mentioned, micro reaction technology can be used to overcome obstacles such as exothermicity and temperature control in general, narrow explosion limits, highly reactive reactants, and influence of residence time distribution. Scale-up is easy in parallel systems, and modular designs can be used. These advantages will lead to broad application of this emerging technology in the synthesis of chemicals.

References

1 K. Weissermel, H.-J. Arpe, *Industrial Organic Chemistry*, 4th edn, Wiley–VCH, Weinheim, 1978–2003
2 *Ullmann's Encyclopedia of Industrial Chemistry*, 6th edn, Wiley, New York, **2003**
3 R.E. Kirk, D.F. Othmer, *Encyclopedia of Chemical Technology*, 4th edn, Wiley, New York, **1998**
4 Montecatini, GB 1000480, **1962**, *Acetylene Production*
5 BASF, DE 879989, **1953**, H. Klein, H. Sachsse, W. Hofditz, F. Haubach, *Verfahren zur Gewinnung von Acetylen*
6 BASF, DRP 296 866, **1912**
7 L. Andrussow, *Angew. Chem.* **1935**, *48*(37), 593–604
8 Degussa, DE 959364, F. Endter, **1954**, *Vorrichtung zur Durchführung einer endothermen Gasphasenreaktion*
9 Shawinigan Chemicals Ltd, US 3097921, D.J. Kennedy and N.B. Shine, **1963**, *Production of Hydrogen Cyanide*
10 M.S. Scurrell, *Appl. Catal.* **1987**, *32*, 1
11 Monsanto Co., J.T. Wrobelski, US 4456764, **1982**, *Process for the production of maleic anhydride*
12 Du Pont, DE 197 42 935, **1998**, R.M. Contractor, H.S. Horowitz, J.D. Sullivan, *Verbessertes Verfahren zur Kalzinierung/Aktivierung eines V-P-O-Katalysators*
13 Hoechst, GB 1 400 855, **1975**, *Process for the manufacture of chloromethanes by thermal chlorination*
14 Chemische Werke Hüls, A. Schmidt, J. Tiesler, DE 1 188 571, **1965**, *Verfahren zur Chlorierung von Kohlenwasserstoffen bzw. Chlorkohlenwasserstoffen*
15 Shell, US 2 403 772, **1946**; Shell, US 2 845 461, **1958**
16 I.M. Towbin, D.M. Boljanowskii, *Maslo. Zhir. Promst.* **1966**, *32*, 29
17 F. Püschel, *Tenside Deterg.* **1971**, *11*, 147
18 Aventis, DE 19905613, A. Hagemeyer und K. Kühlein, **2000**, *Photoinitiierung der Sulfochlorierung oder Sulfoxidation von Alkanen*, and references cited therein
19 Du Pont, US 3530185, **1970**, K. Pugi, *Oxidation Process*
20 Toyo Rayon, DE 1150975, **1960**, A. Yoshikazo Ito, *Verfahren zur Herstellung von Nitrosoverbindungen oder Oximen*
21 Y. Wang, K. Ostsuka, *J. Chem. Soc. Chem. Commun.* **1994**, 2209
22 M. Baerns, G. Grubert, E.V. Kondratenko, D. Linke, U. Rodemerck, DGMK Tagungsbericht 2002-4, 61–78, *Alkanes as substitutes for alkenes in the manufacture of petrochemicals – a continuing challenge of the present and future*
23 O.V. Buyevskaya, M. Baerns, *Catalysis*, **2002**, *16*, 155
24 R.K. Grasselli, *Catal. Today*, **1999**, *49*, 141
25 R.K. Grasselli, G. Centi, F. Trifiro, *Appl. Catal.*, **1990**, *57*, 149
26 Mitsubishi, EP 0 318 295, **1989**, A. Kayo, T. Umezawa, K. Kiyono, I. Sawaki, *Process for the production of nitriles*
27 K.P. DeJong, J.W. Geus, *Catal. Rev.–Sci. Eng.*, **2000**, *42*(4), 481
28 Showa Denko, JP 62078125, **1987**, *Carbon fibre mfr. in vapour phase – by thermally decomposing organic transition metal cpd., feeding into zone at 1000–1300 deg C and adding another organic cpd. Gas*
29 Hyperion Catalysis International, US 4663230, H.G. Tennent, **1986**, *Carbon Fibrils, Method for Producing Same, and Compositions Containing Same*
30 G. Centi, R.K. Grasselli, E. Patane und F. Trifiro, *Stud. Surf. Sci. Catal.*, 55, Ed. G. Centi and F. Trifiro, **1990**, Elsevier Science Publishers, Amsterdam, *Synthesis of Acrylonitrile from Propane over V-Sb-based Mixed Oxides*
31 Standard Oil Company, US Patent US 4,746,641, **1988**, A.T. Guttmann, R.K. Grasselli, J.F Brazdil, *Ammoxidation of paraffins and catalyst therefore*
32 Mitsubishi Chemical Corporation, EP 529 853 B1, **1989**, A. Kayo, T. Umezawa, K. Kiyono, I. Sawaki, *Process for producing nitriles*
33 BP Chemicals Limited, US 4,761,511, M.T. Barlow, **1988**, *Crystalline galloaluminosilicates, steam-modified crystalline gal-*

loaluminosilicates, their preparation and their use as catalysts and catalyst supports
34 Consortium für Elektrochemie, DE 1 049 845, W. Hafner, J. Smidt, R. Jira, R- Rüttinger, J. Sedlmeier, **1957**, *Verfahren zur Herstellung von Carbonylverbindungen*
35 J.L. Seone, P. Boutry, R. Montarnal, *J. Catal.* **1980**, *63*, 182–190
36 Showa Denko, DE 694 12 326 T2, **1999**, T. Suzuki, Y. Toshiro, K. Abe, K. Sano *Verfahren zur Herstellung von Essigsäure*
37 Union Carbide Corporation, EP 2161264 A1, **1986** R.M. Manyik, J.L. Brockwell, J.E. Kendall, *Process for the oxydehydrogenation of ethane*
38 Nippon Kayaku, US 3454630, **1969**, G. Yamaguchi, S. Takenaka, *Process for the oxidation of olefins to aldehydes and acids*
39 Sumitomo Chemical Co., EP 0630879, **1994**, Y. Nagaoka, Y. Nomura, K. Nagai, *Process for the production of an unsaturated carboxylic acid*
40 Mitsubishi Petrochemical Co., EP 239071, **1992**, K. Sarumaru, E. Yamamoto, T. Saito, *Production of composite oxidation catalyst*
41 W.M. Weigert, *Chem. Eng.*, **1973**, *80*, 15, 68
42 Allied, US 2481826, **1947**, J.N. Cosby
43 Standard Oil of Ohio, US 2904580, **1957** J.D. Idol, Jr
44 Du Pont, DE-OS 2 046 007, **1970** O.K. Wayne, J.L. Hatten, *Chlorierung von Butadien*
45 Distillers Co., DE 1 090 652, **1957**, C.W. Capp, H.P. Croker, F.J. Bellinger, *Verfahren zur Herstellung von Dichlorbutenen*
46 Boreskov Institute of Catalysis, WO 95/27650 Panov, G.I., Karithonov, A.S., Shelevea, G.A., **1995**, *Method for producing phenol and its derivatives*
47 Solutia Inc., 6,156,938, Sobolev, V.I., Rodkin, M.A., Urlarte, A.K, Panov, G.I., **2000**, *Process for making phenol or its derivatives*
48 V.I. Sobolev, K.A. Dubkov, O.V. Panna, G.I. Panov, Catalysis Today (**1995**), 24, 251–252, *Selective oxidation of methane to methanol on a FeZSM-5 surface*
49 Wang F, Xu J, Liao S.J., *Chem. Commun.* (Camb). **2002** Mar 21;(6):626–7, *One-step heterogeneously catalytic oxidation of o-cresol by oxygen to salicylaldehyde*
50 A.E. Shilov, G.B. Shul'pin, *Chem. Rev.*, **1997**, 2879
51 D. Baudry, M. Ephritikine, H. Felkin, J. Zakrewski, *J. Chem. Soc., Chem. Commun.*, **1982**, 1235
52 M.J. Burk, R.H. Crabtree, *J. Am. Chem. Soc.* **1987**, *109*, 8025
53 H. Chen, J.F. Hartwig, *Angew. Chem., Int. Ed.* **1999**, *38*, 3391
54 R.H. Crabtree, *Pure Appl. Chem.* **1995**, *67*, 39
55 R.H. Crabtree, *J. Chem. Soc., Dalton Trans.* **2001**, 2437
56 G. Heimann, P. Warneck, *J. Phys. Chem.* **1992**, *96*, 8403
57 K. Tabata, Y. Teng, Y. Yamaguchi, H. Sakurai, E. Suzuki, *J. Phys. Chem.* **2000**, *104*, 2648
58 S.M. Aschmann, A.A. Chew, J. Arey, R. Atkinson, *J. Phys. Chem.* **1997**, *101*, 8042
59 Y. Ishii, S. Kato, T. Iwahama, S. Sakaguchi, *Tetrahedron Lett.* **1996**, *37*, 4993
60 G.B. Shul'pin, Y.N. Kozlov, G.V. Nizova, G. Süss-Fink, S. Stanislas, A. Kitaygorodskij, A.S. Kulikova, *J. Chem. Soc. Perkin, Trans. 2*, **2001**, 1351
61 H. Sun, F. Blatter, H. Frei, *J. Am. Chem. Soc.* **1996**, *118*, 6873
62 S. Sakaguchi, Y. Nishiwaki, T. Kitamura, Y. Ishii, *Angew. Chem. Int. Ed. Engl.* **2001**, *40*, 222
63 P. Krajnik, D. Michos, R.H. Crabtree, *New J. Chem.* **1993**, *17*, 805
64 P.R. Schreiner, O. Lauenstein, E.D. Butova, A.A. Fokin, *Angew. Chem. Int. Ed. Engl.* **1999**, *38*, 2786
65 R.D. Chambers, D. Holling, R.C.H. Spink, G. Sandford, *Lab Chip*, **2001**, *1*, 132
66 V. Hessel, W. Ehrfeld, K. Golbig, C. Hofmann, S. Jungwirth, H. Löwe, T. Richter, M. Storz, A. Wolf, O. Wörz, J. Breysee, in, Proc. 3rd Int. Conf. Microreaction Technology, **1999**, Springer, Berlin, 151
67 Y. Ishii, S. Sakaguchi, T. Iwahama, *Adv. Synth. Catal.* **2001**, *343*, 393
68 K. Jähnisch, M. Baerns, V. Hessel, *Angew. Chem. Int. Ed.* **2004**, *43*, 406

II
C–H Transformation at sp-Hybridized Carbon Atoms

Handbook of C–H Transformations. Gerald Dyker (Ed.)
Copyright © 2005 WILEY-VCH Verlag GmbH & Co. KGaA, Weinheim
ISBN: 3-527-31074-6

1
C–H Transformation at Terminal Alkynes

1.1
Recent Developments in Enantioselective Addition of Terminal Alkynes to Aldehydes

Tobias Ritter and Erick M. Carreira

1.1.1
Introduction

As hydrocarbons, terminal acetylenes enjoy a rich reaction chemistry [1]. This is in no small part because of a unique feature of terminal acetylenes that differentiates them from other hydrocarbons – the acidity of the terminal proton ($pK_a = 25$). It is suggested that the lability of the terminal C–H towards deprotonation results its being bound to an sp-hybridized carbon [2]. This characteristic has been recognized for some time and has led to a diversity of methods for generation of metal acetylides which can participate in coupling reactions.

Alkynylides typically employed in C=O addition reactions are commonly prepared by acetylene deprotonation with strong bases (e.g. BuLi, RMgBr, LDA) [3]. Because aldehydes can undergo nucleophilic additions or are themselves subject to deprotonation under such conditions, generation of metal acetylides is conducted as a separate step before introduction of the electrophilic aldehyde coupling partner. The ability of terminal acetylenes to undergo metalation in the presence of Ag(I) or Cu(I) salts has been recognized for some time [4] and forms the basis of useful processes, for example the Sonogashira reaction. Acetylides derived from Ag(I) and Cu(I), however, do not readily participate in C=O addition reactions. Recent studies have expanded the conditions that can be used to metalate acetylenes to include other metals, in particular Zn(II). This has enabled the development of novel methods for the synthesis of propargylic alcohols. This chapter will cover the latest developments in acetylene activation and deprotonation to alkynylides which have been used in asymmetric aldehyde addition reactions.

Handbook of C–H Transformations. Gerald Dyker (Ed.)
Copyright © 2005 WILEY-VCH Verlag GmbH & Co. KGaA, Weinheim
ISBN: 3-527-31074-6

1.1.2
Background

In 1900 Favorski reported the KOH-mediated addition of acetylene to aldehydes and ketones [5]. Acetylene and KOH have been observed to form a 1:1 complex in liquid ammonia to give a solid. The infra-red spectrum of this solid (in Nujol or KBr disks) gave no signals for the C–C and C–H resonances. The complex was also shown to evolve acetylene when heated at 110 °C for 4 h. Several observations were made in the course of these studies that reveal some intriguing nuances of the process. For example, use of an alkali metal acetylide generated by reaction of K or Na and acetylene proved inferior in catalytic additions of acetylene and acetone, compared with additions employing the alkali metal hydroxides [6]. In this process the use of highly polar solvents, for example DMSO, N-methylpyrrolidone, ammonia, and HMPA, has proven optimum. It been suggested that the high solubility of acetylene in such media and putative hydrogen bonds between solvent and the terminal acetylene are relevant. In 1996, the use of potassium *tert*-butoxide to effect the deprotonation of terminal alkynes in DMSO was reported [7]. Addition of the resulting potassium alkynylides to aldehydes and ketones subsequently was shown to afford propargylic alcohols (Eq. 1).

$$R-\!\!\equiv\!\!-H \;+\; \underset{R^1\;\;\;R^2}{\overset{O}{\|}} \quad\xrightarrow[\text{DMSO, RT, 15 h}]{10-20\text{ mol\% KO}^t\text{Bu}}\quad \underset{R^2}{\overset{OH}{R^1\!\!-\!\!\!-\!\!\!\equiv\!\!-H}} \qquad (1)$$

70 – 90% yield

Corey introduced the use of a 1:1 CsF/CsOH salt to effect the in-situ activation of trimethylsilyl alkyne and subsequent addition to aldehydes (Eq. 2) [8]. The use of catalytic quantities of CsOH to effect the addition of terminal acetylenes to ketones in DMSO/THF or THF has also been documented (Eq. 3) [9].

$$\text{cyclohexyl-CHO} + \text{Me}_3\text{Si}-\!\!\equiv\!\!-\text{Ph} \;\xrightarrow[\text{THF, -20 °C, 8 h}]{\substack{\text{CsF/CsOH}=1:1\\(1.6\text{ equiv F}^\ominus)}}\; \text{cyclohexyl-CH(OH)-}\!\!\equiv\!\!\text{-Ph} \qquad (2)$$

85% yield

$$\text{cyclohexanone} + \text{H}-\!\!\equiv\!\!-\text{R}^3 \;\xrightarrow[\substack{\text{THF/DMSO 1:1}\\23\text{ °C}}]{\text{CsOH·H}_2\text{O (30 mol\%)}}\; \text{1-(alkynyl)cyclohexan-1-ol} \qquad (3)$$

88% yield

Although application and use of acetylenic potassium and organocesium reagents have inherent practical advantages related to cost and safety, stereochemical control with these reagents in additions to C=O electrophiles has, in general, proven elusive. In contrast, the coordination chemistry of the d-block metals is considerably better understood and more facile to control. As such, the development of methods for generation of transition-metal acetylides that are sufficiently reactive to participate in C=O additions would offer advantages in the discovery and development of enantioselective methods for the preparation of propargylic alcohols from terminal acetylenes and aldehydes or ketones.

1.1.3
Enantioselective Addition of Terminal Alkenes to Aldehydes

Methods for the enantioselective addition of metal acetylides to aldehydes can be divided into two categories:
1. addition reactions with stoichiometric, preformed metalated terminal acetylenes, and
2. addition reactions with in-situ formation of the metal alkynylides in substoichiometric amounts.

Asymmetric alkynyl additions to aldehydes by prior, separate generation of the alkynylides (e.g. dialkylzinc reagents) have recently been reviewed and are a topic of current research [10]. They will not be covered in the context of this chapter. Instead, in line with the theme of this book, this chapter will focus on the metalation of terminal alkynes by activation of the terminal C–H and the use of the corresponding metal acetylides in aldehyde and ketone addition reactions.

In 1999, Carreira identified Zn(II) as a metal that, like Ag(I) and Cu(I), is capable of effecting the metalation of terminal acetylenes under mild conditions. Thus, treatment of terminal alkynes with $Zn(OTf)_2$ and NEt_3 at room temperature led to the formation of zinc alkynylides (Eq. 4). The zinc salt and the amine base work in synergy to weaken the acetylenic proton, with the acetylene undergoing complexation to the Zn(II) center and the base effecting subsequent deprotonation (Fig. 1) [11].

$$R\text{\textemdash}\!\!\!\equiv\!\!\!\text{\textemdash}H + Zn(OTf)_2 + Et_3N \rightleftharpoons R\text{\textemdash}\!\!\!\equiv\!\!\!\text{\textemdash}Zn(OTf) + Et_3NH(OTf) \qquad (4)$$

Figure 1. Proposed process for metalation of terminal acetylenes by Zn(II) and amine bases.

The proposed metalation process involving the formation of a putative zinc acetylide was substantiated by a series of infra-red spectroscopic studies (Fig. 2) [12]. The synergistic effect in the deprotonation event of the zinc salt

(Zn(OTf)$_2$) and amine base (e.g. triethylamine, Hünig's base) could be readily monitored by observation of the disappearance of the stretch corresponding to the terminal C–H bond of phenylacetylene. Consistent with a reversible metalation, reappearance of the terminal C–H stretch could be observed on stepwise addition of triflic acid. During treatment of phenylacetylene with either amine base or Zn(OTf)$_2$ alone, no evidence of deprotonation was detected.

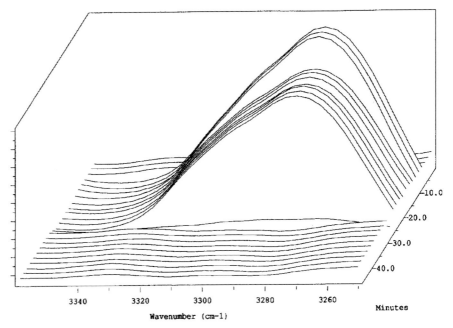

Figure 2. ReactIR spectra of the C–H stretch resonance signal of phenylacetylene in CH$_3$CN. Addition of Et$_3$N and subsequent addition of Zn(OTf)$_2$ resulted in the disappearance of the terminal C–H resonance within 4 min.

The new reactivity mode for the in-situ generation of metal alkynylides was exploited in addition reactions to aldehydes. Stoichiometric quantities of Zn(OTf)$_2$ and NEt$_3$ or Hünig's base effected deprotonation of a number of alkynes which underwent smooth addition to various aldehydes to furnish the corresponding propargylic alcohols (Eq. 5) [13]. Subsequent studies revealed that apart from Zn(OTf)$_2$ other zinc sources such as ZnCl$_2$ and ZnCO$_3$ could be used in this reaction (Eq. 6) [14].

Shortly after this initial success, the isolation of optically active propargyl alcohols in up to 99% ee could be effected by the use of stoichiometric amounts of (+)-*N*-methyl ephedrine (**1**) (Eq. 7). A wide range of aldehydes and acetylenes participate in this addition reaction affording the product alcohols in generally high yields, especially when using aldehydes that are Cα-branched (Eq. 8). Of additional importance, the reaction can be performed with functionalized alkynes, which illustrates the mildness of the metalation conditions (Eq. 9). In further expanding the scope of the addition reaction, ethyne has been shown to yield products with high enantioselectivity (Eq. 10) [15]. Alternatively, 3-hydroxy-3-methylbut-1-yn (**2**) can be used as a convenient ethyne substitute (Eq. 11). Removal of the acetone protecting group on the acetylene can be conveniently accomplished under mild conditions (Eq. 12) [16].

$$\text{R-CHO} + \text{H}\!\!\equiv\!\!\text{H} \xrightarrow[\substack{\text{toluene} \\ 23\,°\text{C, 7-14 d}}]{\substack{1.2\text{ equiv (+)-NME} \\ 1.1\text{ equiv Zn(OTf)}_2 \\ 1.2\text{ equiv EtN}^i\text{Pr}_2}} \text{R}\!-\!\underset{\text{H}}{\overset{\text{OH}}{\text{C}}}\!-\!\!\equiv\!\!\text{H} \quad (10)$$

30-92% yield
91-98% ee

$$\text{R-CHO} + \underset{\substack{\text{Me Me} \\ \mathbf{2}}}{\text{HO}\!-\!\!\equiv\!\!\text{H}} \xrightarrow[\substack{\text{toluene} \\ 23\,°\text{C}}]{\substack{1.2\text{ equiv (-)-NME} \\ 1.1\text{ equiv Zn(OTf)}_2 \\ 1.2\text{ equiv NEt}_3}} \text{R}\!-\!\underset{\text{H}}{\overset{\text{OH}}{\text{C}}}\!-\!\!\equiv\!\!-\!\underset{\text{Me Me}}{\text{C}}\!-\!\text{OH} \quad (11)$$

up to 97% yield
up to 99% ee

$$\text{R}\!-\!\underset{\text{H}}{\overset{\text{OH}}{\text{C}}}\!-\!\!\equiv\!\!-\!\underset{\text{Me Me}}{\text{C}}\!-\!\text{OH} \xrightarrow[\substack{\text{toluene} \\ \text{reflux}}]{\substack{\text{cat. 18-crown-6} \\ 1\text{ equiv K}_2\text{CO}_3}} \text{R}\!-\!\underset{\text{H}}{\overset{\text{OH}}{\text{C}}}\!-\!\!\equiv\!\!\text{H} \quad (12)$$

70-91% yield

Jiang has expanded the Carreira method of alkyne addition to aldehydes to include other ligands and Zn(II) salts (Eq. 13) [17]. Thus use of stoichiometric quantities of Zn(II) difluoromethane sulfonate salt and (1S,2S)-3-(tert-butyldimethylsilyloxy)-2-N,N-dimethylamino-1-(p-nitrophenyl)propane-1-ol (**3**) in the addition reaction can afford propargylic alcohols in high ee. Difluoromethanesulfonic acid is prepared from 3,3,4,4-tetrafluoro[1.2]oxathietane; the amino alcohol has been used in the synthesis of chloramphenicol and is also readily accessible. Application of a combination of this same amino alcohol ligand with Zn(OTf)$_2$ has also been shown to afford products in high yield and ee in addition reactions (Eq. 14) [18].

$$\text{H}\!\!\equiv\!\!\text{Ph} + {}^i\text{Pr-CHO} \xrightarrow[\substack{1.1\text{ equiv Zn(ODf)}_2 \\ 1.1\text{ equiv NEt}_3 \\ \text{toluene} \\ 25\,°\text{C, 2-12 h}}]{\substack{1.2\text{ equiv } \mathbf{3}}} {}^i\text{Pr}\!-\!\underset{}{\overset{\text{OH}}{\text{C}}}\!-\!\!\equiv\!\!\text{Ph} \quad (13)$$

99% yield, 97% ee

1.1 Recent Developments in Enantioselective Addition of Terminal Alkynes to Aldehydes | 37

$$\text{(14)}$$

A catalytic version of the Zn(II)-mediated enantioselective addition of alkynylides to aldehydes was documented after the stoichiometric process [19]. Initially, the reaction was reported to proceed using 22 mol % N-methylephedrine, 20 mol % Zn(OTf)$_2$, and 50 mol % Et$_3$N to furnish the product alcohols in yields and enantioselectivity only marginally lower than in the original stoichiometric version (Eq. 15). The key difference between the stoichiometric and the catalytic procedures is the elevated temperature (60 °C) for the catalytic process. Because the reaction can also be conducted under solvent-free conditions, ensuring a process with a high atom economy and volumetric efficiency (Eq. 16). Under these conditions, the reactions can be conducted with substantially lower catalyst loading (Eq. 17) [13].

$$\text{(15)}$$

$$\text{(16)}$$

$$\text{(17)}$$

1.1.4 Applications

The methods described have already found numerous applications in the synthesis of building blocks and natural products. β-Hydroxy ketones **4** and **5** (Scheme 1) [20] and 1,4-diols **6–8** [21] are conveniently accessed, with high selectivity, by alkynylation then hydroboration (Scheme 2).

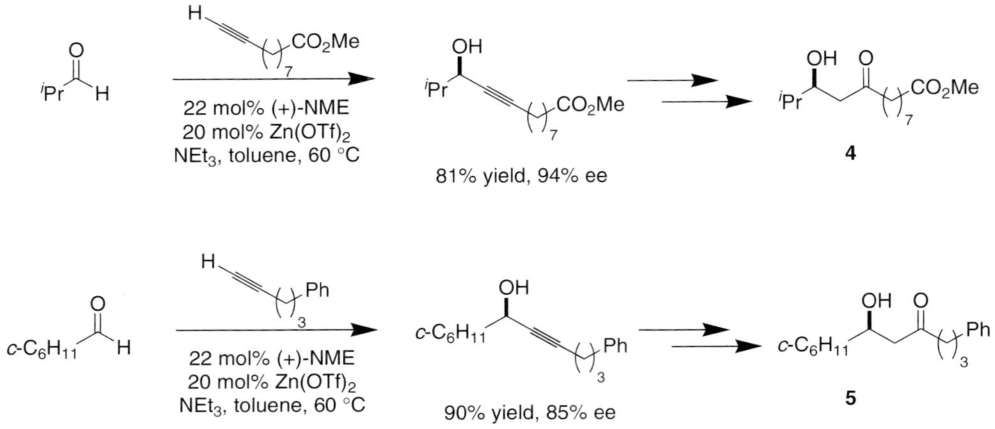

1 C–H Transformation at Terminal Alkynes

Scheme 1

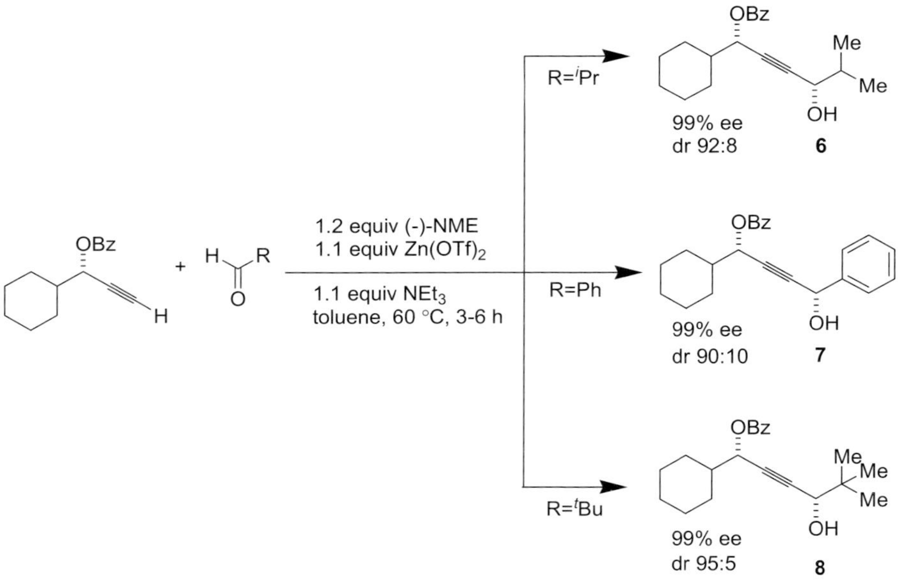

Scheme 2

Trost has exploited the asymmetric alkynylation of an α-unbranched acetaldehyde in an elegant synthesis of the core of the mitomycin analog FR900482 [22]. This is a remarkable example wherein an enolizable aldehyde participates in the addition reaction. It has been suggested that this is critical for the success of these substrates, because the aldehyde undergoes reversible enolization. (Eq. 18).

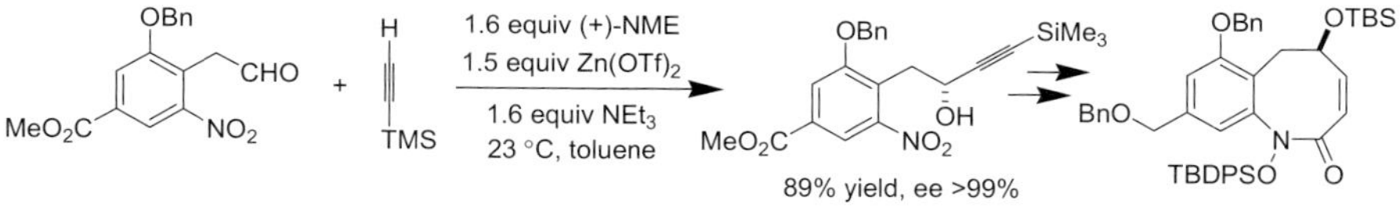

[Eq. (18) reaction scheme: aryl aldehyde with OBn, MeO₂C, NO₂ substituents + H–≡–TMS, 1.6 equiv (+)-NME, 1.5 equiv Zn(OTf)₂, 1.6 equiv NEt₃, 23 °C, toluene → propargyl alcohol product, 89% yield, ee >99% → further transformed to benzazocinone with OBn, OTBS, BnO, TBDPSO, N substituents]

McDonald performed an asymmetric synthesis of D-desosamine, with high selectivity, by diastereoselective addition of TMS-acetylene to an α-unbranched aldehyde obtaining the propargylic alcohol [23]. The reaction proceeded in nearly 100 % diastereoselectivity albeit in moderate (60 %) yield (Eq. 19).

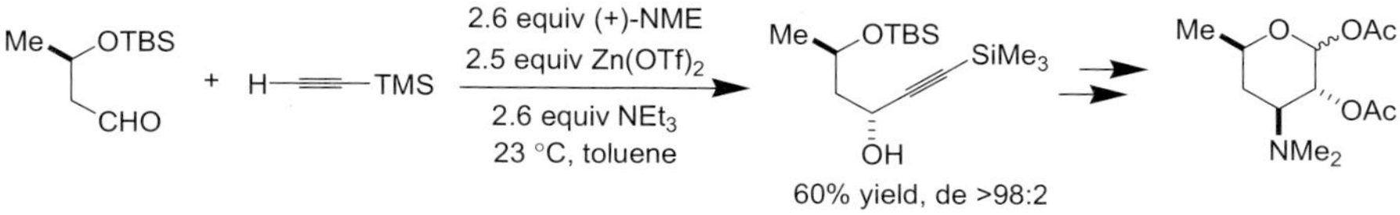

[Eq. (19): Me-CH(OTBS)-CHO + H–≡–TMS, 2.6 equiv (+)-NME, 2.5 equiv Zn(OTf)₂, 2.6 equiv NEt₃, 23 °C, toluene → propargyl alcohol, 60% yield, de >98:2 → pyranose with OAc, OAc, NMe₂ substituents]

Tanaka has shown that in his synthesis of a variety of diastereomeric annonaceous acetogenins (**9**, Scheme 3) the carbinol stereocenter can be generated with predictable selectivity by reagent control of either enantiomer of the chiral ligand [24]. He applied this method in the total synthesis of murisolin (**9**, $n=1$), a member of family of over 350 natural polyketides isolated from various annonaceaes plants (Scheme 4).

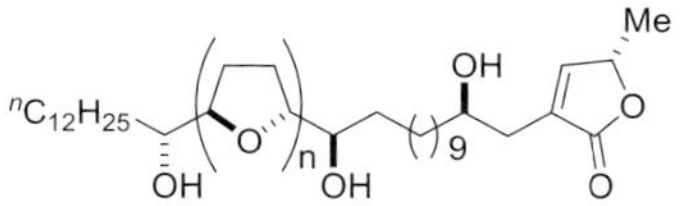

[Structure of acetogenin **9**: $^nC_{12}H_{25}$–CH(OH)–[THF ring]–(CH₂)ₙ–CH(OH)–(CH₂)₉–[butenolide with Me]; $n = 1\text{-}3$, **9** ($n=3$)]

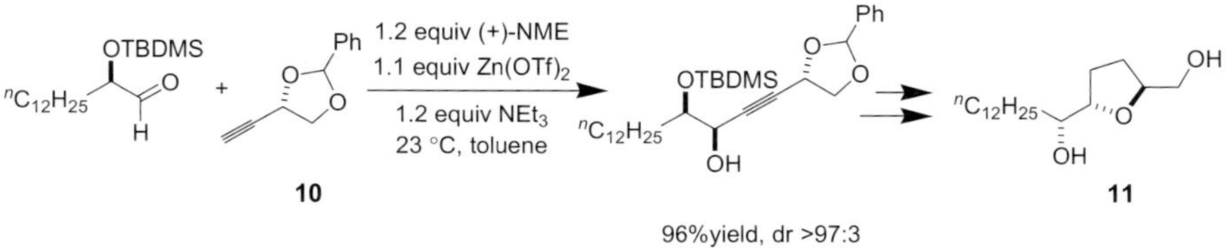

[Scheme 3: $^nC_{12}H_{25}$–CH(OTBDMS)–CHO + dioxolane-alkyne **10**, 1.2 equiv (+)-NME, 1.1 equiv Zn(OTf)₂, 1.2 equiv NEt₃, 23 °C, toluene → propargyl alcohol with OTBDMS, OH, dioxolane(Ph), 96% yield, dr >97:3 → **11**: $^nC_{12}H_{25}$–CH(OH)–[THF]–CH₂OH]

Scheme 3

[Scheme 4: **11** ($^nC_{12}H_{25}$–CH(OH)–[THF]–CHO) + **10**, (+)-NME, equiv Zn(OTf)₂, equiv NEt₃, toluene → $^nC_{12}H_{25}$–CH(OH)–[THF]–CH(OH)–C≡C–dioxolane(Ph), 97% yield, dr >97:3 → bis-THF product with $^nC_{12}H_{25}$, OH, OH substituents]

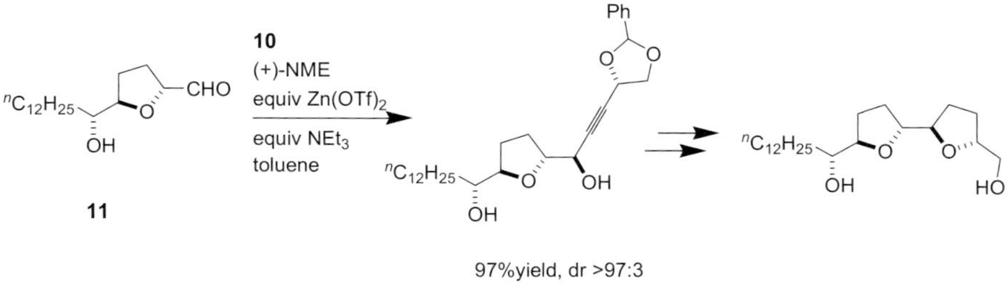

Scheme 4

Rozner published a synthesis of 3′,5′-C-branched nucleosides and used the addition of protected propargyl alcohol to an α-siloxyaldehyde to control absolute stereochemistry at C2′ [25]. He noted that use of the bulky trityl and TBDPS protecting groups were essential for the reaction. In this respect the corresponding product could be isolated in only low yield (>10 %) when changing to MOM-protected propargyl alcohol and TBS-protected oxyaldehyde (Eq. 20).

Smith has employed a related process using the ligand **3** developed by Jiang to obtain a chiral propargyl alcohol. This served as a key building block in an elegant synthesis of (-)-indolizidine 223AB (Eq. 21) [26].

Ariza took advantage of the use of acetylides for preparation of unsymmetrical 1,4-diols, adopting it to the synthesis of the natural product musclide B [27]. He also showed for a number of different aldehydes that by correct choice of the enantiomeric ligand the diastereomeric, protected diols were accessible in 87:13 to 98:2 diastereoselective ratios (Eq. 22).

In a synthesis of (+)-hyptolide Marco exploited the Carreira procedure for alkynylation of a β-silyloxy aldehyde to afford the product proparyl alcohol as a single isomer [28]. Subsequent semireduction of the alkyne furnished the requisite Z double bond to complete the synthesis (Eq. 23).

1.1 Recent Developments in Enantioselective Addition of Terminal Alkynes to Aldehydes | 41

Carreira employed this method in the context of the total syntheses of the complex natural products epothilones A (**12**) and B (**13**) [29] and leucascandrolide (**14**) (Schemes 5 and 6) [30]. The synthesis of strongylodiol B (Scheme 7) expands the scope of the process to include the activation and asymmetric addition of 1,3-diynes such as **15** [31].

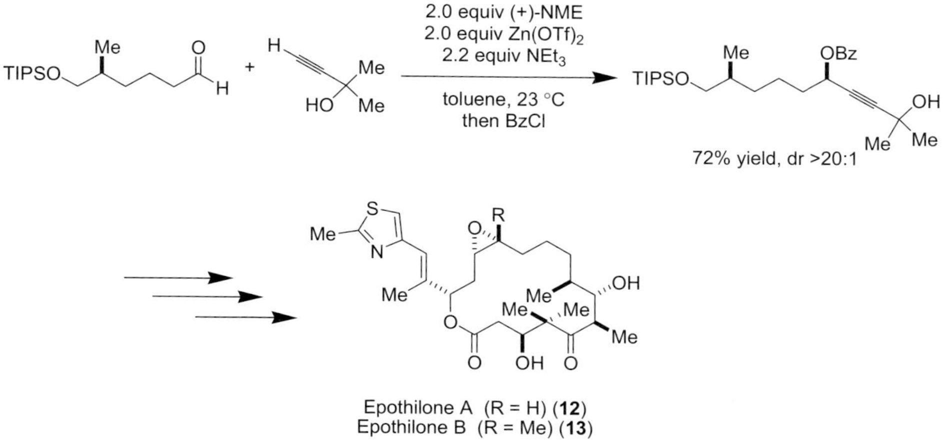

Scheme 5

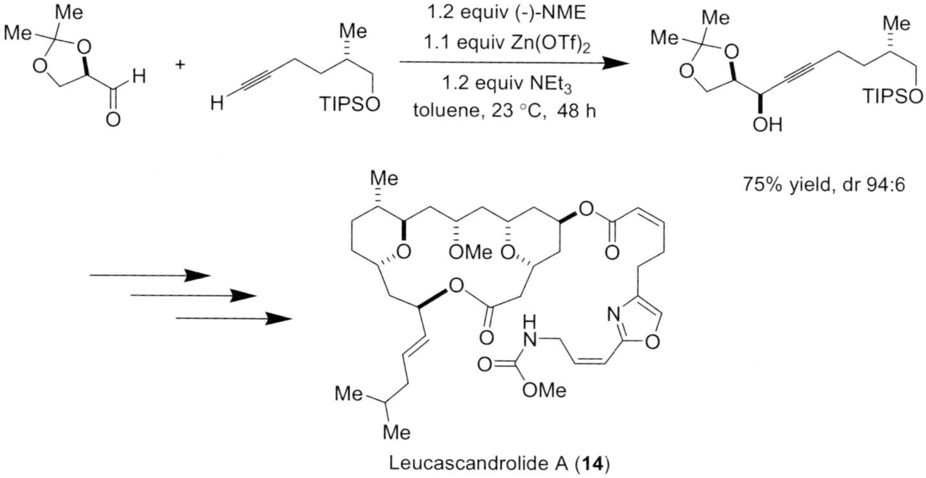

Epothilone A (R = H) (**12**)
Epothilone B (R = Me) (**13**)

Scheme 6

Leucascandrolide A (**14**)

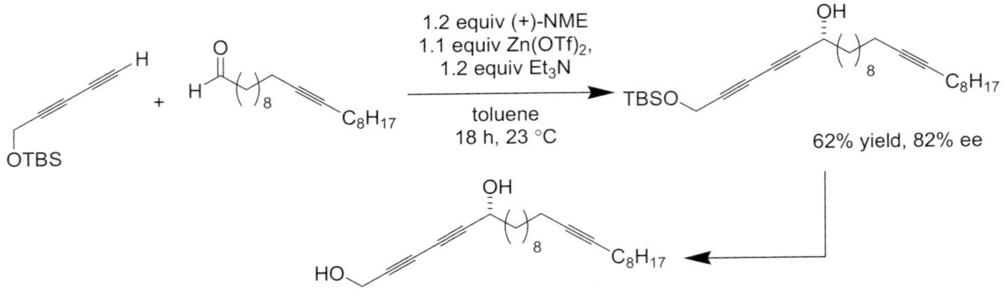

Scheme 7

1.1.5
Conclusion

The use of terminal acetylenes for the synthesis of value-added compounds has several important advantages in contrast to the practice of modern synthetic chemistry. Two of the most prominent of these are that numerous terminal acetylenes are readily available as starting materials at low cost from commercial sources and that the product propargyl alcohols are amenable to a wide range of synthetic transformations. It is thus hardly surprising that addition of acetylenes to aldehydes was examined as early as the turn of the twentieth century. The metalation of terminal acetylenes is one of the simplest C–H-activation processes that can be executed. The discovery that this can be facilitated by a variety of different metal salts under mild conditions has been a boon to the subject. That the reactivity of these can be modulated to effect ligand-accelerated enantioselective additions to C=O also offers new opportunities for enantioselective synthesis of propargylic alcohols. Indeed, the mildness and utility of this new procedure are evident from the various applications in natural product and building block syntheses that have recently appeared.

Experimental

For example 1, see: Anand, N. K.; Carreira, E. M. *J. Am. Chem. Soc.* **2001**, *123*, 9687; Fässler R. Ph.D. Thesis, Swiss Federal Institute of Technology, Diss ETH No. 14936, Zürich, 2003.

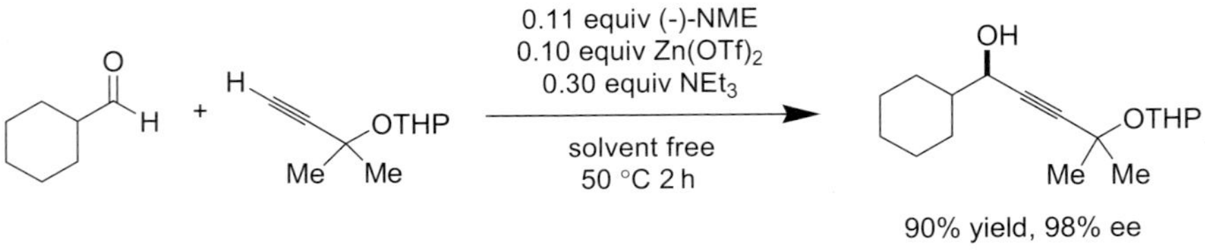

1.1 Recent Developments in Enantioselective Addition of Terminal Alkynes to Aldehydes | 43

An oven-dried *Schlenk* flask was charged with $Zn(OTf)_2$ (545 mg, 1.50 mmol, 0.10 equiv) and evacuated under high vacuum (ca. 0.1 mbar). The salt was dried for 2 h at 120 °C, with stirring. The flask was cooled to ambient temperature, (–)-NME (296 mg, 1.65 mmol, 0.110 equiv) was added, and the solids were mixed with stirring for 30 min at ambient temperature. After purging the flask with argon, 2-(tetrahydropyranyloxy)-2-methyl-3-butyne (2.78 g, 16.5 mmol, 1.10 equiv) and NEt_3 (623 µL, 4.50 mmol, 0.30 equiv) were added, and the resulting mixture was vigorously stirred for ca. 20 min, cyclohexane carboxaldehyde (1.68g, 15.0 mmol, 1.00 equiv) was added and the reaction mixture was warmed to 50 °C and stirred for 2 h at this temperature. After cooling to 23 °C, the reaction mixture was poured on to a stirred mixture of 20.0 mL sat. aq. NH_4Cl solution, 10.0 g ice, and 50.0 mL Et_2O. The phases were separated and the aqueous phase was extracted with diethyl ether (2 × 60 mL). The combined organic layers were washed with sat. aq. NH_4Cl solution and brine, dried over $MgSO_4$, filtered, and concentrated in vacuo. The residue was distilled by Kugelrohr distillation under high vacuum (0.03 mbar) to afford the propargylic alcohol in 90 % yield and 98 % ee (as determined by HPLC analysis of the 3,5-dinitrobenzoate ester; Chiralcel AD, 10 % ^{i}PrOH in hexane, 254 nm) t_r minor= 25.7 min; t_r major= 49.5 min.

$[\alpha]_D^{23}$ +5.2° (c = 1.2, $CHCl_3$); ^{1}H NMR (400 MHz, $CDCl_3$) δ 5.04 (s, 1H), 4.18 (d, J = 5.9 Hz, 1H), 3.98–3.93 (m, 1H), 3.52–3.47 (m, 1H), 2.15 (br s, 1H), 1.89–1.67 (m, 7H), 1.59–1.43 (m, 5H), 1.53 (s, 3H), 1.49 (s, 3H), 1.32–1.01 (m, 5H); ^{13}C NMR (100 MHz, $CDCl_3$) δ 96.1, 88.0, 84.0, 71.0, 67.0, 63.3, 44.2, 32.0, 30.5, 29.9, 28.6, 28.0, 26.4, 25.9, 25.8, 25.3, 20.5; FTIR (thin film) 3604, 3402, 3009, 2931, 2855, 1452, 1381, 1248, 1124, 1074, 1031, 996, 946, 868, 814 cm^{-1}; MS (EI) 280.2 ($[M]^+$, <1), 265.3 ($[M-CH_3]^+$, <1), 161.2 (11), 96.0 (33), 95.1 (30), 85.1 (100), 83.1 (28), 69.1 (56), 55.1 (32); Anal. Calcd. for $C_{17}H_{28}O_3$: C, 72.82 %; H, 10.06 %. Found: C, 72.87 %; H, 9.91 %.

For example 2, see: Trost, B. M.; Ameriks, M. K. *Org. Lett.* **2004**, *6*, 1745.

Triethylamine (492 mg, 677 µL, 4.86 mmol) was added at once to a mixture of zinc trifluromethanesulfonate (1.656g, 4.56 mmol) and (1*S*,2*R*)-(+)-*N*-methylephedrine (871 mg, 4.86 mmol) in toluene (10 mL). The resulting slurry was stirred at room temperature for 2 h. Trimethylsilylacetylene (477 mg, 687 µL, 4.86 mmol) was added at once, and the reaction mixture was stirred for an additional 45 min. The aldehyde (1.00 g, 3.04 mmol) was added at once, and the resulting red mixture was stirred at room temperature for 17.5h. The reaction was quenched with saturated aqueous NH_4Cl, diluted with water, and extracted with EtOAc (×3). The combined organic extracts were dried ($MgSO_4$), filtered, and concentrated under reduced pressure to give an orange oil. The crude product was purified by flash chromatography (SiO$_2$, 10% EtOAc in pet ether to 20% EtOAc in pet ether) to afford 1.158 g (89%) of the target compound as a clear oil. The enantiomeric purity was determined to be >99% by chiral HPLC analysis as compared with the racemic standard. Chiralcel OD at 23 °C, 1.0 mL min^{-1} flow rate, 5% iPrOH in heptane, $\lambda = 254$ nm, retention times for enantiomers: 17.54 (major), 21.26 (minor). $[\alpha]_D^{23}$ +1.21° (c 1.16, CH$_2$Cl$_2$). R_F = 0.47 (25% EtOAc in pet ether). IR (neat): 3486, 1728, 1538, 1293, 1250, 1063 cm^{-1}. ^{1}H NMR (300 MHz, CDCl$_3$): δ 8.09 (d, J = 1.5 Hz, 1H), 7.81 (d, J = 1.2 Hz, 1H), 7.41 (m, 5H), 5.22 (s, 2H), 4.70 (t, J = 6.8 Hz, 1H), 3.95 (s, 3H), 3.49 (d, J = 3.4 Hz, 1H), 3.46 (d, J = 1.7 Hz, 1H), 2.32 (d, J = 7.3 Hz, 1H), 0.13 (s, 9H). ^{13}C NMR (75 MHz, CDCl$_3$): δ 164.9, 157.6, 151.6, 135.3, 130.3, 128.9, 128.5, 127.3, 125.8, 117.8, 115.7, 105.2, 90.2, 71.3, 61.8, 52.8, 33.9, -0.304. Anal. Calcd. for C$_{22}$H$_{25}$NO$_6$Si: C, 61.81; H, 5.89; N, 3.28. Found: C, 61.59; H, 6.00; N, 3.18.

For example 3, see: Smith, A. B., III; Dae-Shik, K. *Org. Lett.* **2004**, *6*, 1493.

83% yield, >99% ee

To a solution of Zn(OTf)$_2$ (4.20 g, 11.6 mmol, 1.10 equiv) and chiral ligand (−)-3 (4.10 g, 11.6 mmol, 1.10 equiv) in toluene (75 mL) was added TEA (1.62 mL, 11.6 mmol, 1.10 equiv) at ambient temperature. After 40 min, 4-butyne (3 mL) was transferred to the reaction mixture from a cold trap (−78 °C) via a cannula. After 15 min, the aldehyde (1.04 mL, 10.5 mmol) in toluene (5 mL) was added over 5 h by means of a syringe pump and stirred for an additional 7 h. The reaction mixture was poured into saturated aqueous NH$_4$Cl (100 mL), extracted with Et$_2$O (3 × 100 mL), washed with brine (50 mL), dried over MgSO$_4$, filtered, and concentrated in vacuo. Flash chromatography on silica gel, using diethyl ether–pentane (1:7) and then ethyl acetate as eluent, provide the target compound (1.20 g, 8.70 mmol, 83% yield) as a colorless oil.

$[\alpha]_D^{23}$ −11.0° (c 0.72, CHCl$_3$); IR (film) 3440 (s, br), 3078 (w), 2977 (s), 2938 (s), 2235 (m), 1641 (m), 1064 (m), 911 (m) cm^{-1}; ^{1}H NMR (500 MHz, CDCl$_3$) d 5.84 (dddd, J = 17.0, 10.2, 6.7 and 6.6 Hz, 1H), 5.07 (dd, J = 17.1 and 1.5 Hz, 1H), 4.99 (dd, J = 10.3 and 1.4 Hz, 1H), 4.40–4.35 (m, 1H), 2.25–2.20 (m, 4H), 1.82–1.71 (m, 3H), 1.14 (t, J = 7.5 Hz, 3H); ^{13}C NMR (125 MHz, CDCl$_3$) δ 137.87, 115.07, 87.15, 80.34, 62.21, 37.22, 29.46, 13.85, 12.35; high resolution mass spectrum m/z 137.0958 [(M - H)$^+$; calcd for C$_9$H$_{13}$O: 137.0966]

1.2
The Sonogashira Coupling Reaction

Herbert Plenio and Anupama Datta

1.2.1
Introduction and Fundamental Examples

The first relatively efficient method available for bond formation between the sp-carbon of terminal alkynes and the sp^2-carbon in aryl iodides was the Stephens–Castro coupling, involving reaction of preformed copper acetylides in pyridine at elevated temperatures to generate the respective aryl acetylenes [1]. In the early 1970s when the power of palladium-catalyzed cross-coupling reactions for the formation of C–C bond started to unfold, Sonogashira, Tohda, and Hagihara demonstrated a much more useful procedure for cross-coupling of terminal acetylenes and aryl halides (Scheme 1) [2, 3]. This reaction, which is now known as the Sonogashira coupling, requires catalytic amounts of Pd and Cu complexes and can be regarded as a major progress, because it removes the difficulties faced in the preparation and handling of copper acetylides and enables a wide range of functionalized 1-alkynes and aryl halides to be coupled. This work was first published in 1975, at the same time as Heck [4] and Cassar [5] reported related processes performed in the absence of copper co-catalyst, both of which require forcing conditions and can in fact be regarded as an extension to terminal acetylenes of what is now known as the Heck reaction.

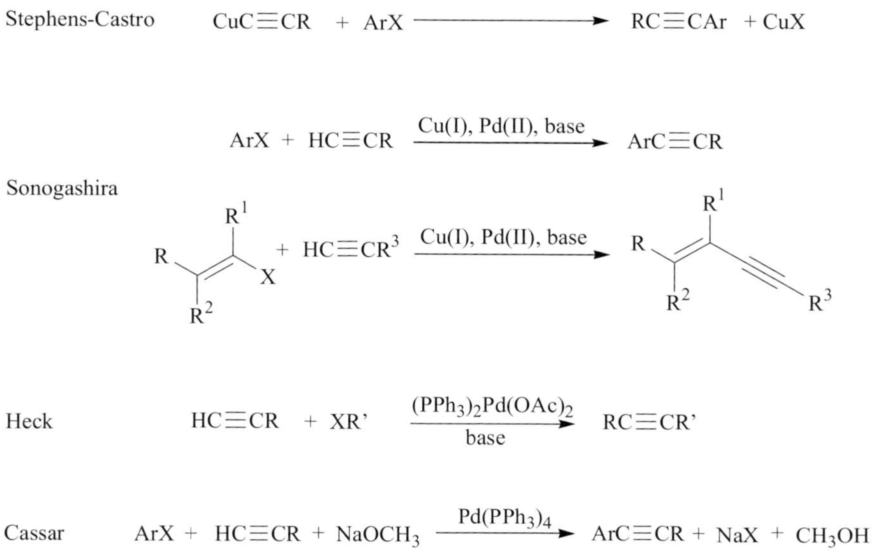

Stephens-Castro $CuC{\equiv}CR + ArX \longrightarrow RC{\equiv}CAr + CuX$

Sonogashira $ArX + HC{\equiv}CR \xrightarrow{Cu(I), Pd(II), base} ArC{\equiv}CR$

$$\underset{R^2}{\overset{R^1}{R}}\!\!\!{>}\!\!\!{=}\!\!\!{<}_X + HC{\equiv}CR^3 \xrightarrow{Cu(I), Pd(II), base} \underset{R^2}{\overset{R^1}{R}}\!\!\!{>}\!\!\!{=}\!\!\!{<}\!\!\!{-}{\equiv}{-}R^3$$

Heck $HC{\equiv}CR + XR' \xrightarrow[base]{(PPh_3)_2Pd(OAc)_2} RC{\equiv}CR'$

Cassar $ArX + HC{\equiv}CR + NaOCH_3 \xrightarrow{Pd(PPh_3)_4} ArC{\equiv}CR + NaX + CH_3OH$

Scheme 1. Alkynylation reactions.

The Sonogashira reaction has finally emerged as the most widely used method for synthesis of substituted alkynes and reflects typical properties of Pd-catalyzed cross-coupling reactions, because it is technically simple, efficient, high yielding and tolerant towards a wide variety of functional groups [6]. In their very early publications Sonogashira et al. demonstrated carbon–carbon coupling between various acetylenes and iodoarenes, bromoalkenes or bromopyridines in the presence of catalytic amounts of $(Ph_3P)_2PdCl_2$ and CuI in diethylamine, leading to the formation of internal acetylenes. Since then, numerous modifications and improvements to this original procedure have been achieved by development of facile procedures, a wider array of substrates, and more active palladium catalysts for cross-coupling with deactivated arenes [7–9].

1.2.2
Mechanism

The mechanism of the Sonogashira reaction has not yet been established clearly. This statement, made in a 2004 publication by Amatore, Jutand and co-workers, certainly holds much truth [10]. Nonetheless, the general outline of the mechanism is known, and involves a sequence of oxidative addition, transmetalation, and reductive elimination, which are common to palladium-catalyzed cross-coupling reactions [6b]. In-depth knowledge of the mechanism, however, is not yet available and, in particular, the precise role of the copper co-catalyst and the structure of the catalytically active species remain uncertain [11, 12]. The mechanism displayed in Scheme 2 includes the catalytic cycle itself, the preactivation step and the copper mediated transfer of acetylide to the Pd complex and is based on proposals already made in the early publications of Sonogashira [6b].

$(PPh_3)_2PdCl_2$, which was originally employed by Sonogashira himself, is still the most commonly used palladium source in Sonogashira coupling. Alternatively, $Pd(OAc)_2$, $(CH_3CN)_2PdCl_2$, or Na_2PdCl_4 with at least two equivalents of a tertiary phosphine can also be used to form the catalytically active species in situ. With all Pd(II) salts the initial step leading to the catalytically active species is preactivation of the catalyst, i.e. reduction of Pd(II) to Pd(0), which is normally effected by reductive elimination from the cis-diacetylide **2** to form the respective butadiyne. Obviously the use of Pd(0) sources such as $Pd_2(dba)_3$, $Pd(dba)_2$, or $Pd(PPh_3)_4$ does not require such a preactivation step. The catalytic cycle probably involves neutral low-coordinate complexes, for example $[L_nPd]$, even though an anionic $[L_nPdX]^-$ species is also considered [13]. In the past $[L_2Pd]$ was believed to be the most likely candidate for the catalytically active species [2], but more recent work from Hartwig has established that [LPd] complexes also need to be considered [14]. The formation of low-coordinate Pd(0) species is favored in the presence of electron-rich and bulky phosphines and explains why very-high-activity catalysts are often formed with such ligands. In this context, the use of Pd(0) sources may not be favorable, because ligands, for example dba, needed to stabilize Pd(0) complexes also tend to coordinate and thus inhibit the catalytically active Pd species [15], as was also shown for the acetylenes (Scheme 2) [10].

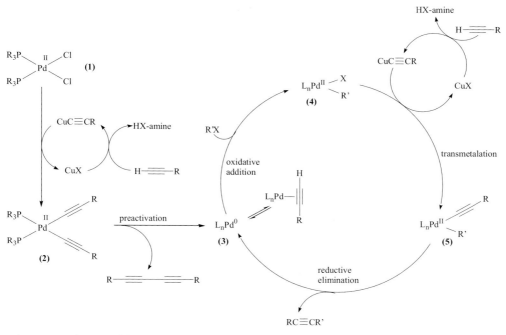

Scheme 2. Mechanism of the Sonogashira reaction for Pd/Cu-catalyzed cross-coupling of sp²-C halides with terminal acetylenes.

Subsequent oxidative addition of aryl or vinyl halides to [L$_n$Pd] results in a Pd(II) intermediate (**4**). The copper-supported alkynylation leads to the alkynylpalladium(II) derivative (**5**), which regenerates the catalytic species (**3**) by reductive elimination of the coupled products. The oxidative addition of the aryl halide to the Pd(0) species is generally regarded as the rate-determining step in reactions with aryl bromides and chlorides. The general order of the overall reactivity of organic halides is: vinyl iodide > vinyl bromide > vinyl chloride = aryl iodide >> aryl bromide >> aryl chloride [6d]. The low reactivity of aryl chlorides is attributed to the strength of the C–Cl bond (bond dissociation energies for Ph–X; X = Cl: 96 kcal mol^{-1}, X = Br: 81 kcal mol^{-1}, X = I: 65 kcal mol^{-1}) which ender them reluctant to add oxidatively to Pd(0) [16]. For a given halide X, substrates carrying electron withdrawing groups ortho or para to the aryl halide will react more readily, because the electron-deficient halides will undergo oxidative addition more easily [17].

1.2.3
Scope and Limitations

The Sonogashira cross-coupling reaction of terminal acetylenes with aryl or vinyl halides is a powerful tool for generation of carbon–carbon bonds between sp^2- and sp-carbon. Numerous molecules of interest can be generated from a wide variety of aryl or vinyl halides (Scheme 3) [18–23].

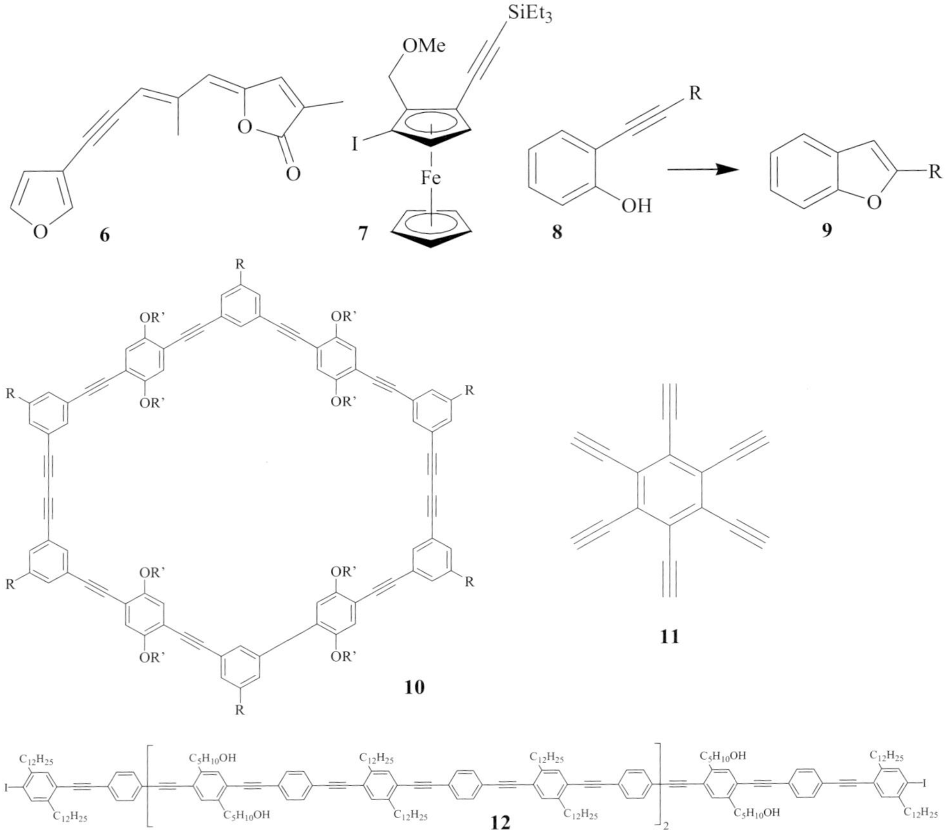

Scheme 3. Examples of molecules generated by Sonogashira coupling reactions.

In the past decade the Sonogashira reaction has found many applications in the synthesis of scaffolds leading to molecular-scale electronic devices [24], dendrimers [25], estradiols [26], enediyne antibiotics [27], carbohydrate sensors [28], liquid crystals [29], and polymers [30]. The antimitotic agent **13** [31, 32], which inhibits the proliferation of human epidermal cells, is being developed for topical treatment of hyperproliferative skin disorders. A key step is a Sonogashira cross-coupling of an aryl halide and a heteroaryl halide (at a later stage the triple bond is reduced to a C_2H_4 unit) (Scheme 4). Terbinafine **14**, which is an active ingredient of the Novartis broad-spectrum antimycotic Lamisil has been realized on an industrial scale by Sonogashira reaction [33]. An important step in the synthesis of terbinafine is the palladium-catalyzed coupling of a substituted vinyl chloride with *tert*-butylacetylene. The Sonogashira reaction is also applied in the Merck synthesis of MK-0462, a 5-HT$_{10}$ receptor agonist and potential antimigraine drug [34], in the DuPont synthesis of a DNA sequencing agent [35], and in the synthesis of a model system of dynemicin A (intermediate **15**) by Nicolaou [36, 37].

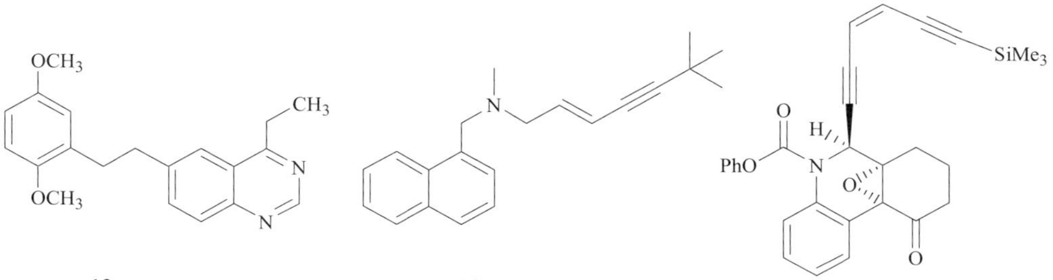

Scheme 4. Sonogashira reactions in the synthesis of 6-[2-(2,5-dimethoxyphenyl)ethyl]-4-ethylquinazoline **13**, terbinafine **14** and dynemicin A **15**.

Until the mid 1990s progress in the use of the Sonogashira coupling was mostly restricted to methods that could readily couple aryl iodides at room temperature [38, 39] and aryl bromides at elevated temperatures [40]. Aryl chlorides would be especially interesting as substrates because of their low price and the large variety of compounds which are commercially available. Hence, the development of catalysts with enhanced activity for cross-coupling reactions of aryl bromides and chlorides had to be addressed [41]. Some success has been reported using bulky, electron-rich phosphine ligands such as P^{t}Bu$_3$, by the groups of Buchwald and Fu [42] and Herrmann and Böhm [43]. These systems display excellent activity in cross-coupling reactions with aryl bromides at room temperature. Until fairly recently, however, only isolated examples were known of Sonogashira coupling of activated aryl chlorides [44, 45]. Effective cross-coupling of alkynes with aryl chlorides was reported by Eberhard et al., who used a phosphonito (PCP) palladium pincer complex **16** in the presence of ZnCl$_2$ [46]. Several aryl chlorides could be coupled with phenyl acetylene but these reactions required reaction for 24 h at 160 °C, a fivefold excess of the aryl chloride, and 10 to 100 mol% ZnCl$_2$ additive. The problems involved in the activation of aryl chlorides have essentially been solved recently through work of Plenio and Köllhofer [47] who reported a general and efficient procedure for Sonogashira coupling of both activated and deactivated aryl chlorides with terminal alkynes by use of a high-yielding catalyst based on Na$_2$[PdCl$_4$]/PR$_3$/CuI (PR$_3$ = (1-Ad)$_2$PBn or PtBu$_3$) (Scheme 5). A variety of activated and deactivated chlorides were coupled with a variety of alkynes in excellent yield in toluene or DMSO at temperatures of 100 or 120 °C. Similarly, in a significant improvement, Buchwald and Gelman for the first time demonstrated the Sonogashira coupling of aryl tosylates with terminal alkynes, using a catalyst prepared from phosphine **17** and [PdCl$_2$(CH$_3$CN)$_2$], a reaction which also works well with aryl chlorides (Scheme 6) [48].

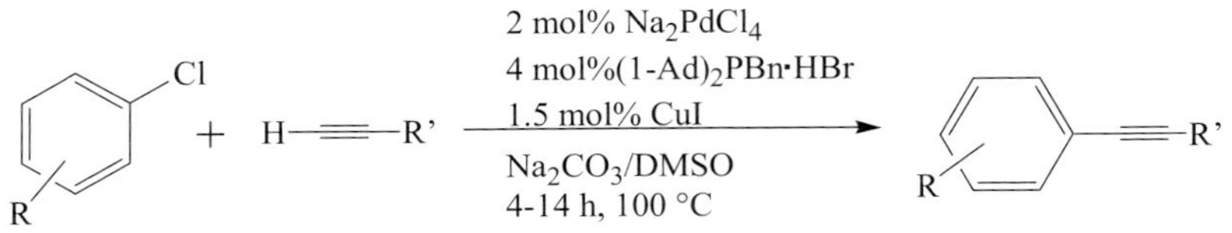

$$\text{Ar-Cl} + \text{H}{\equiv}\text{R'} \xrightarrow[\text{Na}_2\text{CO}_3/\text{DMSO}]{\substack{2\text{ mol\% Na}_2\text{PdCl}_4 \\ 4\text{ mol\%(1-Ad)}_2\text{PBn·HBr} \\ 1.5\text{ mol\% CuI} \\ 4\text{-}14\text{ h, }100\,°\text{C}}} \text{Ar}{\equiv}\text{R'}$$

Scheme 5. Palladium-catalyzed Sonogashira coupling of aryl chlorides according to Plenio and Köllhofer.

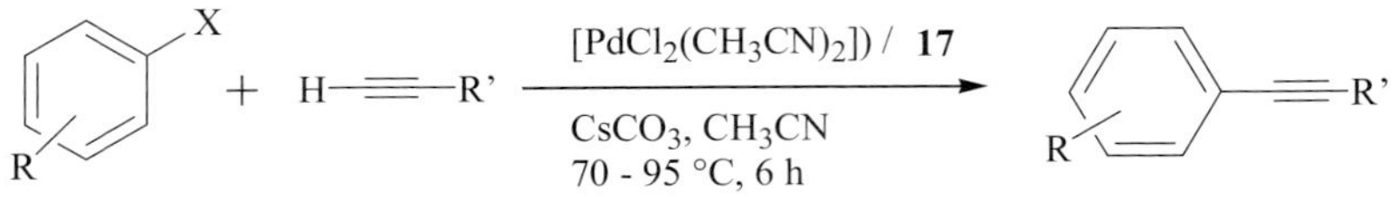

$$\text{Ar-X} + \text{H}{\equiv}\text{R'} \xrightarrow[\substack{\text{CsCO}_3,\text{ CH}_3\text{CN} \\ 70\text{-}95\,°\text{C, }6\text{ h}}]{[\text{PdCl}_2(\text{CH}_3\text{CN})_2])\,/\,\mathbf{17}} \text{Ar}{\equiv}\text{R'}$$

Scheme 6. Palladium-catalyzed Sonogashira coupling of aryl chlorides and tosylates according to Buchwald and Gelman.

Because the Sonogashira process is frequently associated with homo-coupling of the starting alkynes promoted by oxygen impurities in the presence of a copper co-catalyst, some effort has been devoted to the design of suitable catalysts for diminished alkyne dimerization. This was achieved either by using SiMe$_3$-protected terminal acetylenes, which are deprotected during the coupling reaction, or by developing suitable copper-free procedures [49–53].

As an alternative to sterically hindered phosphines in the Sonogashira reaction, N-heterocyclic carbenes (NHC) have recently received increased attention [54–56]. Hermann was first to report the use of chelating ligand complex **18** in the Sonogashira reaction [57]. The effectiveness of NHC ligands in Sonogashira reactions was improved significantly by using the imidazolium system developed by Nolan and co-workers in combination with Pd(OAc)$_2$ (Scheme 7) [58].

Initially, it was assumed that the presence of copper co-catalysts in the Sonogashira reaction is crucial for a highly efficient catalytic conversion. Nonetheless, there are alternatives and several other metal acetylides can be used instead [6a]. Blum et al. demonstrated that tetraalkynyl aluminates, prepared in situ from NaAlH$_4$ and terminal acetylenes, are useful in palladium-catalyzed cross-coupling reactions, because undesired homocoupling is avoided [59]. Fürstner [60] and Soderquist [61] introduced organoboron or trialkyloxyborate acetylides for sp–sp^2 carbon bond formation. Molander et al. [62] have recently demonstrated that cross-coupling with aryl halides and triflates can be efficiently carried out in the presence of potassium alkynyl trifluoroborates. Negishi et al. reported that the Pd-catalyzed alkynylation of alkenyl halides and triflates with zinc acetylides proceeds well even with alkynyl derivatives containing electron-withdrawing groups [63,

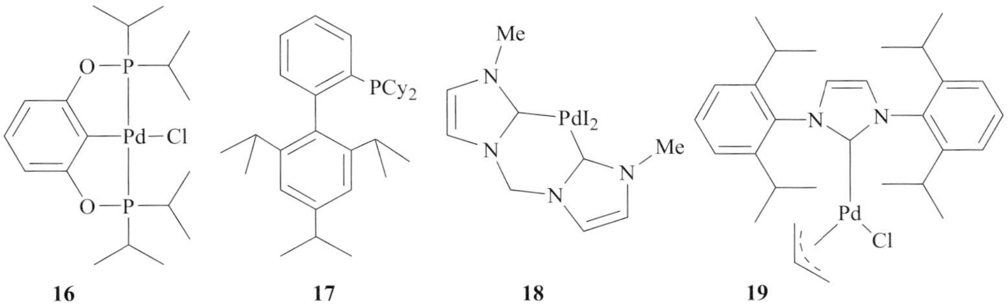

Scheme 7. Ligands and Pd complexes used for Sonogashira coupling.

64]. Wang et al. have discovered that ultrafine Ni powder in the presence of CuI, PPh$_3$, and KOH promotes coupling of terminal alkynes with aryl and alkenyl iodides in high yields [65]. Recent developments have shown, moreover, that the use of co-catalysts (Cu, Zn, Al, etc.) to facilitate the formation of the acetylides is not always required and that cross-coupling reactions of acetylenes and aryl halides can be performed successfully with Pd-based catalysts alone, even with difficult substrates [48, 66]

Microwave heating has emerged as an interesting method to speed up chemical reactions frequently delivering high yields in just a few minutes as opposed to hours with conventional heating. A microwave-enhanced Sonogashira reaction in the presence of Pd(PPh$_3$)$_2$Cl$_2$ and CuI was presented by Erdélyi and Gogoll and applied to the coupling of aryl iodides, bromides, triflates, and activated aryl chlorides with trimethylsilylacetylene [67]. Leadbeater has shown that Sonogashira-type reactions can even be performed with trace amounts of transition metals when applying microwave heating. This approach involves the use of water as solvent, poly(ethylene glycol) as phase-transfer agent, and NaOH, and enables reaction of terminal acetylenes with aryl iodides or bromides [68].

In homogeneously catalyzed reactions loss of the catalyst and contamination of the product with the catalyst metal can be a serious limitation. Developing efficient strategies for separation of the catalysts from the products is now recognized as an important subject in catalysis research [69]. One approach in this respect is the use of liquid–liquid biphasic solvent systems (biphasic catalysis). To enable this strategy a catalyst must be modified with a phase tag such that the solubility properties of this modified catalyst are determined by the phase tag whereas the catalytic properties are not affected [70, 71]. Example of catalysts with polar and nonpolar phase preferences are the Pd/PR$_3$ catalysts prepared from BnP(1-Ad)$_2$ tagged with poly(ethylene glycol) (MeOPEG$_{2000}$) **20** or with soluble poly(*p*-methylstyrene) **21** (Scheme 8). A single batch of such catalysts can be used to efficiently couple aryl bromides or aryl chlorides and acetylenes in DMSO–*n*-heptane and cyclohexane–DMSO, respectively, over at least five consecutive reactions [72, 77]. The same polymer-enlarged catalysts have also been used in nanofiltration experiments and in continuous biphasic catalysis, with the polymer-tagged catalyst immobilized in a stationary solvent, with a flow of reactants passing through it, to enable the conversion of a continuous feed of reactants into a product stream [73, 74]. The Sonogashira reaction has also been applied in a variety of other reaction media, for example aqueous phases [75], polar–nonpolar biphasic solvent mixtures [76–78], ionic liquids [79] and fluorous solvents [80] with the appropriate phase tags.

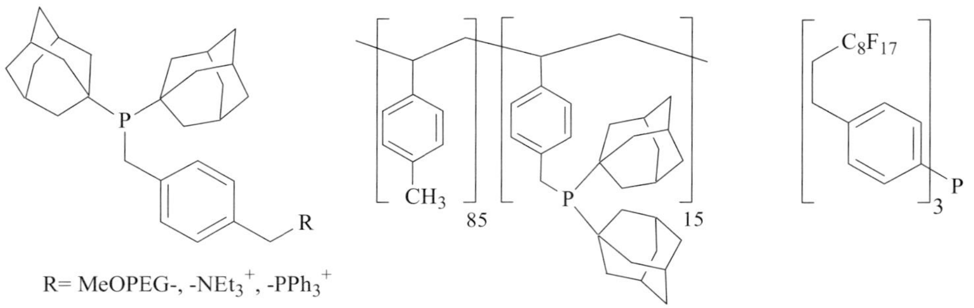

R= MeOPEG-, -NEt$_3^+$, -PPh$_3^+$

Scheme 8. Phase-tagged phosphines as ligands in Pd-catalyzed cross-coupling reactions.

Heterogeneous catalysts for Sonogashira coupling, comprising palladium complexes supported on organic resins [81, 82] or palladium deposited on inorganic supports, are also known. Most notable in this context is work from Choudary et al., who used a ligand-free heterogeneous layered double hydroxide-supporting nanopalladium catalyst which can activate aryl chlorides for Sonogashira reactions and can be reused for up to five consecutive cycles with almost constant activity [83]. A detailed mechanistic study relying on XPS and TGA-MS to elucidate the nature of surface transient organometallic intermediates, giving proof of elementary steps in the catalytic cycle, was carried out in the same group [84]. It is important to note in this respect that many supported Pd catalysts and some molecular catalysts function as a reservoir of soluble palladium nanoclusters, which promote the coupling reactions in solution and not on the support. Such nanoclusters are often redeposited on the solid support after the coupling reaction and can be used again with high activity when the morphology of the deposited material is suitable [85].

Experimental

General Procedure for Sonogashira Coupling of Alkynes and Aryl Chlorides According to Plenio and Köllhofer [47]

To a thoroughly degassed slurry of Na$_2$CO$_3$ (230 mg, 2.1 mmol) in dried toluene (3 mL) under argon atmosphere were added Na$_2$[PdCl$_4$] (8.8 mg, 30 mmol, 2 mol%), PtBu$_3$ (12 mg, 60 mmol), CuI (4.4 mg, 23 mmol, 1.5 mol%), the chloroarene (1.5 mmol), and the alkyne (2 mmol). The mixture was heated to the respective reaction temperature (typically 100 °C) with vigorous stirring. After completion of the reaction (GC monitoring) and cooling to room temperature, the volatiles were removed in vacuo. The crude products are purified by chromatography on silica.

General Procedure for Coupling of Aryl Chlorides with Alkyl Acetylene According to Buchwald and Gelman [48]

An oven-dried Schlenk tube was evacuated and back-filled with argon (the cycle was performed twice) and then charged under a positive pressure of argon with [PdCl$_2$(CH$_3$CN)$_2$] (1.2 mg , 4.62 mmol, 1 mol%), phosphine **17** (6.6 mg, 14 mmol, 3 mol%), Cs$_2$CO$_3$ (391 mg, 1.20 mmol), followed by anhydrous acetonitrile (0.924 mL) and the aryl chloride (0.462 mmol). The slightly yellow suspension was stirred for 25 min. The alkyne (0.6 mmol) was then injected, the Schlenk tube was sealed with a Teflon valve, and the reaction mixture was stirred at the desired temperature for the indicated period of time. The resulting suspension was then left to reach room temperature, diluted with water (3 mL), and extracted with diethyl ether (4×4 mL). The combined organic layers were dried over MgSO$_4$, concentrated, and the residue was purified by flash chromatography on silica gel to provide the desired product.

General Procedure for Coupling of Aryl Tosylates According to Buchwald and Gelman [48]

An oven-dried two-necked flask, equipped with reflux condenser, gas inlet/outlet, and rubber stopper, was evacuated and backfilled with argon (the cycle was performed twice) and then charged under a positive pressure of argon with [PdCl$_2$(CH$_3$CN)$_2$] (7.7 mg, 29.6 mmol, 5 mol%), phosphine **17** (42.4 mg, 89 mmol, 15 mol%), Cs$_2$CO$_3$ (0.87 g, 2.66 mmol), followed by propionitrile (1.8 mL) and the aryl tosylate (0.59 mmol). The slightly yellow suspension was stirred for 25 min at room temperature. (Efficient stirring of the reaction mixture and high purity of the starting tosylate are important for the transformation to be successful.) The reaction mixture was then heated to reflux and the alkyne (0.88 mmol diluted with 1 mL propionitrile) was injected slowly over the course of reaction (8 h) by means of a syringe pump. The reaction mixture was stirred for a further 2 h after addition was complete and the resulting suspension was left to reach room temperature, diluted with water (3 mL), and extracted with diethyl ether (4×4 mL). The combined organic layers were dried over MgSO$_4$, concentrated, and the residue was purified by flash chromatography on silica gel to provide the desired product.

1.3
Glaser Homocoupling and the Cadiot–Chodkiewicz Heterocoupling Reaction

Peter Siemsen and Beatrice Felber

1.3.1
Introduction and Fundamental Examples

Homocoupling of terminal acetylenes first described by Glaser in 1869 [1] occurs in presence of a base, a copper(I) salt (usually CuCl) and dioxygen (Scheme 1a). Because of the widespread occurrence of di- and polyacetylenic structures in natural products and the rapidly growing interest of material sciences in conjugated oligo- and polyacetylenes, the reaction has been investigated intensively and gained far-reaching applicability [2]. Besides progress achieved in the field of oxidative homocoupling processes, Chodkiewicz and Cadiot, in 1957, presented a nonoxidative procedure for achieving efficient copper-mediated heterocoupling of terminal alkynes with 1-haloalkynes in the presence of amines (Scheme 1b) [3].

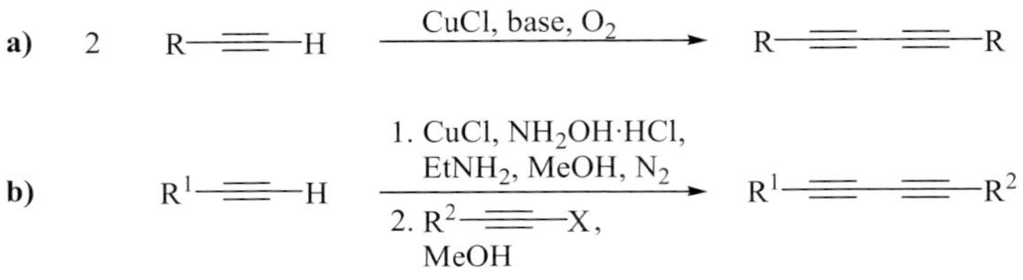
a)
2
R—≡—H
CuCl, base, O2
R—≡—≡—R
b)
R1—≡—H
1. CuCl, NH2OH·HCl,
EtNH2, MeOH, N2
2. R2—≡—X,
MeOH
R1—≡—≡—R2

Scheme 1. General description of (a) the oxidative Glaser-type homocoupling reaction and (b) nonoxidative Cadiot–Chodkiewicz heterocoupling (X = Br, I).

Of various homocoupling procedures derived from the original Glaser process that are still used today, Hay's procedure [4] using catalytic amounts of the bidentate complexing base TMEDA in polar solvents (e.g. acetone, dichloromethane or *o*-dichlorobenzene) is most utilized and is the method favored for preparing linear oligo- and polyacetylenes, as outlined in Scheme 2 [5].

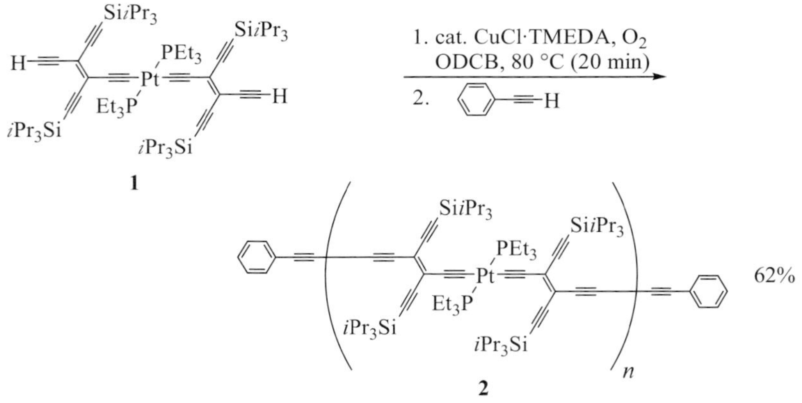
SiiPr3
PEt3
Pt
Et3P
iPr3Si
H
1
1. cat. CuCl·TMEDA, O2
ODCB, 80 °C (20 min)
2.
2
n
62%

Scheme 2. Preparation of polyacetylene **2** ($n \approx 32$) using the coupling procedure described by Hay. (ODCB: *o*-dichlorobenzene, TMEDA: *N*,*N*,*N*′,*N*′-tetramethylethylenediamine.)

Although the Hay coupling has also been successfully used in the synthesis of cyclic oligoacetylenes (shown in the Section 1.3.3, Scheme 8) [6], for extended, macrocyclic oligoacetylenes the modification presented by Eglinton and Galbraith [7a] is quite often the method of choice [2]. Here copper(II) acetate must be used in large excess, most commonly dissolved in pyridine or mixtures of pyridine and methanol (the cosolvent avoids precipitation of copper(I) reaction intermediates [7b]). Annulene syntheses developed by Sondheimer and co-workers [8] are a good example of the use of this procedure, particularly applicable to cyclizations (Scheme 3).

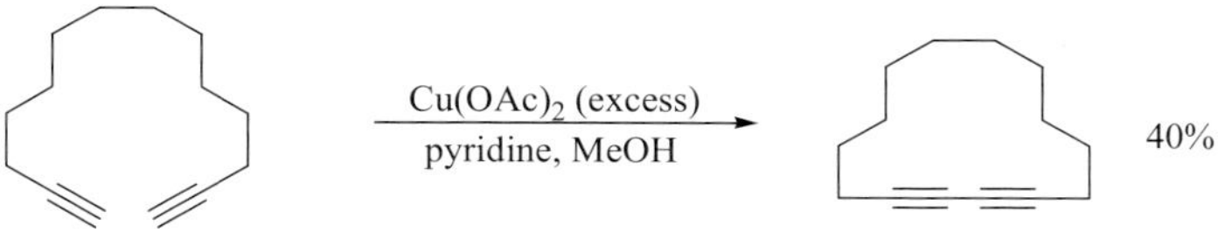

Scheme 3. Intramolecular acetylenic coupling in annulene synthesis described by Sondheimer [8] according to the method of Eglinton and Galbraith [7].

Whereas Glaser-type oxidative coupling opens efficient synthetic pathways toward symmetrical diynes, its performance in heterocoupling is poor. The latter may be accomplished by Cadiot–Chodkiewicz coupling of terminal alkynes with 1-haloalkynes (usually 1-bromoalkynes). The reaction is conducted in the presence of an amine and catalytic amounts of a copper(I) salt. Because, in contrast with the Glaser-type reactions described above, it follows a nonoxidative reaction mechanism, oxygen is not necessary – but needs often not to be excluded (Scheme 4) [9].

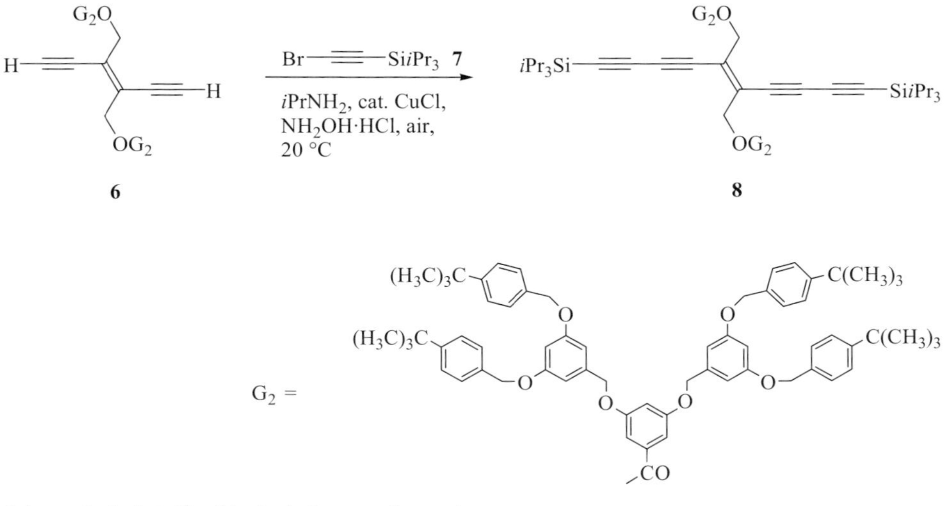

Scheme 4. Cadiot–Chodkiewicz heterocoupling under access of air. Reaction time 1 h; yield of heterocoupling product 64 % [9].

1.3.2
Mechanism

Although acetylenic homo- and heterocouplings have been widely used in different fields of organic synthesis, their exact mechanism is still obscure. Although several mechanistic hypotheses have been postulated for Glaser-type couplings, copper-catalyzed Cadiot–Chodkiewicz heterocoupling reactions have been little studied – largely because of the difficulty of kinetic studies, owing to the rapid reaction rates observed for the bromoalkynes commonly employed [2].

Because all currently known mechanisms of oxidative acetylenic homocouplings are very specific to single reaction conditions, e.g. pH or oxidation state of the used copper salt, this section summarizes the most reasonable mechanistic ideas proposed for the commonly utilized coupling procedures.

1.3.2.1 Oxidative Homocoupling

Studies by different research groups indicate that both Cu^+ and Cu^{2+} ions are involved in the coupling process [2]. Whereas copper(II) serves as the direct oxidizing agent, the role of copper(I) seems more versatile. Kinetic studies of acetylenic couplings in buffered solutions (Et_3N, HOAc, pyridine) in the presence of copper(I) and copper(II) salts showed copper(I) ions to play no role in the reaction, except that of an intermediate electron carrier [10a, b]. For couplings with mixtures of CuCl and $CuCl_2$ in non-buffered pyridine, however, it was found that copper(I) is indeed directly involved in the oxidative coupling process [10c]. The occurrence of higher-order copper(I)–copper(II)–ethynyl complexes in the rate-limiting step of the reaction has been proposed [10e], but not yet proven.

A very reasonable role of copper(I) in the coupling process seems to be intermediate formation of non-reactive copper–π-complexes. Coordination of cuprous ions activates the corresponding alkyne units toward deprotonation (Scheme 5).

Scheme 5. Acetylene activation by π-complex formation between Cu^+ and the triple bond, postulated by Bohlmann et al. [10f].

This activation process can be assumed to be the initial step in the formation of dinuclear copper(II) acetylide complexes, as first proposed by Bohlmann and coworkers 40 years ago (Scheme 6) [10f]. Deprotonated alkyne units **11** (or the corresponding π-complexes **10**) generated therein, stepwise displace the negatively charged counter ions of copper(II) salt dimers (**12**). The dinuclear copper(II) acetylide complex which finally results (**14**) collapses to the coupled product under reductive elimination of copper(I). The existence of higher-order copper acetylide complexes according to **13** and **14** has recently been demonstrated in the solid state [10g], whereas their presence in solution has not yet been proven.

It must be emphasized that current mechanistic understanding of copper-mediated oxidative acetylenic couplings is unsatisfactory. Several studies have shown the strong dependency of the mechanism on the experimental setup, suggesting highly complex coherences and interactions. Nevertheless, the mechanistic idea of Bohlmann et al. described above still provides the most accepted picture for Glaser-type oxidative acetylenic homocouplings.

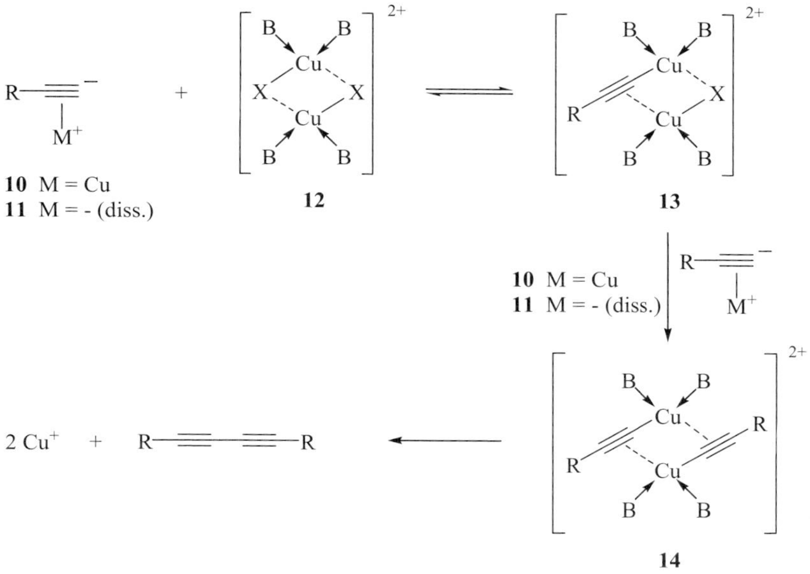

Scheme 6. The mechanistic proposal of Bohlmann et al. shows dimeric copper acetylides as key intermediates in copper-mediated acetylenic coupling (B = N-ligand, e.g. pyridine; X = Cl$^-$, OAc$^-$) [10f].

1.3.2.2 Nonoxidative Heterocoupling

As already mentioned, there have been few mechanistic examinations of the copper-catalyzed Cadiot–Chodkiewicz heterocoupling reaction. Kinetic studies with the less reactive chloroalkynes [11a] have led to the assumption, shown in Scheme 7, that coupling between alkynes and haloalkynes proceeds through initial formation of copper(I) acetylides, probably formed by an acetylenic activation process similar to that described above for oxidative homocouplings. Subsequently, two reaction pathways may be reasonable:

1. Oxidative addition of the copper(I) acetylide to the alkynyl halide with formation of a copper(III) intermediate (**15**), giving the corresponding diacetylene by reductive elimination [11a].
2. Nucleophilic addition without change of the oxidation state, similar to couplings of alkynylmagnesium halides [11b].

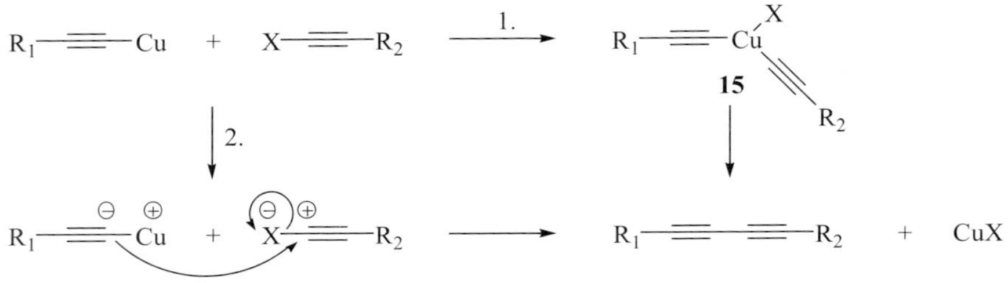

1 C–H Transformation at Terminal Alkynes

$$R_1-\!\!\!\equiv\!\!\!-Cu \;+\; X-\!\!\!\equiv\!\!\!-R_2 \xrightarrow{1.} R_1-\!\!\!\equiv\!\!\!-Cu\!\!\begin{array}{c}X\\ \diagdown\\ \diagdown\\ R_2\end{array}$$

15

$$\downarrow 2.$$

$$R_1-\!\!\!\equiv\!\!\!-Cu^{\ominus\oplus} \;+\; X^{\ominus\oplus}\!\!-\!\!\!\equiv\!\!\!-R_2 \longrightarrow R_1-\!\!\!\equiv\!\!\!\equiv\!\!\!-R_2 \;+\; CuX$$

Scheme 7. Mechanisms assumed for the Cadiot–Chodkiewicz heterocoupling (X = Cl, Br) [11].

Altogether, mechanistic understanding of both coupling processes, oxidative and non-oxidative, still requires much work. Improvements here may be promising for development of new coupling methods, enriching the already wide scope of acetylenic coupling reactions.

1.3.3
Scope and Limitations

Glaser-type homocouplings and Cadiot–Chodkiewicz heterocouplings have been applied to the synthesis of numerous aliphatic and aromatic diynes, as highlighted in a recent review [2]. Both methods are highly tolerant of a large variety of functional groups. Their synthetic capacity has been impressively demonstrated in the synthesis of the conjugated organic oligomers and polymers, presented in this section.

1.3.3.1 Oxidative Homocouplings of Tetraethynylethene Derivatives

Controlled Glaser–Hay coupling of platinum-bridged tetraethynylethene complex **1** rendered the synthesis of mondisperse, soluble oligomers up to the hexamer with a length of 12 nm, and a soluble polymer (**2**) with an average of approximately 32 repeat units and a remarkably narrow molecular weight distribution [5, 12]. Apart from the concentration of starting material, yield and ease of isolation of the single coupling products are strongly dependent on reaction conditions, as solvent, temperature, and time of addition of the end-capping agent phenylacetylene (Table 1).

The high coupling performance, apparent from total yields of isolable coupling products of 60–70 %, has also been found for the dimerization of tetraethinylethane derivative **17** to the cyclic diastereoisomers *syn*-**18** and *anti*-**18** (Scheme 8) [6].

Despite the high turnover rates of Glaser-type couplings, their applicability to the synthesis of macro-molecular structures is limited by two points:

1. Because of the complex mechanism of these reactions, not yet sufficiently understood, directed synthesis of single monodisperse oligomers by end-capping polymerization of bis-deprotected acetylenes is not yet very effi-

cient. Mixtures of oligomers are usually obtained and the isolation of these will always be limited by the performance of current separation techniques.

2. Formation of macrocyclic di- and oligoacetylenes in acceptable yields is rendered more difficult, because of the poor selectivity of Glaser-type couplings.

Table 1. Oligomerization and polymerization of compound **1** by Glaser–Hay coupling. Catalyst formation: CuCl, TMEDA, and O_2 in 1,2-dichlorobenzene [5, 12]. The reaction was performed in the presence of molecular sieves (4 Å). If PhC≡CH was available immediately at beginning of the coupling process (procedure C) an equal amount was added again 1 h before the end of the reaction. (t_r is the total reaction time and t_{add} is the time until addition of the end-capping reagent.)

	Solvent	T [°C]	t_r [h]	t_{add} [min]	n
(A)	o-$C_6H_4Cl_2$	80	2.3	20	~32 (**2**, 62%)
(B)	o-$C_6H_4Cl_2$	80	2	10	1 (**16a**, 17%), 2 (**16b**, 12%), 3 (**16c**, 20%), 4 (**16d**, 13%), 5 (**16e**, 4%), 6 (**16f**, 1%)
(C)	CH_2Cl_2	20	12	0	1 (**16a**, 56%), 2 (**16b**, 9%), 3 (**16c**, 4%)

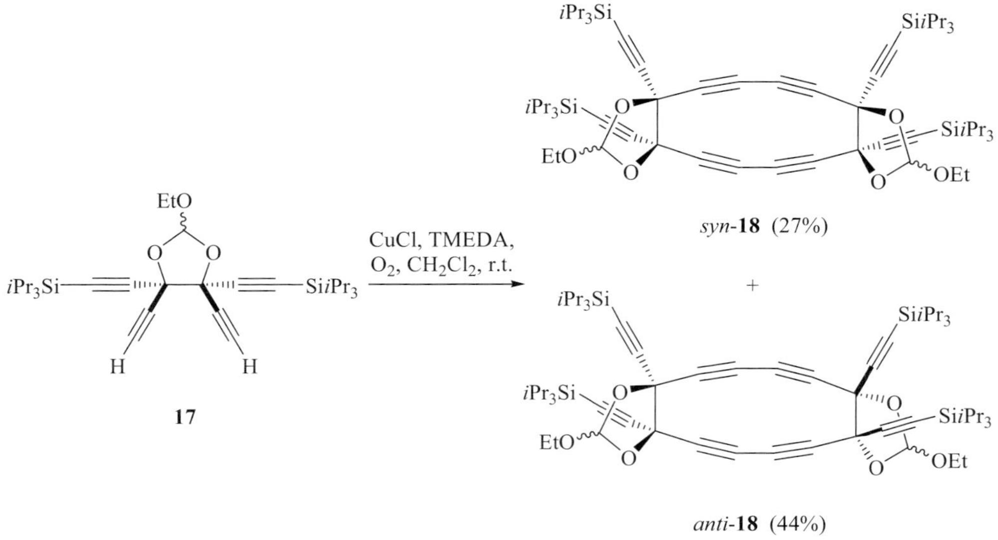

Scheme 8. Glaser–Hay dimerization of tetraethinylethane-1,3-dioxolane **17** yields the cyclic diastereoisomers **18** in 71 % [6].

1.3.3.2 Nonoxidative Heterocoupling of Terminal Alkynes with Haloalkynes: Cadiot–Chodkiewicz Reaction

By applying Cadiot–Chodkiewicz conditions, diethynylethene **6**, shown in Scheme 4, can be coupled with the bromoalkyne **7** to afford the highly conjugated product **8** in 64 % yield [9]. In this type of heterocoupling reaction terminal aromatic and conjugated alkynes usually give higher yields than aliphatic acetylenes. Bromoalkynes are usually used as coupling partners, because of the poor reactivity of the chloro derivatives and the strongly oxidizing character of iodoacetylenes. The latter favors self-coupling of the haloalkyne, initiated by oxidation of copper(I) to copper(II) and simultaneous halogen–metal exchange (Scheme 9) [11a]. To a small extent this concurrent homocoupling reaction also occurs with bromoacetylenes, and thereby considerably reduces the selectivity of the heterocoupling method – especially for more reactive substrates.

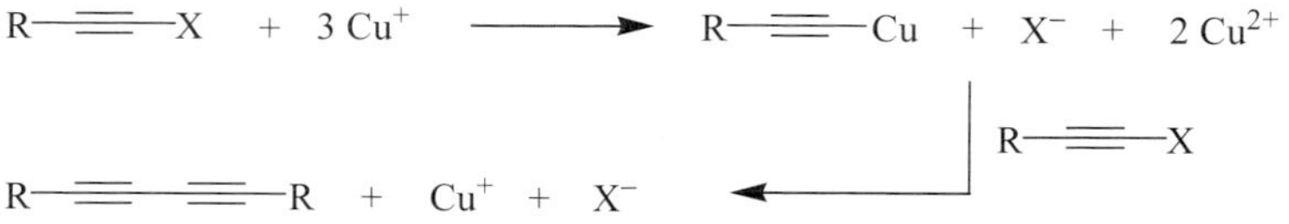

Scheme 9. Side-reaction observed in Cadiot–Chodkiewicz couplings (X = I, Br) [11a].

Where applicable, this disturbing process can be suppressed by addition of amines and employing copper(I) and haloalkyne in reduced concentrations [11a, 13]. Thus, heterocouplings are frequently complicated for acetylenes of poor stability.

Although palladium-catalyzed processes have also achieved improvements in the field of acetylenic coupling [2], the Glaser–Hay and Cadiot–Chodkiewicz copper-mediated methods still remain the most used. Further mechanistic investigations seem to be necessary, to overcome the limitations of the existing procedures, mainly poor selectivity and predictability.

Experimental

Oxidative homocoupling: Glaser–Hay polymerization of compound 1 [5] –
α,ω-Bis[phenylethynyl]poly[trans-bis{(Z)-4-ethynyl-6-(triisopropylsilyl)-3-[(triisopropylsilyl)ethynyl]-hex-3-ene-1,5-diynyl}bis(triethylphosphane)platinum(II)] (2)

A mixture of **1** (100 mg, 77 µmol) and molecular sieves (4 Å) in 1,2-dichlorobenzene was heated to 80 °C. Under a stream of O_2 a solution of Hay catalyst (prepared by stirring CuCl (8.4 mg, 85 µmol) and TMEDA in 1,2-dichlorobenzene (2 mL) under a stream of O_2 for 30 min at 20 °C) was added. After stirring for 20 min at 80 °C, PhC≡CH (176 µL, 1.56 mmol) was added and the mixture stirred for 2 h at that temperature. PhMe (50 mL) was added, and the mixture was extracted with saturated aqueous NH_4Cl solution (2 × 25 mL) and saturated aqueous NaCl solution (50 mL). Filtration over silica gel and evaporation gave a red-brown precipitate which was purified by GPC (Bio-Beads S-X1, PhMe[12]) to give **2** (62 mg, ~62 %) as a solid containing traces of 1,4-diphenylbuta-1,3-diyne which could not be removed by means of three GPC separations. ^{1}H NMR (CDCl$_3$, 500 MHz): δ = 0.83–1.35 (br s), 1.91–2.14 (br m), 7.27–7.44 (m), 7.46–7.49 (m); IR (CHCl$_3$): ν = 3011 (w), 2944 (m), 2867 (m), 2356 (w), 2322 (w), 2067 (s), 1600 (s), 1489 (m), 1461 (m), 1183 (m), 1017 (w), 917 (m), 883 (w), 700 (s), 689 (s); UV–visible (CHCl$_3$): λ_{max} (ε) = 489 (sh, 90 100), 446 (127 300), 423 (105 900), 305 (sh, 77 300), 294 (81 400). (ε calculated for polymer with 32 monomeric repeat units.)

Cadiot–Chodkiewicz heterocoupling: coupling of compounds 6 and 7 [9] –
[(E)-1,10-Bis(triisopropylsilyl)dec-5-ene-1,3,7,9-tetrayne-5,6-diyl]dimethylene Bis{3,5-bis[3,5-bis{[4-(tert-butyl)benzyl]oxy}benzyl)oxy]benzoate} (8)

A solution of **6** (376 mg, 0.18 mmol) and **7** (198 mg, 0.75 mmol) in THF (8 mL) was cooled to 0 °C. Propan-2-amine (2 mL), CuCl (0.2 g, 0.20 mmol), and $NH_2OH \cdot HCl$ (0.02 g, 0.28 mmol) were added and the mixture was stirred for 1 h at 20 °C in the air. Saturated aqueous NH_4Cl solution was added, and the organic phase was extracted with CH_2Cl_2. Column chromatography (SiO$_2$, hexane–CH$_2$Cl$_2$ 1:1 → 1:2) afforded **8** (293 mg, 64 %) as a pale yellow solid with m.p. 84 °C. ^{1}H NMR (CDCl$_3$, 200 MHz): δ = 1.04 (s, 42 H), 1.34 (s, 72 H), 5.02 (s, 24 H), 5.29 (s, 4 H), 6.61 (t, J = 2.3 Hz, 4 H), 6.72 (d, J = 2.3 Hz, 8 H), 6.82 - 6.83 (m, 2 H), 7.35 - 7.46 (m, 36 H); ^{13}C NMR (CDCl$_3$, 50.3 MHz): δ = 11.16, 18.41, 31.23, 34.47, 64.40, 69.95, 70.21, 88.55, 89.41, 95.29, 101.70, 106.43, 107.63, 108.55, 125.56, 127.63,

129.70, 131.63, 133.79, 138.77, 151.13, 159.88, 160.42, 165.88; MALDI–TOF-MS (9-nitroanthracene): 2451 ([M + Na]$^+$). Anal. calc. for $C_{160}H_{192}O_{16}Si_2 \cdot H_2O$ (2445.49): C 78.58, H 8.00, found: C 78.49, H 7.83.

1.4
Dimerization of Terminal Alkynes

Emilio Bustelo and Pierre H. Dixneuf

1.4.1
Introduction and fundamental examples

One of the most challenging topics in organic synthesis is the development of new catalytic systems which efficiently promote the selective formation of carbon–carbon bonds and combine several simple molecules into one useful product only. The simple catalytic dimerization of alkynes is atom-economic and straightforward method for synthesis of a set of versatile functional products. Among these, the preparation of *conjugated enynes* by direct coupling of two terminal alkynes enables easy access to important building blocks in organic synthesis. In particular, the regio- and stereoselective head-to-head dimerization of 1-alkynes is of special importance, because the resulting 1,3-enynes are key units found, for example, in a variety of naturally occurring antibiotics [1, 2]. Furthermore, the dimerization of acetylene itself, catalyzed by an alkynylcopper derivative, enables industrial access to but-1-en-3-yne and to neoprene rubber [3].

1.4.1.1 Simple Dimerization of Alkynes
The synthetic application of the direct coupling of two acetylene units was initially limited by the formation of mixtures of regio (head-to-head and head-to-tail) and stereo (*E*/*Z*) isomers, and preference for trimerization processes [4]. During recent years, however, increased regio and stereochemical control has been achieved by the appropriate choice of catalyst. For example, conjugated polymers with enriched (*E*), (*Z*), or *gem*-vinylene linkages are selectively obtained by the use of suitable Pd, Ru, or Rh catalysts, respectively [5].

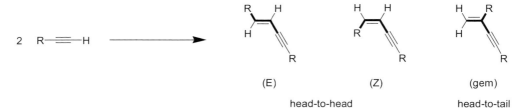

Scheme 1. Linear and branched enyne isomers from dimerization of alkynes.

The catalytic dimerization of alkynes has led to the development of a variety of catalysts by new combinations of transition metals and ligands, and to a better understanding of the processes and mechanism involved, leading to improvement of selectivity and scope. In Table 1 the most relevant catalysts are compared with regard to phenylacetylene dimerization. The nature of the terminal alkyne has also a marked effect on the outcome of this reaction.

Table 1. Selectivity of phenylacetylene dimerization by some transition metal catalysts.

Catalyst[a]	Yield	(E)	(Z)	(gem)
Pd(OAc)$_2$/TDMPP [4, 6]	62%	–	–	100
Pd(OAc)$_2$/IMes·HCl/Cs$_2$CO$_3$ [7]	98%	97	3	–
[(π-allyl)PdCl]$_2$/TDMPP/Et$_2$NH [8]	70%	100	–	–
(PP$_3$)Ru(C≡CPh)$_2$/NH$_4$PF$_6$ [9]	80%	5	84	–
Cp*Ru(PCy$_3$)H$_3$ [10]	86%	10	90	–
Cp*Ru(PMe$_3$)H$_3$ [10]	82%	90	10	–
TpRuCl(PPh$_3$)$_2$ [11]	98%	91	6	–
RuCl$_2$(P^{i}Pr$_3$)$_2$(=C=CHPh) [5]	>99%	1	96	3
[Rh(PMe$_3$)$_2$Cl]$_2$ [12]	>99%	–	–	100
[Me$_2$Si(C$_5$Me$_5$)(NAr)Lu(μ-CCR)]$_2$ [13]	>99%	–	>99	–

[a] TDMPP = P[(2,6-OMe)$_2$C$_6$H$_3$]$_3$, IMes = 1,3-dimesitylimidazolinium, PP$_3$ = P(CH$_2$CH$_2$PPh$_2$)$_3$, Tp = trispyrazolylborate, Cp* = C$_5$Me$_5$

Palladium catalysts enable the dimerization of functional alkynes with selective head-to-tail coupling in the presence of bulky phosphine ligands, for both homo-coupling of terminal alkynes or cross-coupling of mono and disubstituted alkynes [4, 6]. On the other hand, a palladium/imidazolium system gives linear (E)-enynes as the predominant products [7].

Mononuclear ruthenium complexes have become useful catalysts, not only because they can have high regio- and stereoselectivity but also because their catalyzed reactions rely on an elucidated mechanism. This true for the *cis*-dihydride (PP$_3$)RuH$_2$ complex, a catalyst precursor for the selective head-to-head dimerization of phenylacetylene to the corresponding (Z)-enyne, via bis(alkynyl) active species [9]. Ruthenium hydride complexes such as Cp*Ru(PR$_3$)H$_3$ are also versatile catalysts, for which selectivity has been achieved by modulating both the catalyst and the alkyne substrates [10]. On the other hand, TpRuCl(PPh$_3$)$_2$ [11] and Cp*RuCl(=C=CHPh)(PPh$_3$) [14] catalysts lead to the E isomer of 1,3-enynes. A Z-selective head-to-head dimerization of aliphatic alkynes by thiolate-bridged diruthenium complexes has also been reported for aliphatic and functionalized alkynes [15].

The *rhodium–trimethylphosphine system* is remarkable because it catalyzes the dimerization of aryl substituted acetylenes, yielding the scarce branched head-to-tail coupling product [12]. The *iridium* species generated from [Ir(COD)Cl]$_2$ and phosphine selectively yields linear (E) or (Z) enynes from silylalkynes, depending on whether triaryl or tripropylphosphines, respectively, are used [16].

Lanthanide metallocene compounds are also active catalysts for the dimerization of terminal alkynes, giving predominantly the linear head-to-head enyne dimer with a double bond of E configuration [17]. In recent years, however, novel organolanthanide [13] and organoactinide [18, 19] systems have shown their ability to produce, with high selectivity, Z and geminal enyne products, respectively.

1.4.1.1.1 Mechanism

The dimerization of 1-alkynes to enynes by transition metal catalysts occurs either via alkynyl–vinylidene M(C≡CR)(=C=CHR) or alkynyl–alkyne M(C≡CR)(η^2-HC≡CR) coupling, by insertion into the σ Ru–C bond. Selectivity control depends on the previous orientation of the alkyne/vinylidene moiety.

In the alkyne dimerization catalyzed by *palladium systems*, all proposed mechanisms account for an alkynyl/alkyne intermediate with *cis* addition of the alkynyl C–Pd bond to the alkyne in a Markovnikov fashion, in which the palladium is placed at the less-substituted carbon, both to minimize steric hindrance and to provide the most stable C–Pd bond (Scheme 2a). The reverse regioselectivity in the palladium-catalyzed dimerization of aryl acetylenes has been attributed to an agostic interaction between the transition metal and ortho protons of the aromatic ring in the substrate (Scheme 2b) [7, 8].

For *ruthenium catalysts* a detailed study of [(PP$_3$)RuH$_2$] proposes a bis(alkynyl) complex as the real catalyst. The catalytic key step involves the protonation of an alkynyl ligand by external PhC≡CH, allowing subsequent C–C bond formation between *cis* vinylidene and alkynyl groups [9].

Cp*Ru [14] and TpRu [20] complexes have also been studied in depth. As represented in Scheme 2c, the catalytic alkyne dimerization proceeds via coordinatively unsaturated ruthenium alkynyl species. Either a direct alkyne insertion and/or previous vinylidene formation are feasible pathways that determine the selectivity. The head-to-tail dimer cannot be formed by the vinylidene mechanism, whereas the E or Z stereochemistry is controlled by the nature of the alkynyl–vinylidene coupling.

It is noteworthy that alkyne dimerization may also lead to butatriene derivatives RCH=C=C=CHR. This product was first found for *t*-butylacetylene with

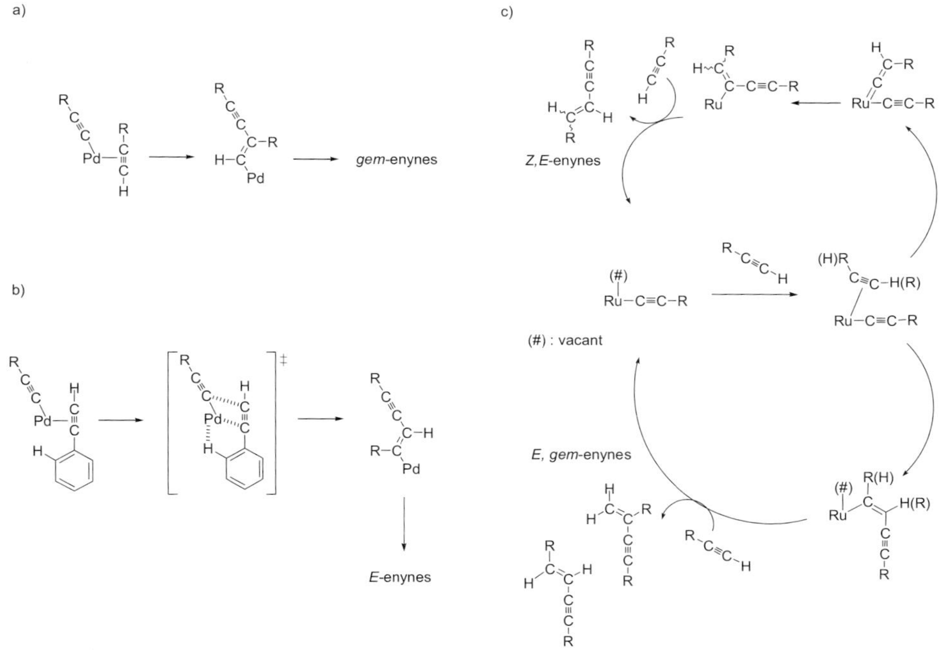

Scheme 2. Pd and Ru-catalyzed alkyne dimerization mechanisms.

[Ru(COD)(COT)]/P^{i}Pr$_3$ [21], but also for benzylacetylene, and with other catalysts such as [Cp*RuH$_3$(PCy$_3$)]/PCy$_3$ [10] and [Ir(COD)Cl]$_2$/P^{n}Pr$_3$ [16]. The [RuH$_2$(CO)(PPh$_3$)$_3$] precatalyst with 3-molar excess of P^{i}Pr$_3$ has the best activity (90 turnover) and selectivity (96 %) for (Z)-1,4-di-t-butylbutatriene [21].

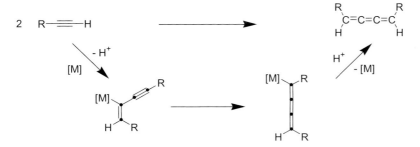

Scheme 3. Catalytic formation of the butatriene isomer.

This process is enhanced by bulky phosphines, thus butatriene formation seems to require a sterically congested environment. Otherwise the 1,3-enyne becomes the major product. Steric interaction between the phosphine ligand and the butenynyl group seems to be the dominant factor promoting 1,3-metal migration to form (Z)-cumulenes (Scheme 3) [21].

1.4.1.2 Dimerization of Alkynes and Propargyl Alcohols into Functional Dienes or Cyclobutenes

The stoichiometric head-to-head oxidative coupling of alkynes with CpRuBr(COD) affords a metallacyclic biscarbene complex [22]. This process has been used to initiate catalytic formation of the RCH=CH-CH=C(Y)R backbone, to produce *functional dienes* from alkynes by addition of H–Y. The complex [Cp*RuCl(COD)] successfully catalyzes this new chemical transformation, involving the combination of two molecules of alkynes and one molecule of carboxylic acid to afford functional conjugated dienes (Scheme 4) [23].

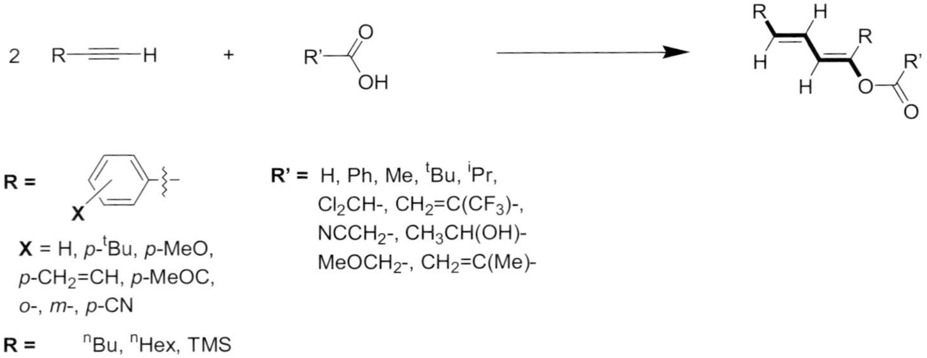

$$2\ R\text{---}\!\!\equiv\!\!\text{---}H\ +\ R'\text{-COOH}\ \longrightarrow\ \text{diene}$$

R = aryl-X

X = H, p-^{t}Bu, p-MeO, p-CH$_2$=CH, p-MeOC, o-, m-, p-CN

R = ^{n}Bu, ^{n}Hex, TMS

R' = H, Ph, Me, ^{t}Bu, ^{i}Pr, Cl$_2$CH-, CH$_2$=C(CF$_3$)-, NCCH$_2$-, CH$_3$CH(OH)-, MeOCH$_2$-, CH$_2$=C(Me)-

Scheme 4. Functional dienes from alkyne/carboxylic acid coupling.

It occurs with stereoselective formal addition of proton and carboxylate at C_1 and C_4 carbon atoms with concomitant C–C, C–H, and C–O bond formation, giving (1E,3E)-1,3-dienyl acetate as the only stereoisomer. This general reaction proceeds for a variety of alkynes and carboxylic acids and is favored by electron-withdrawing groups on arylacetylenes.

On the other hand, when propargyl alcohols are treated with the same catalyst precursor [Cp*RuCl(COD)] in the presence of a carboxylic acid, the catalytic reaction takes a different route, leading to the selective formation of alkenylidenecyclobutenes (Scheme 5) [24, 25].

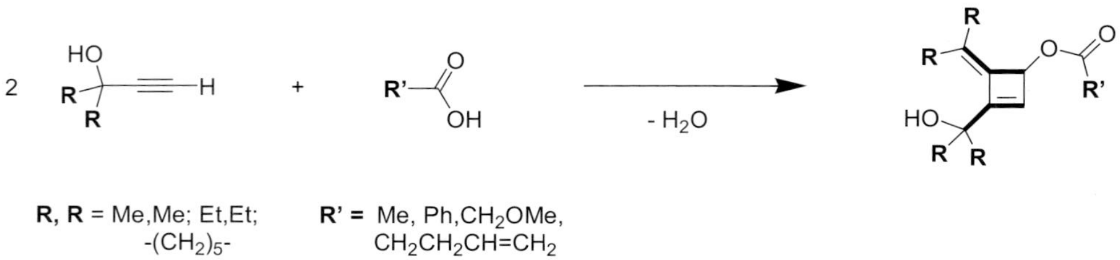

$$2\ \text{HOCR}_2\text{-C}\!\!\equiv\!\!\text{CH}\ +\ R'\text{-COOH}\ \xrightarrow{-H_2O}\ \text{alkenylidenecyclobutene}$$

R, R = Me,Me; Et,Et; -(CH$_2$)$_5$-

R' = Me, Ph, CH$_2$OMe, CH$_2$CH$_2$CH=CH$_2$

Scheme 5. Alkenylidenecyclobutenes from alkynol/carboxylic acid coupling.

This novel catalytic reaction formally corresponds to an initial regioselective [2+2] head-to-head cyclodimerization of the propargyl alcohol with addition of carboxylic acid and elimination of water. Catalytic alkenylidenecyclobutene formation is general for a variety of propargyl alcohols and carboxylic acids and can also be applied to phenol derivatives.

1.4.1.2.1 Mechanism

The stereoselective synthesis of 1,4-disubstituted-1,3-dienes proceeds by head-to-head oxidative coupling of two alkynes with formation of an isolable metallacyclic biscarbene ruthenium complex [23], as shown in Scheme 6. Several key experiments involving labeled reagents and stoichiometric reactions and theoretical studies support the formation of a mixed Fischer–Schrock-type biscarbene complex which undergoes protonation at one carbene carbon atom whereas the other becomes accessible to nucleophilic addition of the carboxylate anion (Scheme 6) [23].

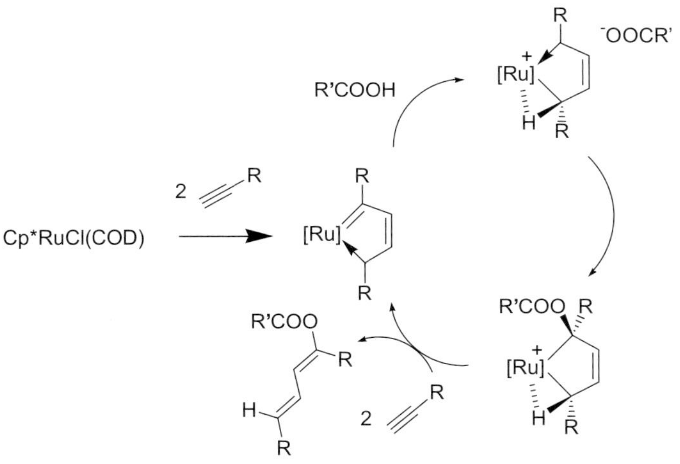

Scheme 6. Catalytic cycle for dienyl ester formation from terminal alkynes and carboxylic acid.

In the synthesis of alkylidenecyclobutenes from propargyl alcohols, stoichiometric experiments show that the first step involves [2+2] oxidative head-to-head coupling of the alkynes, leading to an isolable cyclobutadiene–ruthenium complex. Addition of acid generates a cyclobutenyl metal intermediate which undergoes carboxylate addition on the less substituted allylic carbon atom (Scheme 7).

Computational studies performed on both reactions, leading either to dienyl esters or to alkenylidenecyclobutenes, show that the biscarbene ruthenium intermediate requires a high activation energy to produce the cyclobutadiene complex from terminal alkynes whereas with propargyl alcohols this step occurs readily [25].

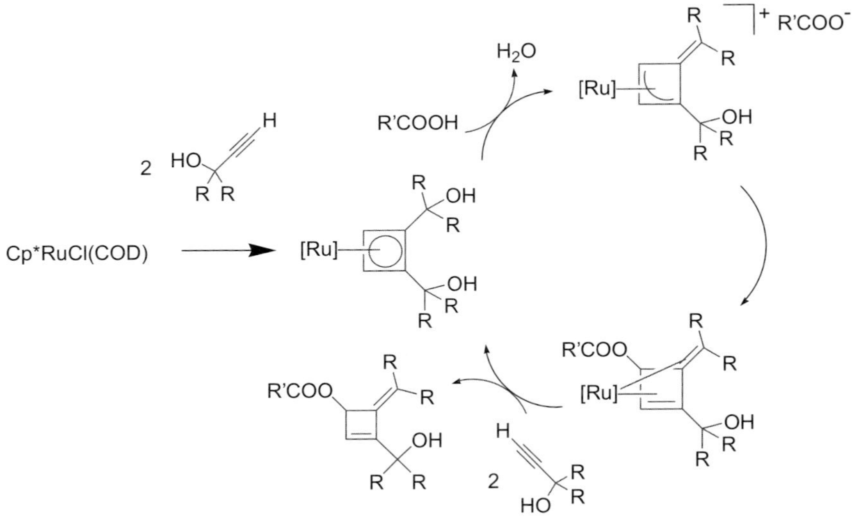

Scheme 7. Catalytic cycle for formation of alkenylidenecyclobutenes from alkynols and carboxylic acid.

1.4.1.2.2 Scope and Limitations

Although alkyne dimerization reactions have been observed with a variety of transition metal catalysts, selectivity is highly dependent not only on the metallic center but also on the ancillary ligands and the substrates.

Palladium catalysts have high tolerance for several functional groups irrespective of their *gem-* or *E*-selectivity [4, 6, 7]. Aldehydes, alcohols, saturated or conjugated ketones, esters, sulfones, malonates and silyl ethers have proved to be compatible. The presence of an additional double bond does not modify the coupling, enabling self-dimerization of non-conjugated enynes as depicted in Scheme 8.

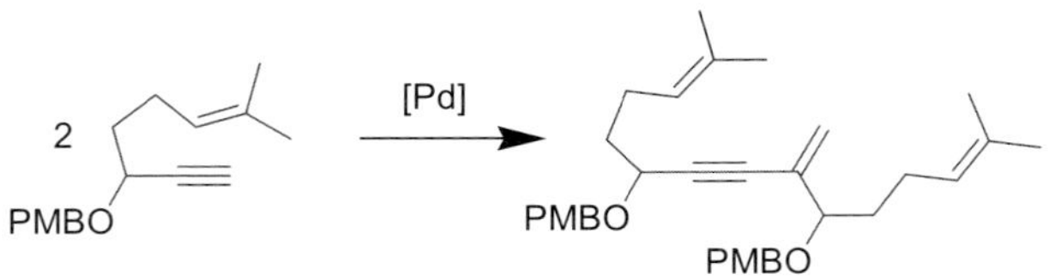

Scheme 8. Catalytic dimerization of non-conjugated enynes with the Pd(OAc)$_2$/TDMPP system.

The palladium–phosphine combination has become a most useful synthetic system, because of the possibility of achieving cross-coupling reactions of terminal and activated internal alkynes. As an example, one-pot alkyne–alkynoate coupling gives dihydropyrans atom-economically and in moderate to good yield (Scheme 9) [26].

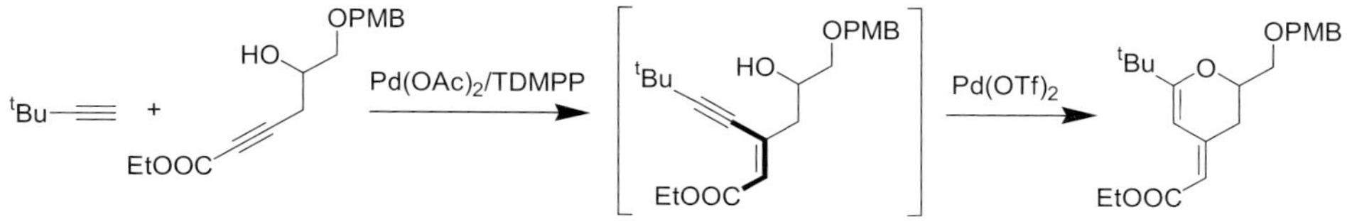

Scheme 9. Oxygen heterocycle by tandem palladium catalysis.

Reports on *ruthenium* catalytic activity focus more on mechanistic consideration of the prototypical phenylacetylene dimerization than in establishing its synthetic applicability. It is not unusual that changing the alkyne substituents results in reversed selectivity (i.e. R = Ph or SiMe$_3$ gave (*E*)- or (*Z*)- isomers, respectively) [27]. Competitive alkyne cyclotrimerization (R = COOMe) [27] or butatriene formation (R = CH$_2$Ph, ^{t}Bu) [10, 21] have occasionally been reported as possible drawbacks in enyne synthesis. The operating mechanism restricts the reaction to terminal alkynes.

Propargyl alcohols are also suitable substrates for dimerization, as reported early on for Wilkinson's catalyst [RhCl(PPh$_3$)$_3$] [28], which provides (*E*)-head-to-head dimers as shown in Scheme 10.

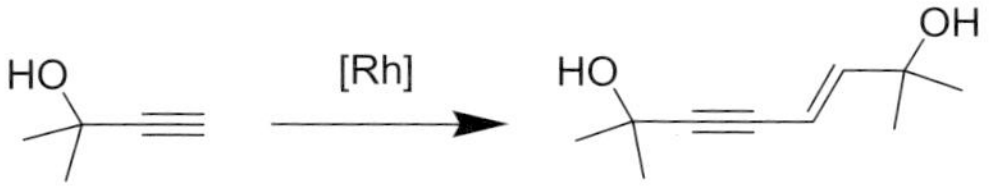

Scheme 10. Head-to-head dimerization of 3-methyl-1-yn-3-ol by [RhCl(PPh$_3$)$_3$] catalyst.

Cyclization of α,ω-diynes in moderate to high yields with high regio- and stereoselectivity has been reported with palladium [29] and binuclear ruthenium catalysts [30], to give *exo*- or *endo*-cyclic (*Z*)-1-en-3-ynes, respectively (Scheme 11).

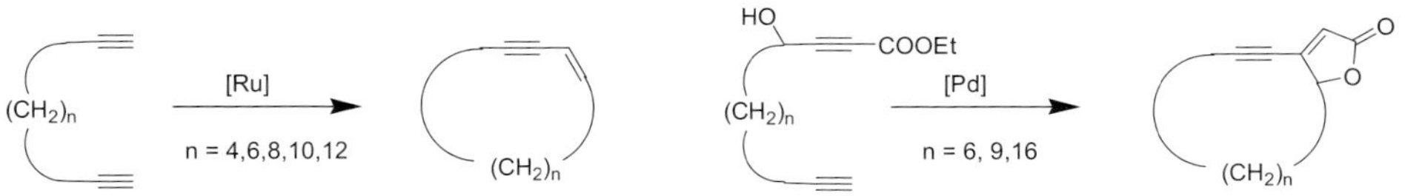

Scheme 11. *Endo*- and *exo*- cyclization of α,ω-diynes.

A recently developed application of this method in the field of polymers involves the synthesis of three geometrical isomers by polyaddition of diethynyl compounds, based on transition-metal-catalyzed dimerization of arylacetylenes [5]. This reaction goes highly regio- and stereoselectively and is controlled by appropriate choice of the catalyst, affording polymers with (*E*)-, (*Z*)- and *gem*-vinylene linkages with selectivity over 92 %, as illustrated in Scheme 12.

The catalytic synthesis of dienyl esters by dimerization of alkynes and addition of carboxylic acids tolerates a large variety of functional groups and carboxylic

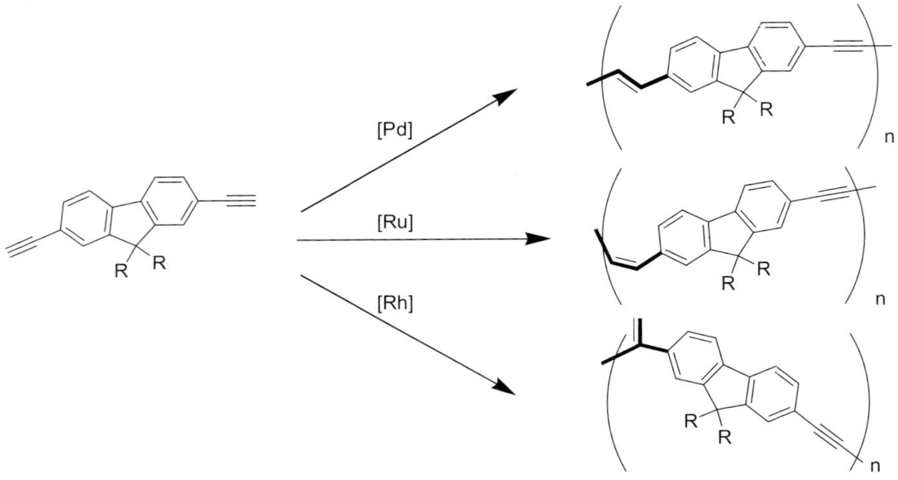

Scheme 12. Conjugated polymers with enriched (E)-, (Z)- or gem-vinylene linkages.

acids, and good yields (60–90%) are obtained from combinations of several arylacetylenes and carboxylic acids. The reaction is, however, inhibited by phenols and strong acids. It can also be extended to alkylacetylenes with moderate yields of dienyl esters. Reaction of arylacetylenes with amino acids leads to dienyl amino esters when the amino group is protected. New monomers for polymerization can be obtained from dicarboxylic acids (Scheme 13) [23].

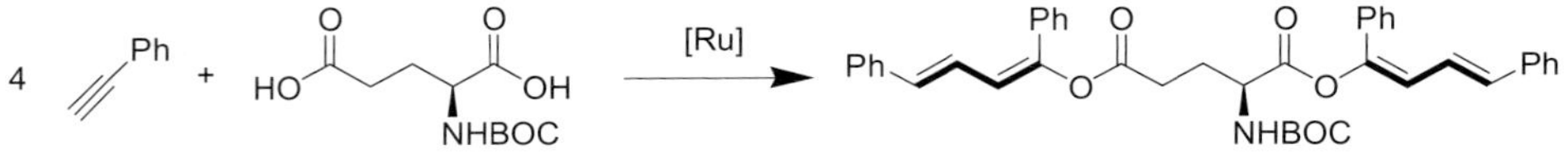

Scheme 13. Alkyne dimerization and coupling with L-glutamic acid catalyzed by [Cp*RuCl(COD)].

The alkylidenecyclobutene synthesis is restricted to propargyl alcohols bearing a terminal triple bond. The scope of the reaction has been established for a variety of carboxylic acids, for example acetic, benzoic, methoxyacetic, and pent-4-enoic acid. It can also be applied to N-protected amino acids, and to phenol derivatives containing reactive sp^2-C–Br bonds (Scheme 14) [24, 25].

Scheme 14. Alkylidenecyclobutenes from propargyl alcohols and acids, with [Cp*RuCl(COD)] as catalyst.

Experimental

Typical Procedure for Ruthenium-Catalyzed Dimerization of Terminal Alkynes with Monocarboxylic Acids [23]

To a solution of terminal alkyne (2.5 mmol, 1 equiv) in degassed dioxane (1 mL) were added Cp*RuCl(COD) [31] (0.125 mmol, 5%) and carboxylic acid (1.25 mmol, 0.5 equiv) under an inert atmosphere at room temperature. The reaction mixture was stirred at room temperature for 15 min to 45 h. The solvent was removed, and the product was purified by silica gel flash column chromatography (eluent pentane–diethyl ether mixtures) to give dimerization adduct as a white solid in 20–98% yield. The compounds were analyzed by NMR (^{1}H and ^{13}C), IR, and mass spectroscopy.

R = Ph, R' = Me

Yield: 90%. ^{1}H NMR (200.131 MHz, CDCl$_3$) δ ppm: 2.21 (s, 3H, MeCO), 6.29 (d, J = 11.1 Hz, 1H, H^1), 6.67 (d, J = 15.5 Hz, 1H, H$_3$), 6.99 (dd, J = 11.1 Hz, J = 15.5 Hz, 1H, H$_2$), 7.21–7.53 (m, 10H, Ph). ^{13}C NMR (50.329 MHz, CDCl$_3$) δ ppm: 169.6, 148.4, 137.2, 134.7, 134.5, 128.9, 128.7, 128.5, 128.4, 127.8, 126.5, 123.3, 120.5, 21.1. MS (EI): m/z 264.1148 (calc for C$_{18}$H$_{16}$O$_2$ 264.1150). FTIR (KBr) ν (cm^{-1}): 3060, 3035, 3022, 1758, 1636, 1594.

Typical procedure for ruthenium-catalyzed dimerization of terminal propargylic alcohols with carboxylic acids [24, 25]

To a solution of terminal alkyne (2.5 mmol, 1 equiv) in degassed isoprene (2 mL) were added Cp*RuCl(COD) [31] (0.125 mmol, 5%) and carboxylic acid (1.25 mmol, 0.5 equiv) under an inert atmosphere at room temperature. The reaction mixture was stirred at 40 °C for 20 h. The solvent was removed and the product was purified by silica gel flash column chromatography (eluent pentane–diethyl ether mixtures) to give dimerization adduct as a viscous oil in 20–77% yield. The compounds were analyzed by NMR (^{1}H and ^{13}C), IR and mass spectrometry.

R = Me; R', R'' = –(CH$_2$)$_5$–

Yield: 66%. ^{1}H NMR (200.131 MHz, CDCl$_3$) δ ppm: 1.4–1.75 (m, 17H, cyclohexyl + OH), 2.01 (s, 3H, MeCO), 1.96–2.08 (m, 2H, cyclohexyl CH$_2$), 2.37–

2.47 (m, 2H, cyclohexyl CH$_2$), 5.59 (d, 1H, $J=0.5$ Hz, H^1), 6.27 (s, 1H, H^2). ^{13}C NMR (50.329 MHz, CDCl$_3$) δ ppm: 171.3, 162.0, 130.7, 129.3, 127.8, 72.4, 69.4, 35.8, 35.6, 31.4, 30.8, 27.9, 27.7, 26.4, 25.5, 21.7, 21.6, 21.2. MS (EI): m/z 290.1879 (calc for C$_{18}$H$_{26}$O$_3$ 290.1882). FTIR (neat) ν (cm^{-1}): 3462, 2928, 1732, 1577.

1.5
anti-Markovnikov Addition to Terminal Alkynes via Ruthenium Vinylidene Intermediates

Christian Bruneau

1.5.1
Introduction

The formation of metal vinylidene complexes directly from terminal alkynes is an elegant way to perform *anti*-Markovnikov addition of nucleophiles to triple bonds [1, 2]. The electrophilic α-carbon of ruthenium vinylidene complexes reacts with nucleophiles to form ruthenium alkenyl species, which liberate this organic fragment on protonolysis (Scheme 1).

Among nucleophilic additions to catalytic metal vinylidene intermediates, tremendous effort was initially devoted to selectively perform unusual *anti*-Markovnikov addition of *O*-nucleophiles such as carbamates and carboxylates to terminal alkynes. The objective was to produce, in one step, useful derivatives such as

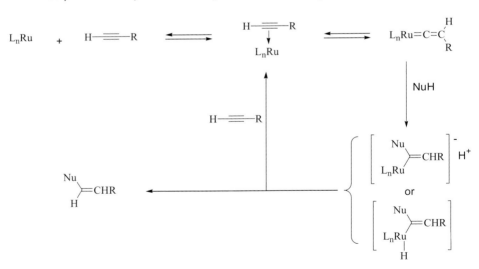

Scheme 1. General mechanism of nucleophilic addition to terminal alkynes *via* ruthenium vinylidene intermediates.

vinylic carbamates and esters without the use of toxic reagents such as phosgene or mercury derivatives. Addition of alcohols and water to alkynes to produce enol ethers, ketones, lactones, and aldehydes are now well controlled. For these reactions, ruthenium catalysts often have excellent activity.

1.5.2
Application to the Synthesis of Vinylcarbamates

Ammonium carbamates are readily and reversibly produced on reaction of secondary amines with carbon dioxide. In the presence of a ruthenium catalyst precursors such as $Ru_3(CO)_{12}$ [3], (arene)$RuCl_2(PR_3)$ [4] or $Ru(methallyl)_2(dppe)$ [5] (dppe = bis(diphenylphosphino)ethane) complexes, the three-component combination of a secondary amine, a terminal alkyne, and carbon dioxide selectively provides vinylcarbamates resulting from addition of carbamate to the terminal carbon of the triple bond (Scheme 2).

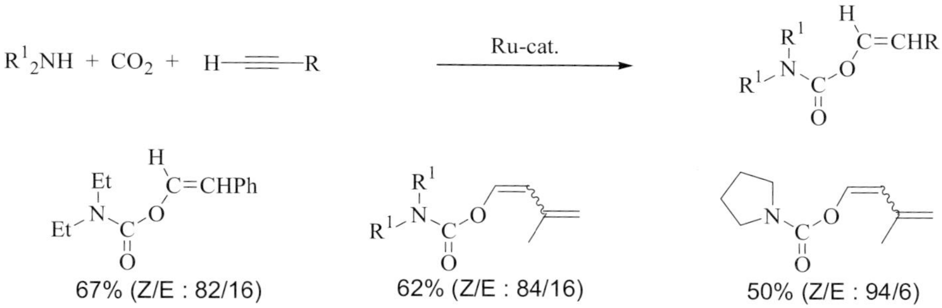

$R^1_2NH + CO_2 + H\text{—}{\equiv}\text{—}R \xrightarrow{\text{Ru-cat.}}$ vinylcarbamate products

67% (Z/E : 82/16)   62% (Z/E : 84/16)   50% (Z/E : 94/6)

Scheme 2. Formation of vinylcarbamates via terminal alkyne C–H bond-activation with ruthenium catalysts.

The major advantage of this one-step catalytic synthesis is that no phosgene derivative such as R^1_2NCOCl is used, which is greener than the traditional stoichiometric processes. It also enables straightforward access to vinylcarbamates, which are useful monomers for functional polymer production.

1.5.3
Application to the Synthesis of Enol Esters

The *anti*-Markovnikov addition of carboxylic acids was attempted with the objective of producing, in one step, alkenyl esters formally produced by acylation of aldehyde enolates. Catalytic addition of carboxylic acids to terminal alkynes usually leads to Markovnikov addition products with classical ruthenium catalysts such as $RuCl_2(phosphine)(arene)$. Efforts were made to modify the regioselectivity of the addition by using more electron-rich catalytic ruthenium moieties. Indeed, as the vinylidene ligand is an electron-withdrawing group, tautomerization of $M(\eta^2\text{-}HC{\equiv}CR)$ into $M(\eta^1{=}C{=}CHR)$ is favored by electron-donating ancillary ligands bonded at the ruthenium site [6]. Success was achieved with the diphosphine ruthenium catalyst Ru(methallyl)$_2$(diphenylphosphinobutane). Thus, when carboxylates are used as nucleophiles, the *anti*-Markovnikov addition of carboxylic acids to terminal alkynes occurs under mild conditions in the presence of catalytic amounts of Ru(methallyl)$_2$(diphosphine), regioselectively providing (Z)-enol esters [7, 8].

Scheme 3. Synthesis of (Z)-enol esters via *anti*-Markovnikov addition of carboxylic acids to terminal alkynes.

The diphosphine of choice for obtaining good regioselectivity is the bis(diphenylphosphino)butane (Scheme 4). It enables addition of a variety of carboxylic acids to phenylacetylene and hexyne. The reaction temperature, which enables complete conversion can be reduced from 80 to 0 °C when the acidity of the carboxylic acid increases in the pK_a range from 5 to 1.5. The milder temperature conditions always lead to higher regioselectivity of the addition. The preparation of functionalized (Z)-enol esters can be conducted in apolar solvents such as toluene or pentane; dienyl esters are selectively produced from conjugated enynes [9].

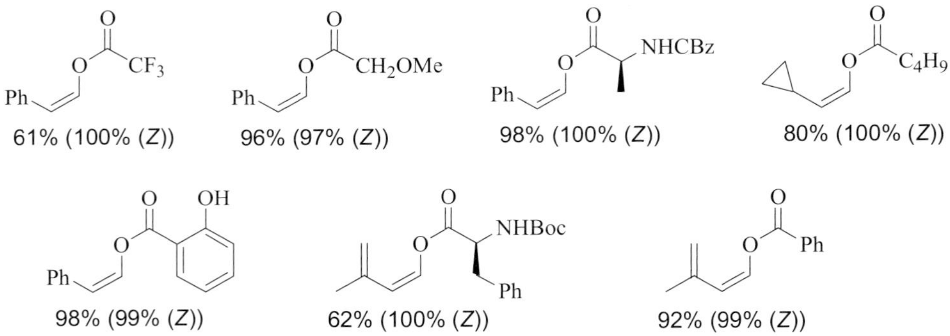

61% (100% (Z))   96% (97% (Z))   98% (100% (Z))   80% (100% (Z))

98% (99% (Z))   62% (100% (Z))   92% (99% (Z))

Scheme 4. Preparation of enol esters and dienyl esters from terminal alkynes and carboxylic acids.

For both reactivity and regioselectivity, however, a compromise must be found between the bulkiness of the reagents (alkyne and carboxylic acid) and the steric hindrance of the diphosphine ligand, all of which are present in the coordination sphere of the ruthenium center during the catalytic process. Thus, with the more bulky trimethylsilylacetylene, the less hindered bis(diphenylphosphino)ethane ligand provides an efficient ruthenium catalyst (Ru(methallyl)$_2$(dppe)) for producing silylated enol esters. Better reactivity is also observed with Ru(methallyl)$_2$(dppe) as catalyst precursor when propargylic ethers are used as acetylenic substrates (Scheme 5) [10].

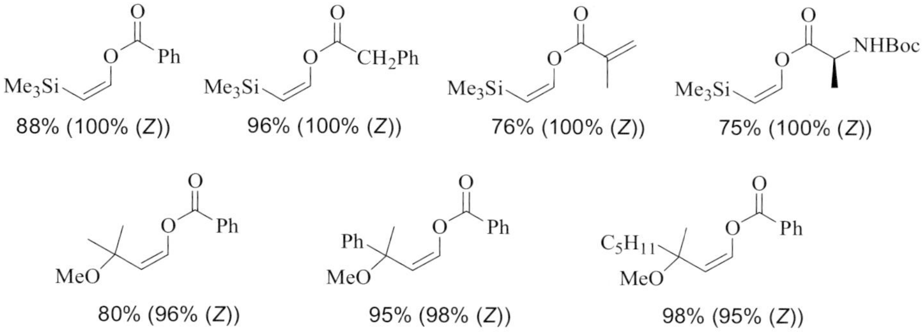

Scheme 5. Preparation of functional enol esters from trimethylsilylacetylene and propargylic ethers.

Very recently, new ruthenium catalysts, for example $RuCl_2$(triazol-5-ylidene)(p-cymene) [11] and the catalytic system generated in situ from $[RuCl_2(p\text{-cymene})]_2$, tris(p-chlorophenyl)phosphine, and 4-dimethylaminopyridine [12], have provided efficient catalysts for synthesis of the same type of enol ester. The regioselective cyclization of acetylenic acids containing a terminal triple bond to give unsaturated lactones was performed in the presence of catalytic amounts of Ru(tris(pyrazolyl)borate)(PhC=C(Ph)C≡CPh)(PMe_2iPr_2) [13].

1.5.4
Application to the Isomerization of Propargylic Alcohols

Propargylic alcohols are also very interesting because *anti*-Markovnikov addition of benzoic acid generates bifunctional 1,3-hydroxy esters with addition of the carboxylate to the terminal carbon of the triple bond [14, 15]. This reaction contrasts

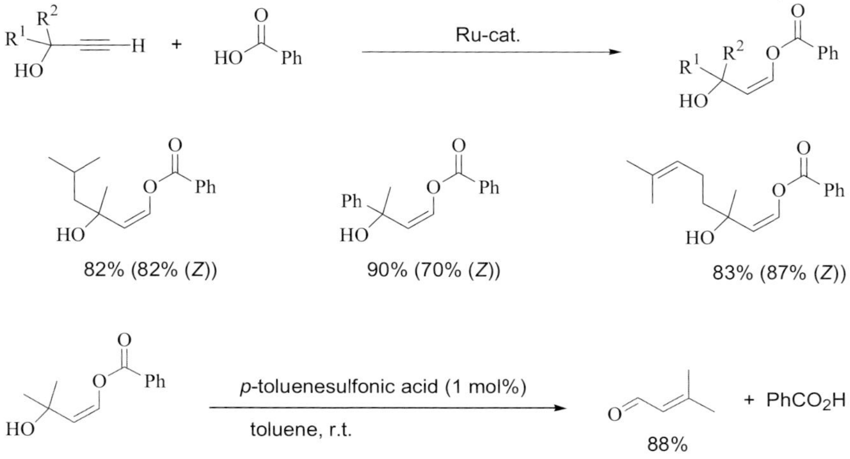

Scheme 6. Preparation of 1,3-hydroxyesters and subsequent cleavage into conjugated enals.

with the Markovnikov addition, which selectively leads to the formation of β-keto esters by intramolecular transesterification [16]. These 1,3-hydroxyesters are readily cleaved under thermal or acidic conditions to afford the conjugated enals, the formal isomerization compounds of the starting propargylic alcohols [14, 15].

1.5.5
Application to the Synthesis of Vinylic Ethers

There are not many examples of addition of alcohols to a triple bond proceeding through an *anti*-Markovnikov reaction. With RuCl(tris(pyrazolyl)borate)(pyridine)$_2$ as catalyst, addition of allyl alcohol to phenylacetylene proceeds in 72 % overall yield in toluene under reflux, and provides a 1:1 mixture of allyl β-styryl ether and 2-phenylpent-4-enal resulting from further Claisen rearrangement (Scheme 7) [17].

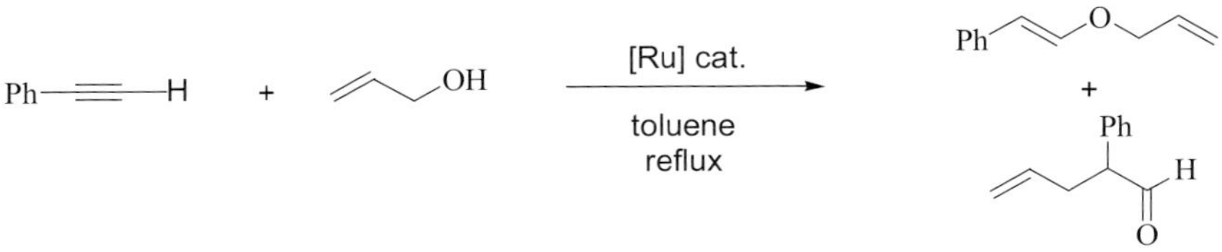

Scheme 7. Ruthenium-catalyzed addition of allyl alcohol to phenylacetylene.

1.5.6
Application to the Synthesis of Unsaturated Ketones

It is well-known that RuCl(cyclopentadienyl)(PPh$_3$)$_2$ stoichiometrically activates terminal alkynes in the presence of a chloride abstractor to produce [Ru(cyclopentadienyl)(=C=CHR)][X] complexes. This reaction has been used with advantage to perform *anti*-Markovnikov addition of allylic alcohols to terminal alkynes followed by intramolecular rearrangement to produce unsaturated ketones according to Scheme 8 [18, 19].

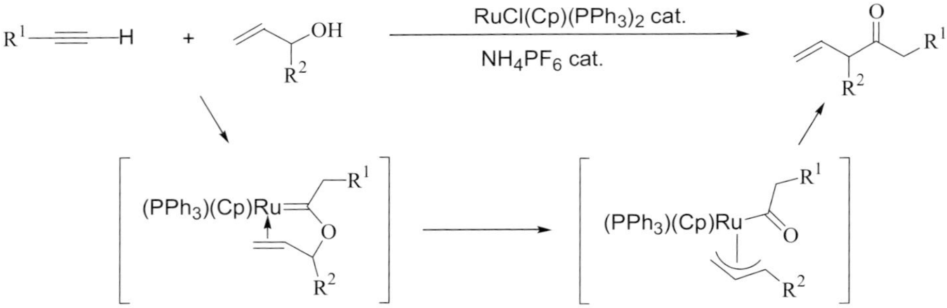

Scheme 8. Ruthenium-catalyzed addition of allyl alcohol to terminal alkynes with skeleton rearrangement.

1.5.7
Application to the Synthesis of Cyclic Enol Ethers and Lactones

Intramolecular addition of a hydroxy group to the terminal sp-carbon of pent-4-yn-1-ols, leading to the corresponding cycloisomerization dihydropyrans, has been successfully achieved with a similar ruthenium catalyst precursor containing the electron-deficient tris(p-fluorophenyl)phosphine ligand, excess phosphine, and sodium N-hydroxysuccinimide as additives (Scheme 9) [20].

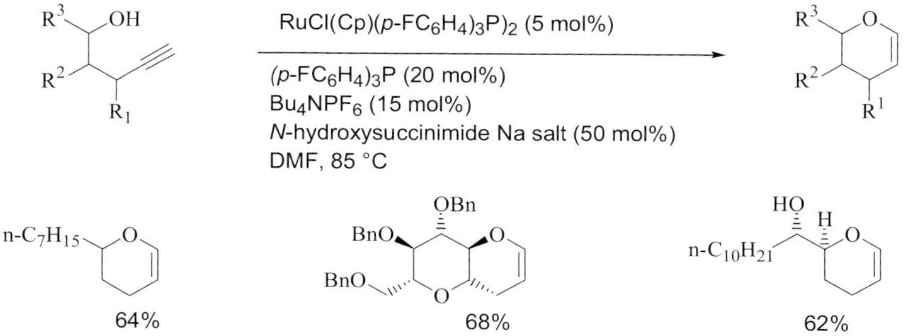

Scheme 9. Cycloisomerization of pent-4-yn-1-ols catalyzed by ruthenium catalysts bearing electron deficient phosphine ligands.

When the ruthenium precursor contains an electron-rich phosphine such as tris(p-methoxyphenyl)phosphine, and if a large excess of this ligand is used in the catalytic reaction, the formation of valerolactones by oxidation of a postulated transient alkoxycarbene ruthenium moiety is favored (Scheme 10) [20].

With N-hydroxysuccinimide as mild oxidant and RuCl(cyclopentadienyl)(cyclooctadiene) as catalyst precursor, in the presence of tris(o-furyl)phosphine and n-butylammonium bromide, a wide range of homopropargylic alcohols were transformed into five-membered γ-butyrolactones via a related cycloisomerization–oxidation reaction [21].

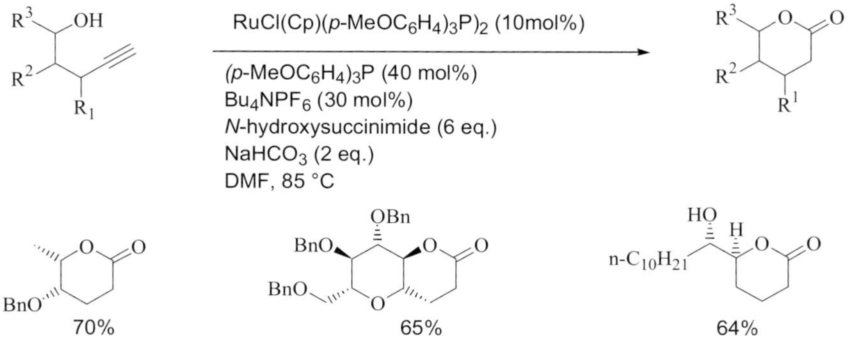

Scheme 10. Oxidative cyclization of pent-4-yn-1-ols involving ruthenium vinylidene intermediates.

It must be pointed out that tungsten [22–26] and molybdenum [26–29] carbonyl precursors also have remarkable catalytic activity in the cycloisomerization of alkynols to produce dihydropyrans and dihydrofurans via intramolecular nucleophilic addition of the hydroxy group to the terminal carbon of the triple bond, activated as a vinylidene metal species.

1.5.8
Application to the Synthesis of Aldehydes

Regioselective addition of water to triple bonds constitutes a straightforward preparation of aldehydes from terminal alkynes (Scheme 11). The *anti*-Markovnikov addition has recently been made possible by use of ruthenium catalysts such as $RuCl(C_6H_6)(PPh_2(C_6F_5)) + 3\,PPh_2(C_6F_5)$ [30], $[RuCl_2(C_6H_6)]_2$ in the presence of an excess of the water-soluble ligand $P(3-C_6H_5SO_3Na)_3$ [30], $RuCl(Cp)(phosphine)_2$ [31], $RuCl(Cp)(diphosphine)$ [31], $RuCl(indenyl)(phosphine)_2$ [32], and $[Ru(Cp)(H_2O)(2\text{-imidazolylphosphine})_2]$ $[CF_3SO_3]$ [33]. Reaction of water with terminal propargylic alcohols has also been used to prepare hydroxy aldehydes by *anti*-Markovnikov addition in the presence of $RuCl(indenyl)(PPh_3)_2$ in an isopropanol–water mixture, or in water containing a surfactant as solvent [32]. It is noteworthy that in the presence of $RuCl(Cp)(PMe_3)_2$ as catalyst precursor, the *anti*-Markovnikov addition is accompanied by a dehydration reaction and conjugated enals are obtained [34].

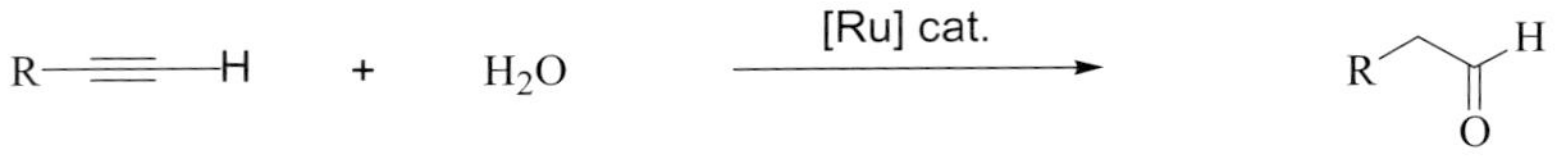

$$R\text{—}\!\!\equiv\!\!\text{—}H + H_2O \xrightarrow{[Ru]\text{ cat.}} R\text{—}CH_2\text{—}CHO$$

Scheme 11. Preparation of aldehydes from terminal alkynes and water in the presence of ruthenium catalyst.

1.5.9
Scope and Limitations

These reactions illustrate the importance of ruthenium vinylidene species, as activated forms of terminal alkynes, in catalysis, because they favor the addition of O-nucleophiles (carbamic and carboxylic acids, alcohols, water) to terminal alkynes and completely reverse the expected regioselectivity of the addition. These examples also show that the activation processes are very sensitive to the nature of the nucleophiles, and the success of the *anti*-Markovnikov addition to terminal alkynes is highly dependent on both the electron richness and steric hindrance of the ancillary ligands coordinated to the active site.

Experimental

Synthesis of Vinylcarbamates – Typical Experiment –
(Z)-β-[(Diethylcarbamoyl)oxy]styrene [4]

Diethylamine (20 mmol), phenylacetylene (10 mmol), and $RuCl_2(C_6Me_6)(PMe_3)$ (0.2 mmol) in 10 mL acetonitrile were placed in a 125-mL stainless steel autoclave. Carbon dioxide was used first to flush out the reactor and then to pressurize it to a starting pressure of 5 MPa. The reaction mixture was stirred at 125 °C for 20 h. After elimination of the solvent under reduced pressure, the product was purified by flash chromatography (1:1 CH_2Cl_2–petroleum ether) and isolated in 67 % yield. ^{1}H NMR (C_6D_6, 300 MHz): δ = 0.84 (t, 3H, J = 7.0 Hz, Me), 0.90 (t, 3H, J = 7.0 Hz, Me), 2.95 (q, 2H, J = 7.0 Hz, NCH_2), 3.06 (q, 2H, J = 7.0 Hz, NCH_2), 6.27 (d, 1H, J = 12.9 Hz, =CHPh), 7.40 (m, 5H, Ph), 8.16 (d, 1H, J = 12.9 Hz, =CHO).

Preparation of the Catalyst

$[RuCl_2(C_6Me_6)]_2$ is prepared by heating a finely powdered mixture of 10.0 g (16.3 mmol) commercially available $[RuCl_2(p\text{-cymene})]_2$ and 30 g (185 mmol) hexamethylbenzene at 205 °C for 20 h. After cooling, the reaction mixture is stirred in hot toluene for 1 h and the remaining solid is isolated by filtration of the hot solution. After washing with toluene and diethyl ether, $[RuCl_2(C_6Me_6)]_2$ is obtained as an orange powder in 96 % yield. Trimethylphosphine (6 mmol, 0.7 mL) is then added to a dichloromethane solution (40 mL) containing 2 g (3 mmol) $[RuCl_2(C_6Me_6)]_2$ under an inert atmosphere and the solution is stirred for 4 h. The resulting solution is evaporated under vacuum and the residue is extracted with boiling ethanol. The hot solution is filtered and dark-red crystals of $RuCl_2(C_6Me_6)(PMe_3)$ precipitate when the solution is cooled (90 %). ^{1}H NMR ($CDCl_3$, 200 MHz) δ = 1.43 (d, 9H, J_{PH} = 10.5 Hz, PMe_3), 2.00 (d, 18H, J_{PH} = 0.7 Hz C_6Me_6). $^{31}P\{^1H\}$ NMR ($CDCl_3$, 81 MHz) δ = 1.96.

Synthesis of Enol Esters – Typical Experiment – (Z)-3-Methylbuta-1,3-dien-1-yl benzoate [9]

A solution of 1.22 g (10 mmol) benzoic acid, 0.86 g (11 mmol) isopropenylacetylene and 64 mg (0.1 mmol) Ru(methallyl)$_2$(diphenylphosphinobutane) in 5 mL toluene was stirred at 50 °C for 18 h. The product was purified by silica-gel chromatography and isolated in 92 % yield. ^{1}H NMR ($CDCl_3$, 300 MHz): δ = 2.18 (t, 3H, J = 1.1 Hz, Me), 4.98 and 5.11 (m, 2H, CH_2), 5.47 (dd, 1H, J = 7.3 Hz, 0.7 Hz, =CH), 7.22 (d, 1H, J = 7.3 Hz, =CHO), 7.45–8.10 (m, 5H, Ph).

Preparation of the Catalyst

Ru(methallyl)$_2$(cyclooctadiene) is prepared by addition of polymeric dichloro(cyclooctadiene)ruthenium to a suspension of (2-methylpropenyl)magnesium chloride in diethyl ether, and stirring at room temperature for 1.5 h. After hydrolysis with cold water at –40 °C, the reaction mixture is extracted twice with diethyl ether. Evaporation of the solvent furnishes a gray powder in 80 % yield. This complex (2.0 g, 6.26 mmol) and 2.6 g (6.10 mmol) 1,3-bis(diphenylphosphino)butane are then heated under reflux for 5 h in 70 mL hexane under argon. After cooling, a yellow powder is isolated by filtration, washed three times with 10 mL hexane and dried in vacuo to give 3.07 g (79 %) Ru(methallyl)$_2$(diphenylphosphinobutane). ^{1}H NMR (CD$_2$Cl$_2$, 300 MHz): δ = 0.56 (dd, 2H, J = 15.9, 4.9 Hz, allylic CH$_2$), 1.02 (d, 2H, J = 13.9 Hz, allylic CH$_2$), 1.05 (s, 2H, allylic CH$_2$), 1.60 (m, 2H, P(CH$_2$)$_4$P), 1.84 (s, 6H, Me), 2.00 (m, 2H, P(CH$_2$)$_4$P), 2.12 (m, 2H, allylic CH$_2$), 2.20–2.40 (m, 2H, P(CH$_2$)$_4$P), 2.70–2.90 (m, 2H, P(CH$_2$)$_4$P), 7.00–7.70 (m, 20H, Ph). ^{31}P{^{1}H} NMR (CD$_2$Cl$_2$, 121 MHz): δ = 44.58.

References to Chapter 1 – C–H Transformation at Terminal Alkynes

References to Section 1.1

1 *Modern Acetylene Chemistry* (Eds.: Stang, P. J.; Diederich, F.), VCH, Weinheim, **1995**.
2 Simandi, L. I. in *The Chemistry of Functional Groups, Supplement C, pt 1* (Eds.: Patai, S.; Rappoport, Z.), Wiley, New York, **1983**, pp. 529–534.
3 (a) Wakefield, B. J. *Organomagnesium Methods in Organic Synthesis*, Academic Press, London, **1995**, Ch. 3, pp. 46–48; (b) Brandsma, L. *Preparative Acetylene Chemistry*, 2nd edn, Elsevier, Amsterdam, **1988**; (c) Wakefield, B. J. *Organolithium Methods*, Academic Press, London, **1988**, Ch. 3, p. 32.
4 (a) Keim, R. *Organosilver Compounds*, Gmelin Handbuch der Anorganischen Chemie, 8th edn, Berlin, **1975**; (b) Breitinger, D. K.; Herrmann, W. A. *Synthetic Methods of Organometallic and Inorganic Chemistry*, Vol. 5, *Copper, Silver, Gold, Zinc, Cadmium, and Mercury*, Herrmann, W. A. (Ed.), Thieme, New York, **1999**; (c) Krause, N. *Modern Organocopper Chemistry*, Wiley–VCH: Weinheim, **2002**.
5 (a) Favorski, A. E.; Skossarewsky, M. *Russ. J. Phys. Chem. Soc.* **1900**, *32*, 652; (b) Favorski, A. E.; Skossarewsky, M. *Bull. Soc. Chim.* **1901**, *26*, 284.
6 (a) Tedeschi, R. J.; Kindley, L. M.; Huckel R. W.; Russell, J. P; Casey, A. W.; Clark, G. S. *J. Org. Chem.* **1963**, *28*, 1740; (b) Tedeshi, R. J. *J. Org. Chem.* **1965**, *30*, 3045.
7 Babler, J. H.; Liptak, V. P.; Phan, N. *J. Org. Chem.* **1996**, *61*, 416.
8 Busch-Petersen, J.; Bo, Y. X.; Corey, E. J. *Tetrahedron Lett.* **1999**, *40*, 2065.
9 Tzalis, D.; Knochel, P. *Angew. Chem. Int. Ed.* **1999**, *38*, 1463.
10 Pu, L. *Tetrahedron* **2003**, *59*, 9873 and references therein.
11 (a) Frantz, D. E.; Fässler, R.; Carreira, E. M. *J. Am. Chem. Soc.* **1999**, *121*, 11245; (b) Frantz, D. E.; Fässler, R.; Tomooka, C. S.; Carreira, E. M. *Acc. Chem. Res.* **2000**, *33*, 373.
12 Fässler, R.; Tomooka, C. S.; Frantz, D. E.; Carreira, E. M. *Proceedings of the National Academy of Sciences* **2004**, *101*, 5843.
13 Frantz, D. E.; Fässler, R.; Carreira, E. M. *J. Am. Chem. Soc.* **2000**, *122*, 1806.
14 Jiang, B.; Si, Y. G. *Tetrahedron Lett.* **2002**, *43*, 8323.
15 Sasaki, H.; Boyall, D.; Carreira, E. M. *Helv. Chim. Acta* **2001**, *84*, 964.
16 Boyall, D.; Lopez, F.; Sasaki, H.; Frantz, D.; Carreira, E. M. *Org. Lett.* **2000**, *2*, 4233.
17 (a) Jiang, B.; Chen, Z. L.; Xiong, W. N. *Chem. Commun.* **2002**, 1524; (b) Jiang B, Si Y. G. *Adv. Synth. Catal.* **2004**, *346*, 669.
18 Jiang, B; Chen, Z.; Tang, X. *Org. Lett.* **2002**, *4*, 3451.
19 Anand, N. K.; Carreira, E. M. *J. Am. Chem. Soc.* **2001**, *123*, 9687.
20 (a) Trost, B. M.; Ball, Z. T.; Jöge, T. *Angew. Chem. Int. Ed.* **2003**, *42*, 3415; (b) El-Sayed, E.; Anand, N. K.; Carreira, E. M. *Org. Lett.* **2001**, *3*, 3017.
21 (a) Amador, M.; Ariza, X.; Garcia, J.; Ortiz, J. *Tetrahedron Lett.* **2002**, *43*, 2691; (b) Diez, R. S.; Adger, B.; Carreira, E. M. *Tetrahedron* **2002**, *58*, 8341.

22 Trost, B. M.; Ameriks, M. K. *Org. Lett.* **2004**, *6*, 1745.
23 McDonald, F. E.; Davidson, M. H. *Org. Lett.* **2004**, *6*, 1601.
24 (a) Maezaki, N.; Tominaga, H.; Kojima, N.; Yanai, M.; Urabe, D.; Tanaka, T. *Chem. Commun.* **2004**, *4*, 406; (b) Kojima, N.; Maezaki, N.; Tominaga, H.; Yanai, M.; Urabe, D.; Tanaka, T. *Chem. Eur. J.* **2004**, *10*, 672; (c) Maezaki, N.; Kojima, N.; Tominaga, H.; Yanai, M.; Tanaka, T. *Org. Lett.* **2003**, *5*, 1411; (d) Maezaki, N.; Kojima, N.; Asai, M.; Tominaga, H.; Tanaka, T. *Org. Lett.* **2002**, *4*, 2977.
25 Rozner, E.; Xu, Q. *Org. Lett.* **2003**, *5*, 3999.
26 Smith, A. B., III; Dae-Shik, K. *Org. Lett.* **2004**, *6*, 1493.
27 Amador, M.; Ariza, X.; Garcia, J.; Ortiz, J. *Tetrahedron Lett.* **2002**, *43*, 2691.
28 Murga, J.; Garcia-Fortanet, J.; Carda, M.; Marco, J. A. *Tetrahedron Lett.* **2003**, *44*, 1737.
29 (a) Bode, J. W.; Carreira, E. M. *J. Am. Chem. Soc.* **2001**, *123*, 3611; (b) Bode, J. W.; Carreira, E. M. *J. Org. Chem.* **2001**, *66*, 6410.
30 (a) Fettes, A.; Carreira, E. M. *Angew. Chem. Int. Ed.* **2002**, *41*, 4098; (b) Fettes, A.; Carreira, E. M. *J. Org. Chem.* **2003**, *68*, 9274.
31 Reber, S.; Knöpfel, T. F.; Carreira, E. M. *Tetrahedron* **2003**, *59*, 6813.

References and Notes to Section 1.2

1 R. D. Stephens, C. E. Castro, *J. Org. Chem.* **1963**, *28*, 3313–3315.
2 K. Sonogashira, Y. Tohda, N. Hagihara, *Tetrahedron Lett.* **1975**, 4467–4470.
3 The name of Sonogashira (or Sonogashira–Hagihara) is commonly employed to identify this reaction, while even for the copper-free procedures Heck and Cassar are rarely mentioned.
4 H. A. Dieck, R. F. Heck, *J. Organomet. Chem.* **1975**, *93*, 259–263.
5 L. Cassar, *J. Organomet. Chem.* **1975**, *93*, 253–257.
6 Recent reviews on Pd-catalyzed alkynylation: (a) E. Negishi, L. Anastasia, *Chem. Rev.* **2003**, *103*, 1979–2017;
(b) K. Sonogashira, p. 493–549, in *Handbook of Organopalladium Chemistry for Organic Synthesis*, (E. Negishi, A. deMeijere, eds.), Wiley–Interscience, New York **2002**.
7 R. R. Tykwinski, *Angew. Chem. Int. Ed.* **2003**, *42*, 1566–1568.
8 *Modern Acetylene Chemistry*, P. J. Stang, F. Diederich (Eds.), VCH, Weinheim, **1995**.
9 W. A. Herrmann, V. P. W. Böhm, C.-P. Reisinger, *J. Organomet. Chem.* **1999**, *576*, 23–41.
10 C. Amatore, S. Bensalem, S. Ghalem, A. Jutand, Y. Medjour, *Eur. J. Org. Chem.* **2004**, 366–371.
11 The role of copper acetylide is well established, while it is not clear whether copper also acts as a scavenger for phosphines.
12 K. Sonogashira, T. Yatake, Y. Tohda, S. Takahashi, N. Hagihara, *Chem. Commun.* **1977**, 291–292.
13 V. Grosshenny, F. M. Romero, R. Ziessel, *J. Org. Chem.* **1997**, *62*, 1491–1500.
14 J. P. Stambuli, M. Bühl, J. F. Hartwig, *J. Am. Chem. Soc.* **2002**, *124*, 9346–9347.
15 M. G. Andreu, A. Zapf, M. Beller, *Chem. Commun.* **2001**, 2745–2746.
16 A. F. Littke, G. C. Fu, *Angew. Chem. Int. Ed.* **2002**, *41*, 4176–4211.
17 P. Fitton, E. A. Rick, *J. Organomet. Chem.* **1971**, *28*, 287–291.
18 F. Liu, E. Negishi, *J. Org. Chem.* **1997**, *62*, 8591–8594.
19 H. Plenio, J. Hermann, A. Sehring, *Chem. Eur. J.* **2000**, *6*, 1820–1826.
20 (a) N. G. Kundu, M. W. Khan, *Tetrahedron* **2000**, *56*, 4777–4783; (b) G. W. Kabalka, L. Wang, R. M. Pagni, *Tetrahedron* **2001**, *57*, 8017–8022.
21 S. Höger, *Chem. Eur. J.* **2004**, *10*, 1321–1329.
22 R. Diercks, K. P. C. Vollhardt, *J. Am. Chem. Soc.* **1986**, *108*, 3150–3151.
23 J. M. Tour, *Acc. Chem. Res.* **2000**, *33*, 791–804.
24 S. H. Chanteau, J. M. Tour, *Tetrahedron Lett.* **2001**, *42*, 3057–3060.
25 V. J. Pugh, Q.-S. Hu, X. Zuo, F. D. Lewis, L. Pu, *J. Org. Chem.* **2001**, *66*, 6136–6140.

26 J. B. Arterburn, K. Venkateswara, R. Perry, M. C. Perry, *Tetrahedron Lett.* **2000**, *41*, 839–842.
27 K. C. Nicolaou, W.-M. Dai, *Angew. Chem. Int. Ed. Engl.* **1991**, *30*, 1387–1416.
28 M. Inouye, K. Takahashi, H. Nakazumi, *J. Am. Chem. Soc.* **1999**, *121*, 341–345.
29 J. Wu, M. D. Watson, L. Zhang, Z. Wang, K. Müllen, *J. Am. Chem. Soc.* **2004**, *126*, 177–186.
30 J. N. Wilson, S. M. Waybright, K. McAlpine, U. H. F. Bunz, *Macromolecules* **2002**, *35*, 3799–3800.
31 P. Nussbaumer, WO 9628430, **1996**.
32 K. Königsberger, G.-P. Chen, R. R. Wu, M. J. Girgis, K. Prasad, O. Repic, T. J. Blacklock, *Org. Proc. Res. Dev.* **2003**, *7*, 733–742.
33 U. Beutler, J. Mazacek, G. Penn, B. Schenkel, D. Wasmuth, *Chimica* **1996**, *50*, 154–156.
34 C. Y. Chen, D. R. Liebermann, R. D. Larsen, R. A. Reamer, T. R. Verhoeven, P. J. Reider, J. F. Cottrell, P. G. Houghton, *Tetrahedron Lett.* **1994**, *35*, 6981–6983.
35 G. W. Parshall, W. A. Nugent, *Chemtech* **1988**, 376–378.
36 *Classics in Total Synthesis*, K. C. Nicolaou, E. J. Sorensen (eds.), Wiley–VCH, Weinheim, **1996**.
37 *Chemistry and Biology of Naturally-Occurring Acetylenes and Related Compounds*, J. Lam, H. Breteler, T. Arnason, L. Hansen (eds.), Elsevier, Amsterdam, **1988**.
38 R. Singh, G. Just, *J. Org. Chem.* **1989**, *54*, 4453–4457.
39 K. Hirota, Y. Kitade, Y. Isobe, Y. Maki, *Heterocycles* **1987**, *26*, 355–358.
40 K. Sonogashira in *Comprehensive Organic Synthesis* B. M. Trost, I. Fleming (Eds.), Pergamon Press, Oxford, p. 521–548, Vol. 3, **1991**.
41 S. Thorand, N. Krause, *J. Org. Chem.* **1998**, *63*, 8551.
42 T. Hundertmark, A. F. Littke, S. L. Buchwald, G. C. Fu, *Org. Lett.* **2000**, *2*, 1729–1731.
43 V. P. W. Böhm, W. A. Herrmann, *Eur. J. Org. Chem.* **2000**, 3679–3681.
44 M. Ansorge, T. J. J. Müller, *J. Organomet. Chem.* **1999**, *585*, 174–178.
45 M. R. Buchmeiser, T. Schareine, R. Kempe, K. Wurst, *J. Organomet. Chem.* **2001**, *634*, 39–46.
46 M. R. Eberhard, Z. Wang, C. M. Jensen, *Chem. Commun.* **2002**, 818–819.
47 A. Köllhofer, T. Pullmann, H. Plenio, *Angew. Chem. Int. Ed.* **2003**, *42*, 1056–1058.
48 D. Gelman, S. L. Buchwald, *Angew. Chem. Int. Ed.* **2003**, *115*, 6175–6178.
49 Z. Xu, J. S. Moore, *Angew. Chem., Int. Ed. Engl.* **1993**, *32*, 246–248.
50 A. S. Hay, *J. Org. Chem.* **1960**, *25*, 1275–1276.
51 D. Méry, K. Heuzé, D. Astruc, *Chem. Commun.* **2003**, 1934–1935.
52 X. Fu, S. Zhang, J. Yin, D. Schumacher, *Tetrahedron Lett.* **2002**, *43*, 6673–6676.
53 D. A. Alonso, C. Nájera, M. C. Pacheco, *Tetrahedron Lett.* **2002**, *43*, 9365–9368.
54 A. C. Hillier, G. A. Grasa, M. S. Viciu, H. M. Lee, C. Yang, S. P. Nolan, *J. Organomet. Chem.* **2002**, *653*, 69–82.
55 M. J. Mio, L. C. Kopel, J. B. Braun, T. L. Gadzikwa, K. L. Hull, R. G. Brisbois, R. A. Batey, M. Shen, A. J. Lough, *Org. Lett.* **2002**, *4*, 1411–1414.
56 Y. Ma, C. Song, W. Jiang, Q. Wu, Y. Wang, X. Liu, M. B. Andrus, *Org. Lett.* **2003**, *5*, 3317–3319.
57 W. A. Herrmann, C.-P. Reisinger, M. Spiegler, *J. Organomet. Chem.* **1998**, *557*, 93–96.
58 C. Yang, S. P. Nolan, *Organometallics* **2002**, *21*, 1020
59 D. Gelman, D. Tsvelikhovsky, G. A. Molander, J. Blum, *J. Org. Chem.* **2002**, *67*, 6287–6290.
60 A. Fürstner, G. Seidel, *Tetrahedron* **1995**, *51*, 11165–11176.
61 J. A. Soderquist, K. Matos, A. Rane, J. Ramos, *Tetrahedron Lett.* **1995**, *36*, 2401–2402.
62 G. A. Molander, B. W. Katona, F. Machrouhi, *J. Org. Chem.* **2002**, *67*, 8416–8423.
63 L. Anastasia, E. Negishi, *Org. Lett.* **2001**, *3*, 3111–3113.
64 E. Negishi, M. Qian, F. Zeng, L. Anastasia, D. Babinski, *Org. Lett.* **2003**, *5*, 1597–1600.
65 L. Wang, P. Li and Y. Zhang, *Chem. Commun.* **2004**, 514–515.

66 A. Soheili, J. Albaneze-Walker, J. A. Murry, P. G. Dormer, D. L. Hughes, *Org. Lett.* **2003**, *5*, 4191–4194.
67 M. Erdélyi, A. Gogoll, *J. Org. Chem.* **2001**, *66*, 4165–4169.
68 N. E. Leadbeater, M. Marco, B. J. Tominack, *Org. Lett.* **2003**, *5*, 3919–3022.
69 *Thematic Issue: Recoverable Catalysts and Reagents*, J. A. Gladysz (ed.), *Chem. Rev.* **2002**, *102*.
70 C. C. Tzschucke, C. Markert, W. Bannwarth, S. Roller, A. Hebel, R. Haag, *Angew. Chem. Int. Ed.* **2002**, *41*, 3964–4000.
71 D. E. Bergbreiter, *Chem. Rev.* **2002**, *102*, 3345–3384.
72 H. Remmele, A. Köllhofer, H. Plenio, *Organometallics* **2003**, *22*, 4098–4103.
73 A. Datta, K. Ebert, H. Plenio, *Organometallics* **2003**, *22*, 4685–4691.
74 J. Hillerich, H. Plenio, *Chem. Commun.* **2003**, 3024–3025.
75 *Aqueous-Phase Organometallic Chemistry*, B. Cornils, W. A. Herrmann (eds.), Wiley–VCH, Weinheim, **1998**.
76 D. E. Bergbreiter, P. L. Osburn, A. Wilson, E. M. Sink, *J. Am. Chem. Soc.* **2000**, *122*, 9058–9064.
77 A. Köllhofer, H. Plenio, *Chem. Eur. J.* **2003**, *9*, 1416–1425.
78 A. Datta, H. Plenio, *Chem. Commun.* **2003**, 1504–1505.
79 T. Fukuyama, M. Shinmen, S. Nishitani, M. Sato, I. Ryu, *Org. Lett.* **2002**, *4*, 1691–1694.
80 C. Markert, W. Bannwarth, *Helv. Chim. Acta*, **2002**, *85*, 1877–1882.
81 E. Gonthier, R. Breinbauer, *Synlett.* **2003**, 1049–1051.
82 Y. Liao, R. Futhi, M. Reitman, Y. Zhang, *Tetrahedron Lett.* **2001**, *42*, 1815–1818.
83 B. M. Choudary, S. Madhi, N. S. Chowdari, M. L. Kantam, B. Sreedhar, *J. Am. Chem. Soc.* **2002**, *124*, 14127–14136.
84 B. M. Choudary, S. Madhi, M. L. Kantam, B. Sreedhar, Y. Iwasawa, *J. Am. Chem. Soc.* **2004**, *126*, 2292–2293.
85 M. Moreno-Manas, R. Pleixats, *Acc. Chem. Res.* **2003**, *36*, 638–643.

References and Notes to Section 1.3

1 C. Glaser, *Ber. Dtsch. Chem. Ges.* **1869**, *2*, 422–424; C. Glaser, *Ann. Chem. Pharm.* **1870**, *154*, 137–171.
2 P. Siemsen, R. C. Livingston, F. Diederich, *Angew. Chem.* **2000**, *112*, 2740–2767; *Angew. Chem., Int. Ed.* **2000**, *39*, 2632–2657.
3 W. Chodkiewicz, P. Cadiot, *C. R. Hebd. Seances Acad. Sci.* **1955**, *241*, 1055–1057; W. Chodkiewicz, *Ann. Chim. (Paris)* **1957**, *2*, 819–869.
4 A. S. Hay, *J. Org. Chem.* **1962**, *27*, 3320–3321.
5 P. Siemsen, U. Gubler, C. Bosshard, P. Günter, F. Diederich, *Chem. Eur. J.* **2001**, *7*, 1333–1341.
6 S. Kammermeier, R. R. Tykwinsky, P. Siemsen, P. Seiler, F. Diederich, *Chem. Comm.* **1998**, 1285–1286.
7 (a) G. Eglinton, A. R. Galbraith, *Chem. Ind. (London)* **1956**, 737–738; (b) G. Eglinton, A. R. Galbraith, *J. Chem. Soc.* **1959**, 889–896.
8 F. Sondheimer, *Pure Appl. Chem.* **1963**, *7*, 363–388.
9 A. P. H. J. Schenning, J.-D. Arndt, M. Ito, A. Stoddard, M. Schreiber, P. Siemsen, R. E. Martin, C. Boudon, J.-P. Gisselbrecht, M. Gross, V. Gramlich, F. Diederich, *Helv. Chim. Acta* **2001**, *84*, 296–334.
10 (a) L. G. Fedenok, V. M. Berdnikov, M. S. Shvartsberg, *J. Org. Chem. USSR* **1973**, *9*, 1806–1809; (b) L. G. Fedenok, V. M. Berdnikov, M. S. Shvartsberg, *J. Org. Chem. USSR* **1976**, *12*, 1385–1387; (c) L. G. Fedenok, V. M. Berdnikov, M. S. Shvartsberg, *J. Org. Chem. USSR* **1978**, *14*, 1328–1333; (d) L. G. Fedenok, V. M. Berdnikov, M. S. Shvartsberg, *J. Org. Chem. USSR* **1978**, *14*, 1334–1337; (e) H. M. Hoan, S. M. Brailovskii, O. N. Temkin, *Kinet. Catal. (Engl. Transl.)* **1994**, *35*, 242–246; (f) F. Bohlmann, H. Schönowsky, E. Inhoffen, G. Grau, *Chem. Ber.* **1964**, *97*, 794–800; (g) B. M. Mykhalichko, *Russ. J. Coord. Chem. (Engl. Transl.)* **1999**, *25*, 336–341.

11 (a) P. Cadiot, W. Chodkiewicz in *Chemistry of Acetylenes* (Ed.: H. G. Viehe), Dekker, New York, **1969**, chap. 9, pp. 597–647; (b) H. G. Viehe, *Chem. Ber.* **1959**, *92*, 3064–3075.

12 For detailed description of syntheses, isolations and characterizations, see: P. Siemsen, ETH Dissertation 13788, Zürich **2000**, pp. 96–138.

13 L. Brandsma, *Preparative Acetylenic Chemistry*, Elsevier, Amsterdam, **1988**, chap. 10, pp. 219–227.

References and Notes to Section 1.4

1 K. C. Nicolaou, W. M. Dai, S.C. Tsay, V. A. Estevez, W. Wrasidlo, *Science* **1992**, *256*, 1172–1178.

2 B. M. Trost, *Angew. Chem., Int. Ed. Engl.* **1995**, *34*, 259–281.

3 G. W. Parshall, S: D. Ittel, *Homogeneous Catalysis*, 2nd ed.; John Wiley & Sons: New York, **1992**.

4 B. M. Trost, C. Chan, G. Rühter, *J. Am. Chem. Soc.* **1987**, *109*, 3486–3487.

5 H. Katayama, M. Nakayama, T. Nakano, C. Wada, K. Akamatsu, F. Ozawa, *Macromolecules* **2004**, *37*, 13–17.

6 B.M. Trost, M. T. Sorum, C. Chan, A. E. Harms, G. Rühter, *J. Am. Chem. Soc.* **1997**, *119*, 698–708.

7 C. Yang, S. P. Nolan, *J. Org. Chem.* **2002**, *67*, 591–593.

8 M. Rubina, V. Gevorgyan, *J. Am. Chem. Soc.* **2001**, *123*, 11107–11108.

9 C. Bianchini, P. Frediani, D. Masi, M. Peruzzini, F. Zanobini, *Organometallics* **1994**, *13*, 4616–4632.

10 C. S. Yi, N. Liu, *Organometallics* **1996**, *15*, 3968–3971.

11 C. Slugovc, K. Mereiter, E. Zobetz, R. Schmid, K. Kirchner, *Organometallics* **1996**, *15*, 5275–5277.

12 W. T. Boese, A. S. Goldman, *Organometallics* **1991**, *10*, 782–786.

13 M. Nishiura, Z. Hou, Y. Wakatsuki, T. Yamaki, T. Miyamoto, *J. Am. Chem. Soc.* **2003**, *125*, 1184–1185.

14 C. S. Yi, N. Liu, A. L. Rheingold, L. M. Liable-Sands, *Organometallics* **1997**, *16*, 3910–3913.

15 J.-P. Qü, D. Masui, Y. Ishii, M. Hidai, *Chem. Lett.* **1998**, 1003–1004.

16 T. Ohmura, S. Yorozuya, Y. Yamamoto, N. Miyaura, *Organometallics* **2000**, *19*, 365–367.

17 Heeres, H. J.; Teuben, J. H. *Organometallics* **1991**, *10*, 1980–1986.

18 den Haan, K. H.; Wielstra, Y.; Teuben, J. H. *Organometallics* **1987**, 6, 2053–2060.

19 A. K. Dash, I. Gourevich, J. Q. Wang, J. Wang, M. Kapon, M. S. Eisen *Organometallics* **2001**, *20*, 5084–5104.

20 S. Pavlik, C. Gemel, C. Slugovc, K. Mereiter, R. Schmid, K. Kirchner, *J. Organomet. Chem.* **2001**, *617–618*, 301–310.

21 Y. Wakatsuki, H. Yamazaki, N. Kumegawa, T. Satoh, J. Y. Satoh, *J. Am. Chem. Soc.* **1991**, 113, 9604–9610.

22 M. O. Albers, D. J. A. de Waal, D. C. Liles, D. J. Robinson, E. Singleton, M. B. Wiege, *J. Chem. Soc., Chem. Commun.* **1986**, 1680–1682.

23 J. Le Paih, F. Monnier, S. Dérien, P. H. Dixneuf, E. Clot, O. Eisenstein, *J. Am. Chem. Soc.* **2003**, *125*, 11964–11975.

24 J. Le Paih, S. Derien, C. Bruneau, B. Demerseman, L. Toupet, P. H. Dixneuf, *Angew. Chem. Int. Ed.* **2001**, *40*, 2912–2915.

25 J. Le Paih, S. Dérien, B. Demerseman, C. Bruneau, P. H. Dixneuf, L. Toupet, G. Dazinger, K. Kirchner to be submitted.

26 B. M. Trost, A. J. Frontier, *J. Am. Chem. Soc.* **2000**, *122*, 11727–11728.

27 C. Gemel, G. Kickelbick, R. Schmid, K. Kirchner, *J. Chem. Soc., Dalton Trans.* **1997**, 2113 - 2117.

28 H. Singer, G. Wilkinson, *J. Chem. Soc. (A)* **1968**, 849–853.

29 B. M. Trost, S. Matsubara, J.J. Caringi, *J. Am. Chem. Soc.* **1989**, *111*, 8745–8746.

30 Y. Nishibayashi, M. Yamanashi, I. Wakiji, M. Hidai, *Angew. Chem., Int. Ed.* **2000**, *39*, 2909–2911.

31 P. J. Fagan, W. S. Mahoney, J. C. Calabrese, I. D. Williams, *Organometallics* **1990**, *9*, 1843–1852.

References and Notes to Section 1.5

1. C. Bruneau, P. H. Dixneuf, *Acc. Chem. Res.*, **1999**, *32*, 311–323.
2. C. Bruneau, *Ruthenium vinylidenes and allenylidenes in catalysis* in: *Ruthenium catalysts and fine chemistry*, Topics in Organomet. Chem., C. Bruneau, P. H. Dixneuf (Eds), Springer, 2004, vol. 11, pp 125–153.
3. Y. Sasaki, P. H. Dixneuf, *J. Chem. Soc., Chem. Commun.*, **1986**, 790–791.
4. R. Mahé, Y. Sasaki, C. Bruneau, P.H. Dixneuf, *J. Org. Chem.*, **1989**, *54*, 1518–1523.
5. J. Höfer, H. Doucet, C. Bruneau, P. H. Dixneuf, *Tetrahedron Lett.*, **1991**, *32*, 7409–7410.
6. J. Ipaktschi, J. Mohsseni-Ala, S. Uhlig, *Eur. J. Inorg. Chem.*, **2003**, 4313–4320.
7. H. Doucet, J. Höfer, C. Bruneau, P. H. Dixneuf, *J. Chem. Soc., Chem. Commun.*, **1993**, 850–851.
8. H. Doucet, B. Martin-Vaca, C. Bruneau, P. H. Dixneuf, *J. Org. Chem.*, **1995**, *60*, 7247–7255.
9. H. Doucet, J. Höfer, N. Derrien, C. Bruneau, P. H. Dixneuf, *Bull. Soc. Chim. France*, **1996**, *133*, 939–944.
10. H. Doucet, N. Derrien, Z. Kabouche, C. Bruneau, P. H. Dixneuf, *J. Organomet. Chem.*, **1997**, *551*, 151–157.
11. K. Melis, P. Samulkiewicz, J. Rynkowski, F. Verpoort, *Tetrahedron Lett.*, **2002**, *43*, 2713–2716.
12. L. J. Goossen, J. Paetzold, D. Koley, *Chem. Commun.*, **2003**, 706–707.
13. M. Jimenez Tenorio, M. C. Puerta, P. Valerga, F. J. Moreno-Dorado, F. M. Guerra, G. M. Massenet, *Chem. Commun.*, **2001**, 2324–2325.
14. M. Picquet, C. Bruneau, P. H. Dixneuf, *J. Chem. Soc., Chem. Commun.*, **1997**, 1201–1202.
15. M. Picquet, A. Fernandez, C. Bruneau, P. H. Dixneuf, *Eur. J. Org. Chem.*, **2000**, 2361–2366.
16. C. Bruneau, Z. Kabouche, M. Neveux, B. Seiller, P. H. Dixneuf, *Inorg. Chim. Acta*, **1994**, *222*, 155–163.
17. C. Gemel, G. Trimmel, C. Slugovc, S. Kremel, K. Mereiter, R. Schmid, K. Kirchner, *Organometallics*, **1996**, *15*, 3998–4004.
18. B. M. Trost, G. Dyker, R. J. Kulaviec, *J. Am. Chem. Soc.*, **1990**, *112*, 7809–7811.
19. B. M. Trost, R. J. Kulawiec, *J. Am. Chem. Soc.*, **1992**, *114*, 5579–5584.
20. B. M. Trost, Y. H. Rhee, *J. Am. Chem. Soc.*, **2002**, *124*, 2528–2533.
21. B. M. Trost, Y. H. Rhee, *J. Am. Chem. Soc.*, **1999**, *121*, 11680–11683.
22. F. E. McDonald, K. S. Reddy, *J. Organomet. Chem.*, **2001**, *617/618*, 444–452.
23. M. H. Davidson, F. E. McDonald, *Org. Lett.*, **2004**, *6*, 1601–1603.
24. W. W. Cutchins, F. E. McDonald, *Org. Lett.*, **2002**, *4*, 749–752.
25. P. Wipf, T.H. Graham, *J. Org. Chem.*, **2003**, *68*, 8798–8807.
26. F. E. McDonald, *Chem. Eur. J.*, **1999**, *5*, 3103–3106.
27. F. E. McDonald, M. M. Gleason, *J. Am. Chem. Soc.*, **1996**, *118*, 6648–6659.
28. F. E. McDonald, M. M. Gleason, *Angew. Chem. Int. Ed. Engl.*, **1995**, *34*, 350–352.
29. F. E. McDonald, C. B. Connolly, M. M. Gleason, T. B. Towne, K. D. Treiber, *J. Org. Chem.*, **1993**, *58*, 6952–6953.
30. M. Tokunaga, Y. Wakatsuki, *Angew. Chem. Int. Ed.*, **1998**, *37*, 2867–2868.
31. T. Suzuki, M. Tokunoga, Y. Wakatsuki, *Org. Lett.*, **2001**, *3*, 735–737.
32. P. Alvarez, M. Bassetti, J. Gimeno, G. Mancini, *Tetrahedron Lett.*, **2001**, *42*, 8467–8470.
33. D. B. Grotjahn, C. D. Incarvito, A. L. Rheingold, *Angew. Chem. Int. Ed.*, **2001**, *40*, 3884–3887.
34. T. Suzuki, M. Tokunoga, Y. Wakatsuki, *Tetrahedron Lett.*, **2002**, *43*, 7531–7533.

2
Asymmetric Hydrocyanation of Alkenes
Jos Wilting and Dieter Vogt

2.1
Introduction

The hydrocyanation of alkenes [1] has great potential in catalytic carbon–carbon bond-formation because the nitriles obtained can be converted into a variety of products [2]. Although the cyanation of aryl halides [3] and carbon-hetero double bonds (aldehydes, ketones, and imines) [4] is well studied, the hydrocyanation of alkenes has mainly focused on the DuPont adiponitrile process [5]. Adiponitrile is produced from butadiene in a three-step process via hydrocyanation, isomerization, and a second hydrocyanation step, as displayed in Figure 1. This process was developed in the 1970s with a monodentate phosphite-based zerovalent nickel catalyst [6].

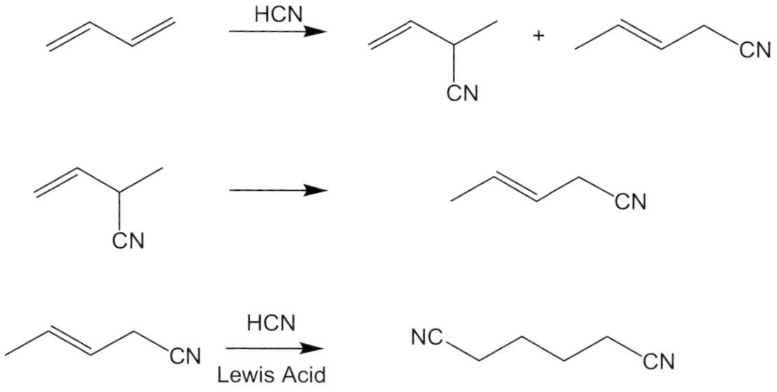

Figure 1. The three-step adiponitrile process.

Although several enzymes can enantioselectively catalyze the hydrocyanation of $R_2C=O$ and $R_2C=NR$ bonds [7], (asymmetric) hydrocyanation of C=C double bonds has no precedents in biology. In homogeneous catalysis asymmetric hydrocyanation is still underdeveloped, as is apparent from the relatively few reports in the literature. In the following paragraphs a short overview will be given divided into the two major substrate classes investigated, cyclic (di)enes and vinylarenes.

Handbook of C–H Transformations. Gerald Dyker (Ed.)
Copyright © 2005 WILEY-VCH Verlag GmbH & Co. KGaA, Weinheim
ISBN: 3-527-31074-6

2.1.1
Cyclic (Di)enes

In 1979 Elmes and Jackson [8–10] described the asymmetric hydrocyanation of norbornene and norbornadiene, with Pd(DIOP)$_2$ as catalyst, resulting in *exo*-2-cyanonorbornane. Parker et al. [11, 12] investigated this reaction in more detail and studied the coordination chemistry of (DIOP)Pd(C$_2$H$_4$) in the presence of HCN. Under these conditions a hydrido cyano and an alkyl cyano palladium complex were detected by NMR. Takaya et al. [13] applied the phosphine–phosphite ligand binaphos with palladium and nickel and obtained ee (enantiomeric excesses) of 48 % (Pd) and 40 % (Ni) with yields of up to 50 %. Pringle and Baker [14, 15] investigated the phosphite ligand 1, based on (R)-2,2'-binaphthol, with nickel and obtained isolated yields of 40–70 % with ee of 28–38 % (M/L/S/ACH [metal/ligand/substrate/acetone cyanohydrin] 1:2:700:350) at different temperatures. Chan et al. [16] used the diphosphite ligand 2, based on (S)-binaphthol, with nickel and obtained the highest ee, 55.0 %, for hydrocyanation of norbornene; the yield was 89 % (M/L/S/ACH 1:7:100:110).

Figure 2. Chiral ligands used in the asymmetric hydrocyanation of norbornene.

2.1.2
Vinylarenes

Chan et al. also tested the Ni(ligand 2) system with other substrates, and obtained 50 % ee (89 % yield) in the hydrocyanation of styrene [16]. With Ni(cod)$_2$ (nickel(0)-bis-[1,5-cyclooctadiene]) in the presence of ligand 3 Babin et al. obtained an ee of 64 % for styrene as substrate, although the regioselectivity of the system was modest (~65 %) [17]. Vogt et al. investigated diphosphonite ligands with an achiral xanthene backbone and binaphthoxy substituents [18]. These resulted in ee of 42 % (styrene), 63 % (4-isobutylstyrene), and 29 % (MVN [6-methoxy-2-vinylnaphthalene]) with ligand 4.

RajanBabu and Casalnuovo [19, 20] tested diphosphinite ligand systems (5 and 6 in Figure 4) based on carbohydrate backbones. The steric and electronic properties depended on the substituents on the aryl groups on the phosphorus atoms. The use of different chlorophosphine precursors led to the electronically asymmetric ligand 6. This approach resulted in both enantiomers of naproxen nitrile from MVN in 91 % ee (S)-nitrile (ligand 5) and 95 % ee (R)-nitrile (ligand 6) at 0 °C.

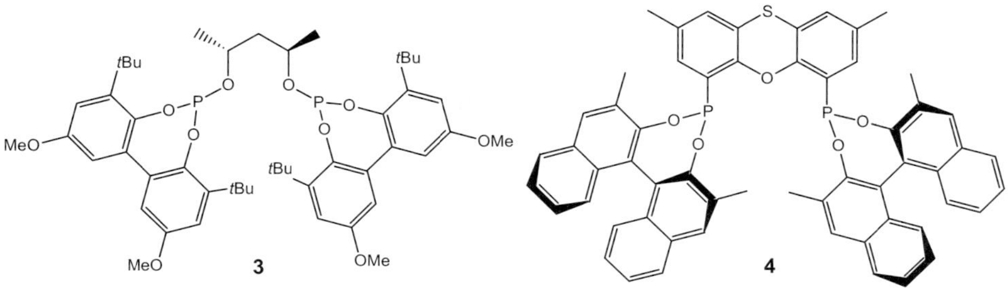

Figure 3. Ligands studied in the asymmetric hydrocyanation of vinylarenes.

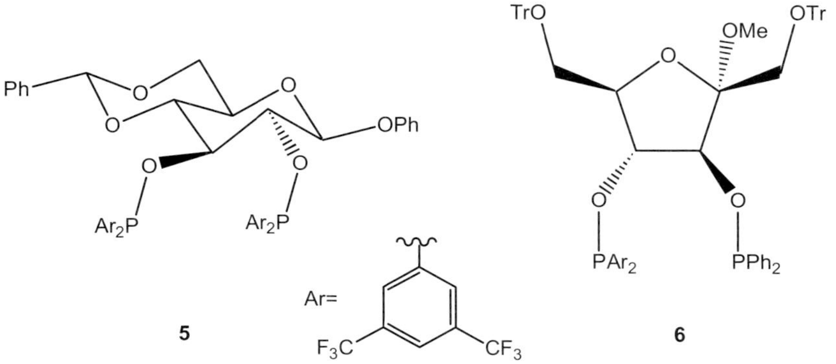

Figure 4. Ligands used by RajanBabu and coworkers.

2.2
Mechanism

Tolman, Druliner, and McKinney [5] were pioneers in nickel-catalyzed hydrocyanation; they used monodentate phosphites mainly to understand and improve the adiponitrile process. Although bidentate ligands give better results in the adiponitrile process [21], mechanistic studies with these systems are rare; bidentate phosphinites have been studied in the asymmetric hydrocyanation of MVN [19].

The hydrocyanation of ethylene with a monodentate ligand has been shown to have zero-order rates in substrate and in HCN but second-order in catalyst [22]. The possible intermediate $[(L)Ni(CN)(C_2H_5)(C_2H_4)]$ has been characterized by NMR, leading to the conclusion that reductive elimination is the rate-determining and most important step. During this step an additional ligand would need to recombine with this nickel species resulting in a second-order rate law in catalyst. This is based on the fact that five-coordinated nickel complexes are more effective than the four-coordinated complexes in the reductive elimination [23].

Accordingly, it seems plausible that the reductive elimination step in the catalytic cycle with bidentate ligands (**F** to **E** in Figure 5) follows an associative mechanism in which first an alkene or nitrile (displayed as S) coordinates to the nickel, thereby stabilizing the zero-valent complex **E**, which is formed during the reductive elimination. Complex **E** can then to exchange S for HCN (resulting in **B**) or the substrate (resulting in **D**). Labeling studies with deuterated MVN [19] and 2M3BN (2-methyl-3-butenenitrile) [24] revealed an equilibrium between species **F** and **C** in the catalytic cycle. Proton migration (**C** to **F** or **H**) is favored to the branched species **F** by the η^1–η^3 equilibrium (**F**–**G**), which is not possible for the linear species **H**. Although several η^3-benzylic complexes have been characterized [25, 26], the (α-methyl)benzylic cyano complex (**G**) could not be isolated or even observed. The η^3–η^1 equilibrium has been investigated for an aliphatic allyl cyano nickel(II) complex [27].

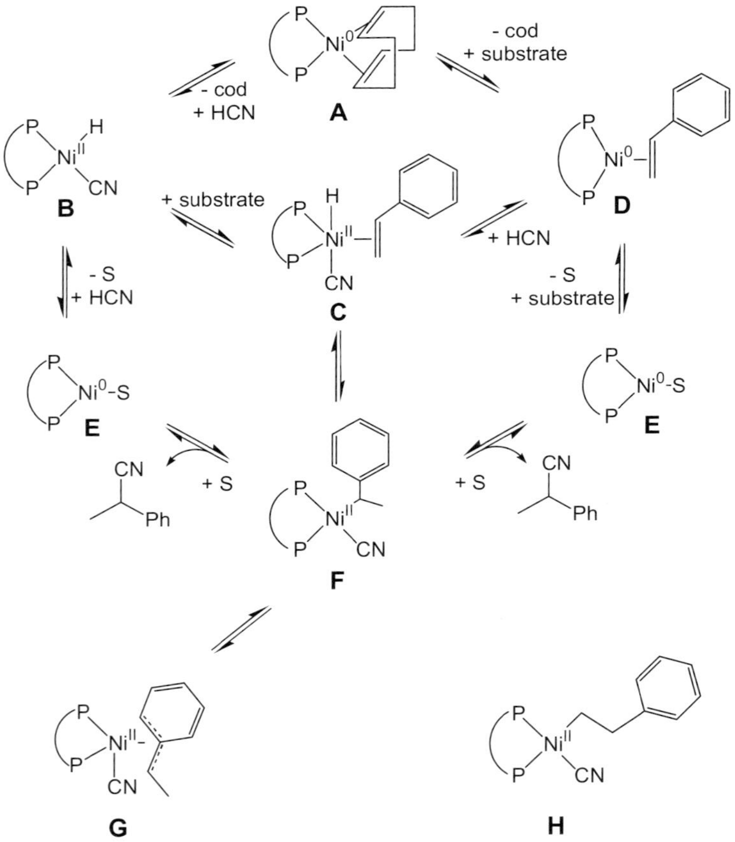

Figure 5. The catalytic cycle of hydrocyanation of styrene as example of the vinylarenes.

For hydrocyanation of MVN with bidentate phosphinites zero-order in substrate and in HCN was again observed. Yet first-order in catalyst was obtained because of the bidentate ligand. The hydrocyanation followed in time should, in theory, fit a simple zero-order model (d[RCN]/dt = k_1[Ni]). The catalyst is, however, deactivated in time, by formation of an insoluble dicyanonickel complex $(PP)Ni(CN)_2$, making deactivation of the catalyst an important issue. The kinetic model is, therefore, more consistent with the experimental data when both zero-order product formation ($d[RCN]/dt = k_1[Ni]$) and deactivation ($-d[Ni]/dt = k_2[HCN][Ni]$) are used, as is displayed in Figure 6.

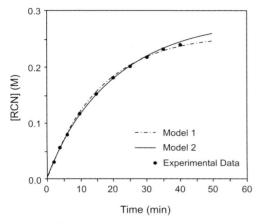

Figure 6. The hydrocyanation of MVN, kinetic model 1: $d[RCN]/dt = k_1[Ni]$ and $-d[Ni]/dt = k_2[HCN][Ni]$ with $k_1 = 27$ min^{-1} and $k_2 = 0.15$ M^{-1} min^{-1}. This figure is reprinted from Ref. 19. Copyright 1994 American Chemical Society.

Several other interesting effects have been observed in the reductive elimination and hydrocyanation reactions. The bite angle [28] has a significant effect in the reductive elimination. A 10^4 fold rate increase was observed progressing from a small bite angle (~85°) to a larger bite angle in (DIOP)Pd(CH$_2$TMS)CN (~100°, TMS = trimethylsilyl) [29].The rate depends on the alkyl group, causing it to vary by orders of magnitude (CH$_2$CMe$_3$ ~10 × CH$_2$TMS >> CH$_3$). Addition of the Lewis acid AlPh$_3$ increases the rate by a factor of 50 [30].

The electronic properties of the phosphorus groups affect the activity and stability of the catalyst during the hydrocyanation. Diphosphine and diphosphinite catalyst systems benefit greatly from electron-withdrawing substituents on the aryl groups [19, 31]. Benzene, toluene, and hexane [32] are usually the solvents of choice, coordinating solvents, for example CH$_3$CN, have been shown to have a negative effect, resulting in a dramatic drop in yield and ee. The effect of the HCN source (either directly as HCN or in situ via ACH), has not been studied carefully. The concentration of free HCN is much lower when ACH is used, leading to less deactivation. The role of the acetone formed during the reaction has not been determined, however. On the other hand a low concentration of HCN can also be achieved by slowly bubbling HCN through the reaction mixture or by adding an HCN solution by syringe pump [33].

2.3
Scope and Limitations

In asymmetric hydrocyanation reactions the desired isomers are the chiral branched products only. Good regioselectivity toward the branched product (>98 %) is limited to vinylarenes. Hydrocyanation of 1,3-dienes gives a variety of mixtures depending on the catalyst and conditions; 1-alkenes give the linear nitrile as major product [34]. Both are seen in the adiponitrile process in which the unwanted branched 2M3BN (hydrocyanation product from 1,3-butadiene) is isomerized to the linear product 3-pentenenitrile, which is then hydrocyanated by in-situ isomerization to 4-pentenenitrile, resulting in the linear adiponitrile. Thus vinylarenes and cyclic alkenes (mainly norbornene) are usually the substrates of choice for the asymmetric hydrocyanation. Hopefully 1,3-dienes will become feasible substrates in the near future.

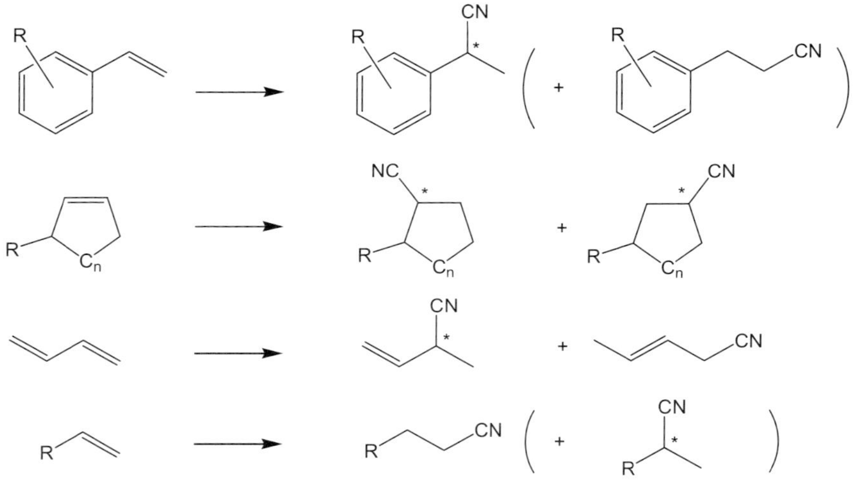

Figure 7. Substrate variations in asymmetric hydrocyanation (minor products between brackets).

The poor turn-over numbers (TON) in the hydrocyanation reactions are another limitation, because of degradation of the catalyst, which still has to be overcome. The maximum TON reached so far have been in the order of 500–750, which is extremely low compared with other homogenous catalytic reactions.

Experimental

The preparative synthesis of the (S) nitrile described here was originally reported by RajanBabu and Casulnuovo [19]; the synthesis of the (R) nitrile is reported elsewhere [20].

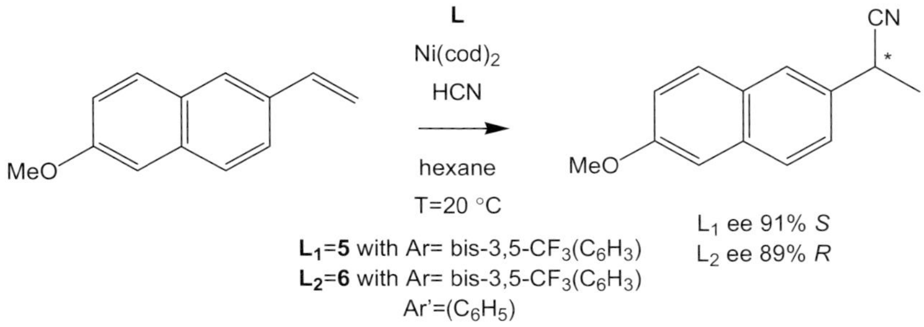

Figure 8. The asymmetric hydrocyanation of MVN.

Caution! HCN is a highly toxic, volatile liquid (bp 27 °C) that is also susceptible to explosive polymerization in the presence of base catalysts. It should be handled only in a well-ventilated fume hood and by teams of at least two technically qualified persons who have received appropriate medical training for treating HCN poisoning. Sensible precautions include having available proper first aid equipment and HCN monitors. Uninhibited HCN should be stored at a temperature below its melting point (-13 °C). Excess HCN may be disposed by addition to aqueous sodium hypochlorite, which converts the cyanide to cyanate.

All reactions are carried out under inert atmospheres using standard Schlenk techniques. All solvents are dried and distilled before use. Ni(cod)$_2$ is extremely air-sensitive, especially in solution, and thermally unstable. Ni(cod)$_2$ [35], (Et$_2$N)PCl$_2$ [36], 3,5-bis-(trifluoromethyl)bromobenzene, and (–)-phenyl-4,6-O-benzylidene-β-D-glucopyranoside are commercially available.

Bis[3,5-bis(trifluoromethyl)phenyl]chlorophosphine

A 1.0 M solution of [3,5-bis(trifluoromethyl)phenyl]magnesium bromide is prepared by slow addition of 18.5 g (60 mmol) 3,5-bis-(trifluoromethyl)bromobenzene in 40 mL THF to a slurry of Mg turnings in 20 mL THF. This is stirred for 1 h at 20 °C and, after filtration, slowly added to a solution of 5.0 g (29 mmol) (Et$_2$N)PCl$_2$ in 30 mL THF at 0 °C. After stirring for 2 h the mixture obtained is concentrated in vacuo. Cyclohexane (100 mL) is added and the mixture is filtered through celite resulting in a solution of bis[3,5-bis(trifluoromethyl)phenyl](diethylamino)phosphine. Dry HCl is passed through this solution for 1 h (a solution of HCl in diethyl ether, which is commercially available, can also be used). After filtration and concentration, 12.4 g (88 %) of the chlorophosphine can be collected as a white solid. ^{31}P{^{1}H} NMR (C$_6$D$_6$): d 70.4 (s)

Preparation of Ligand 5

A solution of (3,5-(CF$_3$)$_2$C$_6$H$_3$)$_2$PCl (1.246 g, 2.53 mmol) in 5 mL CH$_2$Cl$_2$ is cooled to 0 °C and added dropwise to a solution of phenyl 4,6-O-benzylidene-β-D-glucopyranoside (0.397 g, 1.15 mmol) and dimethylaminopyridine (0.030 g, 0.25 mmol) in 15 mL CH$_2$Cl$_2$/pyridine (1:1) at 0 °C. The reaction mixture is warmed to room temperature, stirred overnight, and concentrated to dryness in vacuo. Hot benzene is added to the resulting solids and the mixture is filtered. The filtrate obtained is concentrated in vacuo to a dry white solid. This white solid is then washed with 2 mL cold hexane and dried in vacuo to give 1.135 g (79 %) ligand **5**. ^{31}P{^{1}H} NMR (C$_6$D$_6$): *d* 108.4 (s, 1P), 107.5 (s, 1P).

Hydrogen Cyanide (HCN) [37]

Iron(II)sulfate heptahydrate (1.9 g) is added to 125 mL sulfuric acid. This solution is heated under an argon stream to 80 °C. Subsequently a solution of 96.5 g KCN (1.49 mol) in 200 mL H$_2$O is added by syringe pump during 2 h and the gaseous HCN produced is led over CaCl$_2$ after which it is condensed in a thick glass bottle at -78 °C. The argon, which might contain traces of HCN, is led through a solution of 100 mL 35 % H$_2$O$_2$ and 60 g KOH in 200 mL H$_2$O, which neutralizes the cyanide. Yield: ~20 mL (35 %). Instead of condensing the HCN it is also possible to use the HCN–argon mixture immediately in the catalytic hydrocyanation reaction.

Preparative Asymmetric Hydrocyanation of MVN

To a solution of Ni(cod)$_2$ (0.059 g, 0.22 mmol) and **5** (0.271 g, 0.22 mmol) in 10 mL hexane is added 4.00 g MVN (21.7 mmol) and 110 mL hexane. An approximately 2 M solution of HCN in toluene (11 mL, 22 mmol) is added to the resulting slurry within 2.5 h by using an addition funnel or syringe pump. The initially heterogeneous solution becomes an orange–brown homogenous solution about halfway through the addition after which the product will precipitate as a white powder. Conversion can be monitored by GC analysis and if the reaction does not go to completion additional HCN solution can be added slowly. The original report showed ~94 % conversion. The reaction is stirred overnight, purged with argon for 5 min to remove traces of HCN, and then benzene is added to dissolve all of the solids.

After concentration of the reaction mixture in vacuo, hexane (200 mL) is added and the product is collected by filtration (yield 3.5 g of an off-white solid). The filtrate is then concentrated in vacuo to about 50 mL to isolate a second crop of product by filtration, (yield 0.5 g). A third crop of product can be isolated by flash chromatographic work-up (silica) of the filtrate (90:10 hexane–Et$_2$O), yield 0.4 g, total yield 4.4 g (96 %). Recrystallization from boiling 10 % ether–hexanes results in optically pure (*S*)-nitrile (occasionally it may be necessary to perform a second recrystallization). Mp 99–100 °C, $[\alpha]_D^{23}$ = -29.4 ± 0.8 ° (*c* = 1) Enantiomeric excess can be measured by HPLC (Chiracel OJ column with 5 % isopropanol–hexane as eluent (flow rate 1 mL min^{-1}) at 40 °C, retention times (min): MVN 28.52, (*S*)-nitrile 32.87, (*R*)-nitrile 40.60).

References and Notes

1 Huthmacher, K.; Krill, S. Reactions with hydrogen cyanide (hydrocyanation). In *Applied Homogeneous Catalysis with Organometallic Compounds*, 2nd edn; Cornils, B., Hermann, W. A., Eds.; Wiley–VCH: Weinheim, 2002; Vol. 1 pp 465–486.

2 Pollak, P.; Romeder, G.; Hagedorn, F.; Gelbke, H. Nitriles. In *Ullman's Encyclopedia of Industrial Chemistry*, 5th edn; Wiley–VCH: Weinheim, 1985–1996; Vol. A17 pp 363–376.

3 Sundermeier, M.; Zapf, A.; Beller, M. *Eur. J. Inorg. Chem.* **2003**, (19), 3513–3526.

4 Groger, H. *Chem. Eur. J.* **2001**, *7* (24), 5246–5251.

5 Tolman, C. A.; McKinney, R. J.; Seidel, W. C.; Druliner, J. D.; Stevens, W. R. *Adv. Catal.* **1985**, *33*, 1–46.

6 Drinkard, W. C. Hydrocyanation of olefins using selected nickel phosphite catalysts. US 3,496,215, **1970**.

7 Gregory, R. J. H. *Chem. Rev.* **1999**, *99* (12), 3649–3682.

8 Elmes, P. S.; Jackson, W. R. *J. Am. Chem. Soc.* **1979**, *101* (20), 6128–6129.

9 Elmes, P. S.; Jackson, W. R. *Aust. J. Chem.* **1982**, *35* (10), 2041–2051.

10 Jackson, W. R.; Lovel, C. G. *Aust. J. Chem.* **1982**, *35* (10), 2053–2067.

11 Hodgson, M.; Parker, D. *J. Organomet. Chem.* **1987**, *325* (1/2), C27–C30.

12 Hodgson, M.; Parker, D.; Taylor, R. J.; Ferguson, G. *Organometallics* **1988**, *7* (8), 1761–1766.

13 Horiuchi, T.; Shirakawa, E.; Nozaki, K.; Takaya, H. *Tetrahedron: Asymmetry* **1997**, *8* (1), 57–63.

14 Baker, M. J.; Pringle, P. G. *J. Chem. Soc., Chem. Commun.* **1991**, (18), 1292–1293.

15 Baker, M. J.; Harrison, K. N.; Orpen, A. G.; Pringle, P. G.; Shaw, G. *J. Chem. Soc., Chem. Commun.* **1991**, (12), 803–804.

16 Yan, M.; Xu, Q. Y.; Chan, A. S. C. *Tetrahedron: Asymmetry* **2000**, *11* (4), 845–849.

17 Babin, J. E.; Whiteker, G. T. Asymmetric syntheses using optically active metal–ligand complex catalysts. US 5,360,938, **1994**.

18 Goertz, W.; Kamer, P. C. J.; Van Leeuwen, P. W. N. M.; Vogt, D. *Chem. Eur. J.* **2001**, *7* (8), 1614–1618.

19 Casalnuovo, A. L.; Rajanbabu, T. V.; Ayers, T. A.; Warren, T. H. *J. Am. Chem. Soc.* **1994**, *116* (22), 9869–9882.

20 Rajanbabu, T. V.; Casalnuovo, A. L. *J. Am. Chem. Soc.* **1996**, *118* (26), 6325–6326.

21 Kreutzer, K. A.; Tam, W. Hydrocyanation process and multidentate phosphite and nickel catalyst composition therefore. US 5,663,369, **1997**.

22 McKinney, R. J.; Roe, D. C. *J. Am. Chem. Soc.* **1985**, *107* (1), 261–262.

23 Tatsumi, K.; Nakamura, A.; Komiya, S.; Yamamoto, A.; Yamamoto, T. *J. Am. Chem. Soc.* **1984**, *106* (26), 8181–8188.

24 Druliner, J. D. *Organometallics* **1984**, *3* (2), 205–208.

25 Gatti, G.; Lopez, J. A.; Mealli, C.; Musco, A. *J. Organomet. Chem.* **1994**, *483* (1/2), 77–89.

26 Albers, I.; Alvarez, E.; Campora, J.; Maya, C. M.; Palma, P.; Sanchez, L. J.; Passaglia, E. *J. Organomet. Chem.* **2004**, *689* (4), 833–839.

27 Brunkan, N. M.; Jones, W. D. *J. Organomet. Chem.* **2003**, *683* (1), 77–82.

28 Casey, C. P.; Whiteker, G. T. *Isr. J. Chem.* **1990**, *30* (4), 299–304.

29 Marcone, J. E.; Moloy, K. G. *J. Am. Chem. Soc.* **1998**, *120* (33), 8527–8528.

30 Huang, J.; Haar, C. M.; Nolan, S. P.; Marcone, J. E.; Moloy, K. G. *Organometallics* **1999**, *18* (3), 297–299.

31 Goertz, W.; Keim, W.; Vogt, D.; Englert, U.; Boele, M. D. K.; Van der Veen, L. A.; Kamer, P. C. J.; Van Leeuwen, P. W. N. M. *J. Chem. Soc., Dalton Trans.* **1998**, (18), 2981–2988.

32 The catalyst precursor is not very soluble in hexane and gives a slurry which becomes homogenous during catalysis

33 The slow addition of HCN to the reaction mixture is very important;

we obtained good results using a syringe pump to add the HCN solution

34 Goertz, W.; Kamer, P. C. J.; Van Leeuwen, P. W. N. M.; Vogt, D. *Chem. Commun.* **1997**, (16), 1521–1522.

35 Schunn, R. A. *Inorg. Synth.* **1974**, *15*, 5–9.

36 Whitaker, C. M.; Kott, K. L.; McMahon, R. J. *J. Org. Chem.* **1995**, *60* (11), 3499–3508.

37 Slotta, K. H. *Berichte d. D. Chem. Gesellschaft* **1934**, *67*, 1028–1030.

III
C–H Transformation at sp²-hybridized Carbon Atoms

1
C–H Transformation at Arenes

1.1
Direct Oxidation of Arenes to Phenols and Quinones
Vsevolod V. Rostovtsev

Dedicated to Professor John E. Bercaw on the occasion of his 60th birthday.

1.1.1
Introduction

Direct oxidation of arenes to phenols is a difficult reaction with enormous potential for both synthetic and industrial chemistry. Although metallation of arenes, whether with main group or transition metals, to give compounds with metal–aryl bonds is very common, a one-step transformation of benzene C–H bonds to C–OH bonds is not nearly as widespread. Many complexes capable of arene C–H bond activation do not survive under the oxidizing conditions required for hydroxylation. When metal–carbon bonds are actually formed during catalysis, transformation of Ar–M bonds to Ar–OH bonds requires not only a robust catalyst but also a selective oxidant as well as acceptable rates of ligand exchange. Although the aromatic π-system is an attractive target for an electrophilic attack, it is with rare exceptions less reactive towards radicals and nucleophiles, limiting potential approaches to arene hydroxylation. Furthermore, since phenols are more electron-rich than the starting materials, overoxidation is a problem. Several approaches have been used to overcome these problems and the solutions are detailed below.

1.1.2
Radical Hydroxylations

The first example of the direct hydroxylation of arenes to phenols was reported in 1900 and described oxidation of benzene with a mixture of iron(II) sulfate and hydrogen peroxide (Fenton's reagent) [1]. Low yields of phenol (21 % yield based on H_2O_2) are typically obtained in such cases, in addition to significant amounts of biphenyl (24 % yield based on H_2O_2) [1–3]. Both selectivity to phenol and overall

Handbook of C–H Transformations. Gerald Dyker (Ed.)
Copyright © 2005 WILEY-VCH Verlag GmbH & Co. KGaA, Weinheim
ISBN: 3-527-31074-6

yield can be improved by using oxidants stronger than iron(III) [2]. With copper(II), for example, the yield of phenol increases to 57% and the ratio of phenol to biphenyl improves to 140:1. Similar results are obtained with the peroxydisulfate/iron(II) system [4]. Electrocatalytic hydroxylation coupled with continuous extraction also gives better yields of phenols [5]. Udenfriend's reagent is another iron(II)-based system capable of radical arene hydroxylation. Molecular oxygen in combination with ascorbic acid is used as the oxidant in Udenfriend's system [6, 7]. Molecular oxygen can also be used directly in combination with copper(I) in aqueous sulfuric acid, although the yield of phenol is low (24%) and a significant amount of hydroquinone is formed at the same time (8% yield) [8, 9]. Photochemically generated hydroxyl radicals can also be used for arene hydroxylation [10].

The distinguishing feature of all these systems is the generation of a highly reactive hydroxyl radical which is responsible for arene hydroxylation (Scheme 1) [11–13]. The radical is produced in the reaction between hydrogen peroxide and an iron(II) ion, or any other transition metal ion with a suitable redox potential. Next, the hydroxyl radical adds to an arene molecule and produces a hydroxycyclohexadienyl intermediate, which has been observed experimentally under the reaction conditions [14]. Several pathways are available for further transformation of the radical – oxidation to phenol with iron(III) or other oxidant; dimerization followed by dehydration to afford biphenyl and loss of HO$^-$ to give an aromatic radical cation, which can be reduced back to starting material by iron (II); or dimerization followed by dehydration to afford biphenyl. With some substrates (*e.g.* phenylacetic acid) elimination or rearrangement of the aromatic side chains is also observed [2]. The exact fate of the hydroxycyclohexadienyl radical depends on the acidity of the reaction media and on the nature of the oxidants present in the system. Typically, a mixture of phenols and biphenyls is formed. With copper(II) or excess of iron(III) present, the oxidation step is favored and higher yields of phenol are observed, along with better selectivities. The acidity of the reaction mixture also plays a role – the phenol/biphenyl ratio increases with decreasing pH. The same mechanistic manifold can also be accessed in the hydroxylation of arenes with peroxydisulfates.

In more general terms, a review of product yields and distributions in radical hydroxylations shows that a great variety of yields and selectivities can be achieved by careful manipulation of experimental conditions. Occasionally, formation of biphenyl, which is often regarded as indicating the presence of a radical-based process, can be completely suppressed. Therefore, it is very risky to make statements about the mechanism of hydroxylation based solely on product distributions and without thorough mechanistic investigations [11].

Compared with the starting arenes, phenols are more reactive toward the hydroxyl radical, resulting in overoxidation to quinones and decreasing selectivity and yield of the reaction. Physical separation of products from the starting materials by means of a biphasic reaction mixture can be successfully used in this case [15]. Yields as high as 80% (based on benzene) have been reported with cetyl-

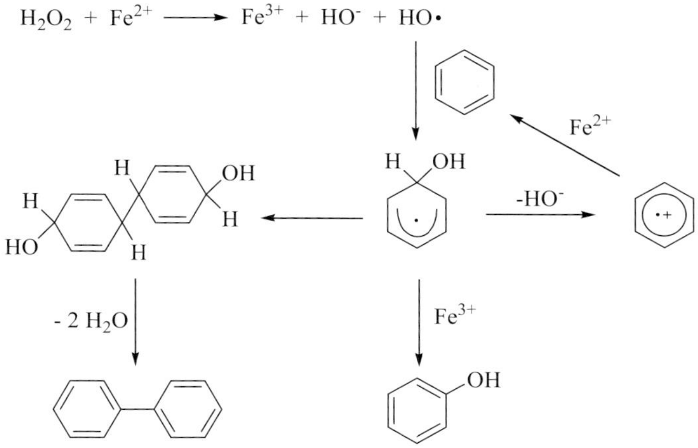

$H_2O_2 + Fe^{2+} \longrightarrow Fe^{3+} + HO^- + HO\cdot$

Scheme 1

trimethylammonium bromide phase-transfer catalyst (0.67 mol%) and iron(III) nitrate (0.33 mol%), using 0.5 M hydrogen peroxide (50 mol%) in a water–benzene mixture at pH 3 and 50 °C. No overoxidation was observed, and the source of extra oxidizing equivalents (0.8 equiv phenol formed with 0.5 equiv H_2O_2) was not discussed in the original work.

Further improvement of the Fe(II)/H_2O_2 system was recently reported by Bianchi [16–18]. Selectivity and yield were greatly improved by using a pyrazine-carboxylate ligand (Eq. 1). Physical separation of the oxidant (H_2O_2) from the products, achieved by using a biphasic system, with careful optimization of the auxiliary pyrazine-carboxylate ligand and of reaction conditions, greatly improved selectivity (78 % based on H_2O_2 and 85 % based on benzene) and conversion (up to 8.6 % based on benzene) of the phenol synthesis. The best results were obtained with 3-carboxypyrazine-N-oxide.

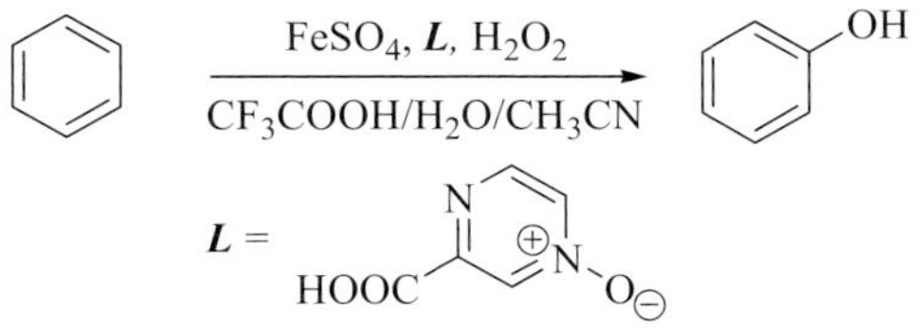

Although significant improvements have been made in the synthesis of phenol from benzene, the practical utility of direct radical hydroxylation of substituted arenes remains very low. A mixture of *ortho*-, *meta*- and *para*-substituted phenols is typically formed. Alkyl substituents are subject to radical H-atom abstraction, giving benzyl alcohol, benzaldehyde, and benzoic acid in addition to the mixture of cresols. Hydroxylation of phenylacetic acid leads to decarboxylation and gives benzyl alcohol along with phenolic products [2]. A mixture of naphthols is produced in radical oxidations of naphthalene, in addition to diols and hydroxyketones [19].

In 1983, Mimoun and co-workers reported that benzene can be oxidized to phenol stoichiometrically with hydrogen peroxide in 56 % yield, using peroxo-vanadium complex **1** (Eq. 2) [20]. Oxidation of toluene gave a mixture of *ortho*-, *meta*- and *para*-cresols with only traces of benzaldehyde. The catalytic version of the reaction was described by Shul'pin [21] and Conte [22]. In both cases, conversion of benzene was low (0.3–2 %) and catalyst turned over 200 and 25 times, respectively. The reaction is thought to proceed through a radical chain mechanism with an electrophilic oxygen-centered and vanadium-bound radical species [23].

$$\text{complex } \mathbf{1} + C_6H_6 \xrightarrow{CH_3CN,\ 2h,\ rt} [\text{intermediate}] + C_6H_5OH \quad (2)$$

56 %

Hydrogen peroxide can be replaced by molecular oxygen in vanadium-catalyzed hydroxylations of arenes, as first reported by Mukaiyama (Eq. 3) [24]. Improved versions were later described by Battistel (VCl$_3$/ascorbate) [25] and Shul'pin ((*n*-Bu)$_4$NVO$_3$/pyrazine-2-carboxylic acid/ascorbic acid) [26]. It is noteworthy that a molecular oxygen/sacrificial reductant (crotonaldehyde, zinc, or ascorbate) combination was used (*cf.* Udenfriend's reagent above) and that the mechanism is most likely to involve hydroxyl radicals.

$$C_6H_6 \xrightarrow[CH_3C(O)CH_2CH_3/CH_3CO_2H]{\text{cat. } VO(\mathbf{L})_2,\ O_2,\ rt,\ \text{CH}_2=\text{CHCHO}} C_6H_5OH + C_6H_5OC(O)CH=CHCH_3 \quad (3)$$

$$\mathbf{L} = \text{3-}n\text{-butyl-2,4-pentanedionate}$$

21 %   1 %

1.1.3
Electrophilic Hydroxylations

Electrophilic hydroxylation of arenes with hydrogen peroxide proceeds in low yield and is hampered by phenol overoxidation. The yield and selectivity improve when either a strong Broensted (HF, FSO$_3$H–SO$_2$ClF) [27] or Lewis (AlCl$_3$) [28] acid is used. These reagents associate with phenolic oxygen and discourage further hydroxylation. In the former case, benzene was converted to phenol in 54–67 % yield. With substituted arenes, mostly ortho and para isomers were obtained in 43–82 % yields. With xylenes, migration of the methyl group during hydroxylation was observed (Eq. 4), in line with the proposed formation of hydroxyarenium intermediates.

Electron-rich arenes, for example phenols, anisoles, and arylamines, can be hydroxylated with potassium persulfate in basic media [29]. Hydroxylation of phenols and anisoles is known as the Elbs oxidation (Eq. 5) [30]. The process is performed at high pH and results in the formation of an aryl sulfate ester which is hydrolyzed in a separate step. The yields are moderate and para-substitution is usually preferred. Alcohols, aldehydes, and olefins are tolerated in this process. The reaction is thought to proceed via electrophilic attack of persulfate on the phenolate ion.

Hydroxylation of arylamines with persulfate ion, or Boyland–Sims oxidation, gives *ortho*-substituted aminophenols in good yields [29]. As with the Elbs oxidation, the procedure is also carried out in two steps – first, treatment with the oxidant to obtain an aminophenyl sulfate ester and, second, hydrolysis to obtain the final product. Primary, secondary and tertiary amines can all be used in this reaction. The ortho product is formed, except when no *ortho*-positions are available, which leads to *para*-substitution. Electrophilic attack on the *ipso*-carbon is believed to be the most likely mechanism, although minor radical pathways also seem to be present.

Hydroxylations of electron-rich arenes (phenols and hydroxyarenes) can be achieved catalytically with (P–P)Pt(CF$_3$)X and H$_2$O$_2$ (where P–P is dppe and X is (CH$_2$Cl$_2$)ClO$_4$) [31]. The ortho- and para-isomers are formed, with ortho substitution preferred. The authors propose that the higher yields of *ortho*-hydroxylated products are a manifestation of catalyst pre-coordination.

Palladium-catalyzed hydroxylation of benzene to phenol with oxygen can be accomplished by use of a sacrificial co-reductant, for example carbon monoxide [32, 33]. The reaction is carried out in acetic acid under 15 atm dioxygen and 15 atm carbon monoxide at 180 °C giving phenol in 25 % yield. The catalyst is a mixture of $Pd(O_2CCH_3)_2$ and 1,10-phenanthroline. Carbon monoxide serves as an oxygen atom acceptor and can be replaced with dihydrogen [34]. Carbon dioxide and water, respectively, are formed as by-products. The use of a co-reductant can be avoided altogether if heteropolyacids are used as co-catalysts [35]. The scope of these processes is still mainly limited to benzene, although a rare example of hydroxylation-carboxylation of biphenyl with catalytic $Pd(O_2CCF_3)_2$, giving a mixture of 3′- and 4′-hydroxybiphenylcarboxylic acids, was recently reported [36]. Formic acid was used as a carbonyl source and potassium peroxysulfate as oxidant (Eq. 6).

$$\text{biphenyl} \xrightarrow[\text{K}_2\text{S}_2\text{O}_8,\ \text{CF}_3\text{CO}_2\text{H},\ 50\ °\text{C},\ 2\ \text{d}]{\text{cat. Pd(O}_2\text{CCF}_3)_2;\ \text{HCO}_2\text{H}} \text{4'-hydroxybiphenyl-3-carboxylic acid}\quad 45\,\% \tag{6}$$

Although the exact mechanism of palladium-catalyzed hydroxylations is unclear at the moment, it is very likely that palladium–aryl intermediates, formed by electrophilic arene activation, are on the reaction pathway [36–38]. In some cases, it has been shown that molecular oxygen is the ultimate source of the element in phenol. Formation of the carbon–oxygen bond might happen in several different ways: (1) insertion of an oxygen atom into a Pd–Ar bond then ligand exchange; (2) reductive elimination from Ar–Pd^{II}–OAc species, then hydrolysis; or (3) oxidation to Ar–Pd^{IV}–OR then reductive elimination and hydrolysis. It is conceivable that the presence of carbon monoxide or hydrogen ensures that palladium is present in the catalytically active oxidation state. Clearly, further mechanistic work is needed in this area.

1.1.4
Nucleophilic Hydroxylations

Arenes activated with an electron-withdrawing group may react with alkyl peroxides directly, by nucleophilic attack at the *ortho*- or *para*-position. Nitroarenes can be hydroxylated with alkyl hydroperoxides in strongly basic media in moderate to good yields (Eq. 7) [39].

$$\text{3-chloronitrobenzene} \xrightarrow[\text{NaOH, liq. NH}_3]{t\text{-BuOOH}} [\text{Meisenheimer intermediate}] \longrightarrow \text{2-chloro-4-nitrophenol}\quad 73\,\% \tag{7}$$

Electron-poor arenes can be converted into phenols by anodic oxidation [40]. A mixture of *ortho*-, *meta*- and *para*-substituted phenols is obtained in 18–89 % yields (Eq. (8); TFA is trifluoroacetic acid). Aryl trifluoroacetate is formed first and then hydrolyzed to phenol in a separate step. The oxidation potential of phenyl trifluoroacetate is higher than that of phenol, which explains the absence of overoxidation.

$$\text{Br-C}_6\text{H}_5 \xrightarrow[\text{2. H}_2\text{O}]{\text{1. -2e Et}_3\text{N, TFA/CH}_2\text{Cl}_2} \text{Br-C}_6\text{H}_4\text{-OH} \qquad (8)$$

1.1.5
Direct Synthesis of Quinones from Arenes

The direct oxidation of arenes to quinones is a reaction with a limited scope [41]. Only substrates that form stable quinones give good yields. For example, oxidation of anthracene to stable 9,10-anthraquinone with chromic acid is practiced on industrial scale. Such oxidations are believed to proceed through a series of one-electron oxidation/solvolysis steps. Yields and selectivity may be improved by using a strong one-electron oxidant such as cerium ammonium nitrate (CAN), as in the oxidation of phenanthrene to phenanthrenequinones (Eq. 9) [42].

$$\text{phenanthrene} \xrightarrow[\text{25 °C, 6 h}]{\text{CAN, H}_2\text{SO}_4} \text{9,10-phenanthrenequinone (60\%)} + \text{1,4-phenanthrenequinone (15\%)} \qquad (9)$$

Use of transition metal catalysts opens up previously unavailable mechanistic pathways. With hydrogen peroxide and catalytic amounts of methyl trioxorhenium (MTO), 2-methylnaphthalene can be converted to 2-methylnaphtha-1,4-quinone (vitamin K_3 or menadione) in 58 % yield and 86 % selectivity at 81 % conversion (Eq. 10) [43, 44]. Metalloporphyrin-catalyzed oxidation of 2-methylnaphthalene with $KHSO_5$ can also be used to prepare vitamin K_3 [45]. The MTO-catalyzed process can also be applied to the synthesis of quinones from phenols [46, 47]. In particular, several benzoquinones of cardanol derivatives were prepared in this manner [48]. The oxidation is thought to proceed through the formation of arene oxide intermediates [47].

Recently, a copper-catalyzed synthesis of trimethyl-1,4-benzoquinone, a key intermediate in the industrial synthesis of vitamin E, has been reported (Eq. 11) [49]. In the proposed mechanism, a tetranuclear cluster $[Cu_4(\mu^4\text{-O})Cl_{10}]^{4+}$, isolated from the reaction mixture, deprotonates phenol and oxidizes it to a copper-bound phenolate radical, which reacts with dioxygen.

Experimental

2-Methylnaphtha-1,4-quinone

H_2O_2 (85%, 14.3 mL, 0.49 mmol) was added dropwise at room temperature to a solution of 10.0 g (70.3 mmol) 2-methylnaphthalene and 0.349 g (1.4 mmol) methyltrioxorhenium in 150 mL acetic acid. The reaction mixture was stirred at 40 °C for 4 h under a N_2 atmosphere and then 100 mL water was added. The mixture was extracted with CH_2Cl_2 (3 × 100 mL). The combined organic phases were washed with 150 mL water, dried over Na_2SO_4, and concentrated at 20–40 °C/10–20 Torr. The conversion was 81 % (GC–MS). The quinone products were separated from the starting material by column chromatography (silica gel 32–63 μm mesh). Recrystallization of the chromatographed product from ethanol afforded the quinone in >98 % purity and 58 % yield (5.65 g).

1.2
Metalation of Arenes

1.2.1
Directed ortho and Remote Metalation (DoM and DreM)

Victor Snieckus and T. Macklin

1.2.1.1 Introduction and Fundamental Concepts

For the chemist practicing polysubstituted aromatic and heteroaromatic synthesis, methods steeped in classical electrophilic (**1**, Scheme 1) [1] and nucleophilic substitution [2] and $S_{RN}1$ (**2**) [3] reactions have been joined and, not infrequently superceded, by vicarious substitution (**3**) [4] and by DoM (**4**) [5] processes. The Murai ortho CH activation (**5**) [6] is a recently evolving and potential competitive method to the DoM tactic. The 60 years since its discovery by Wittig and Gilman, and 40 years since its systematic study by the school of Hauser, the DoM reaction has advanced by the contributions of Christensen, Beak, Meyers, and many other groups [5] of fundamental and dependable directed metalation groups (DMG) (Table 1) which provide this chemistry with broad carbo- and heteroatom-based functional group scope for synthetic manipulation of aromatic structures.

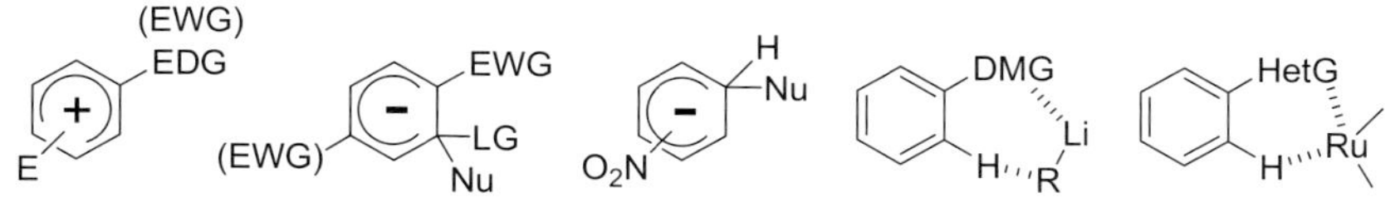

Scheme 1

Table 1. Hierarchal order of DMGs and test conditions for their DoM.

DMG	Base	Solvent	Additive	Temp., °C
OCONR$_2$	s-BuLi	THF	TMEDA	−78
SO$_{(1,2)}$t-Bu	n-BuLi	THF	none	−78
CONR$_2$	s-BuLi	THF	TMEDA	−78
CONLiR	n-BuLi	THF or Et$_2$O	TMEDA or none	−78 to reflux
SO$_2$NR$_2$	n-BuLi	THF	none	−78 to 0
SO$_2$NLiR	n-BuLi	THF	none	−10 to 25
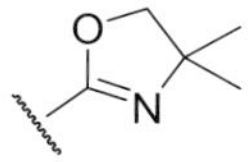 (oxazoline)	n-BuLi or s-BuLi	THF or Et$_2$O	none	−45 to 0
CO$_2$Li	s-BuLi	THF	TMEDA	−90
OMOM	n-BuLi or t-BuLi	Et$_2$O	none	−20 to 25
NLi(BOC)	t-BuLi	THF	none	−50 to −20
NLi(Piv)	n-BuLi	THF	none	−40 to 0
F	n-BuLi	Et$_2$O	none	< −50

Increasing Directing Ability ↑

In comparison with E$^+$ and Nu$^-$ aromatic substitution methods, mild conditions, incontestable ortho regioselectivity, and, perhaps most significant, post-DoM synthetic potential are the hallmarks of the DoM reaction. E$^+$ reactions are controlled by the normal EDG/EWG ortho, para/meta directing effects, occur under harsh acidic conditions [7], and invariably require separation of the resulting mixtures of isomers; Nu$^-$ S$_{RN}$1 processes are highly dependent on EWG situated ortho/para to the displaceable LG. The potential of the vicarious Nu$^-$ substitution in aromatic synthetic chemistry remains to be broadly implemented [4]. Although complementary to E$^+$ and Nu$^-$ substitution, the DoM process has features and advantages for adoption and application in synthesis as follows:
- The sequential, sometime one-pot, introduction of two similar or different electrophiles enables the preparation of contiguously substituted aromatics which are rare in catalogs and difficult to prepare by alternative methods (**6**, Scheme 2) [8].

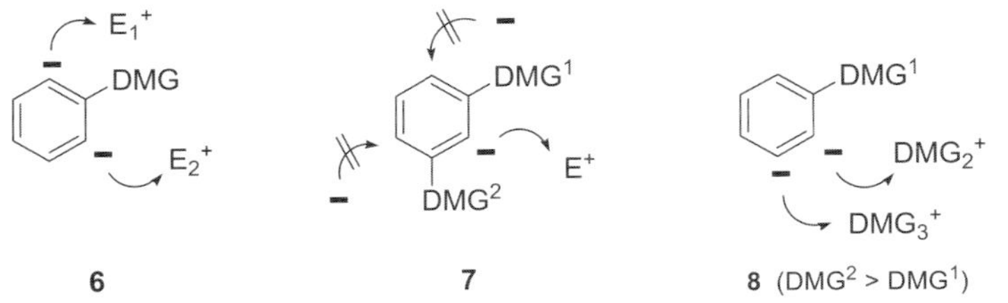

Scheme 2

- The cooperative nature of two similar or different meta-related DMG enables DoM at the in-between site thus constituting a different route to continuous 1,2,3-substitution patterns (**7**) [9]. The analogous *meta* and *para*-related DMG isomers provide routes, most of which have not been tested, to other patterns which are a function of their relative hierarchy qualitatively established by inter- or intramolecular competition experiments (Scheme 3) [5].

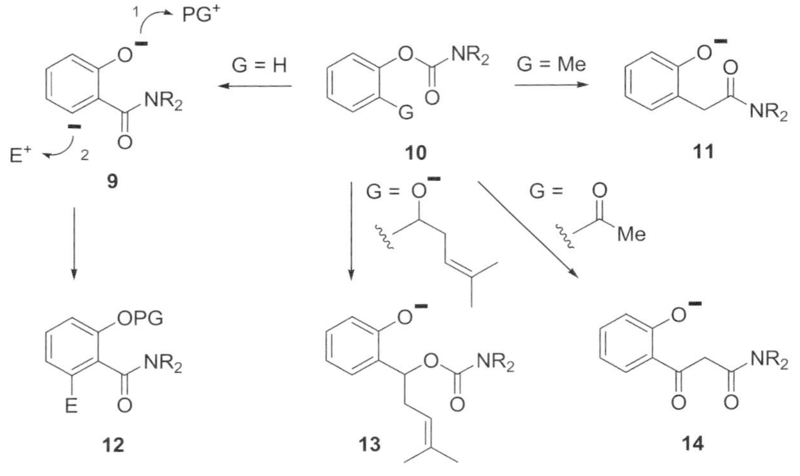

Scheme 3

- The DoM-mediated introduction of a DMG^2 to the existing DMG^1 offers the opportunity to take advantage of the more powerful (e.g. DMG^2) for a walk-around-the-ring metalation excursion (**8**) [10].
- Aside from its summit position as a DMG (**10**, Scheme 3) the OCONEt$_2$ provides versatile anionic aromatic chemistry: anionic ortho-Fries rearrangement (**9**, Scheme 3) [11], a useful synthetic step by itself but also one that provides a path, after incipient phenol protection, to further DoM, resulting in a 1,2,3-substituted aromatic derivative (**9 → 12**, see also Scheme 13) [12]; homo-ortho-Fries rearrangement to aryl acetamides (**10 → 11**) providing a route to benzofuranones which are difficult to prepare by Friedel–Crafts-based strategies [9]; intra-

molecular carbamoyl transfer in an ortho-acetyl system, **10 → 14**, constituting a modern variant of the Baker–Venkataraman reaction [13]; and **10 → 13**, a key step in general chromanone preparation which has also found application in total synthesis (see Scheme 15) [14].

- Directed remote metalation (DreM) of biaryl amides and O-carbamates, conceptually based on the complex-induced proximity effect (CIPE) [15] provides, especially in view of their link to transition metal-catalyzed cross coupling regimens [16], general and versatile routes to fluorenones (**16 → 15**, Scheme 4) [5, 17] and biaryl amides (**16 → 17**) [18] whose features are overriding Friedel–Crafts reactivity and yield enhancement in comparison to Suzuki–Miyaura coupling routes for highly hindered biaryls, respectively. Additional features of the O-carbamate DreM result is potential further DoM of **17** with appropriate phenol protection and cyclization to dibenzopyranones [18].

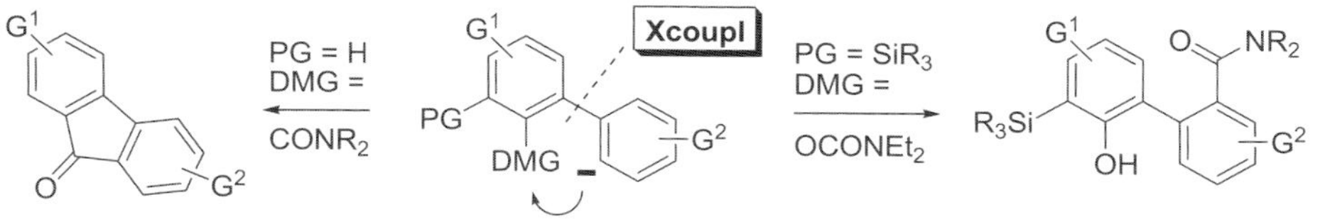

Scheme 4

- Introduction of RLi-unreactive silicon substituents has advantages in protection of Ar-C-H and Ar-CH$_3$ sites. Thus taking advantage of the cooperativity of amide and methoxy DMG, metalation–silylation followed by metalation–E$^+$ quench affords, after fluoride-mediated desilylation and amide hydrolysis, a route 1,2,5-substituted benzoic acids, **18 → 19** (Scheme 5). Lateral metalation, of considerable utility in post-DoM chain extension [19], followed by double silylation and further DoM–E$^+$ quench and the same fluoride and acid treatment steps, furnishes 1,2,3,4-tetrasubstituted aromatic compounds, **20 → 21** [10, 20].

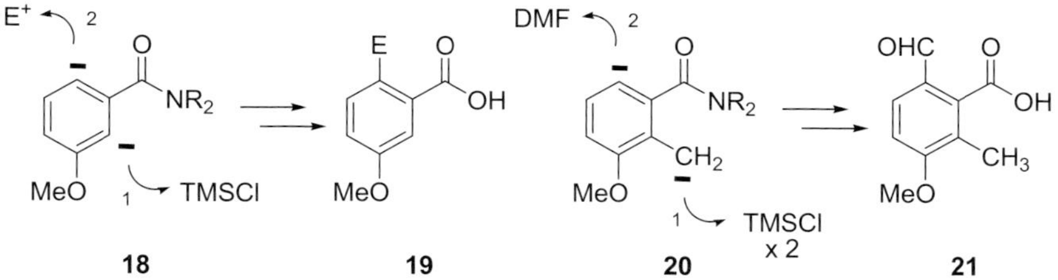

Scheme 5

- Once introduced, a methyl group has enhanced acidity on many DMG systems and is amenable, by LDA-mediated deprotonation to chain extension and heteroannulation processes [21]. In a route as yet not investigated for potential, DoM followed by magnesium transmetalation and allylation enables convenient access to derivatives which on LDA-induced cyclization provide routes to oxygenated naphthalenes, **23** (Scheme 6) [22].

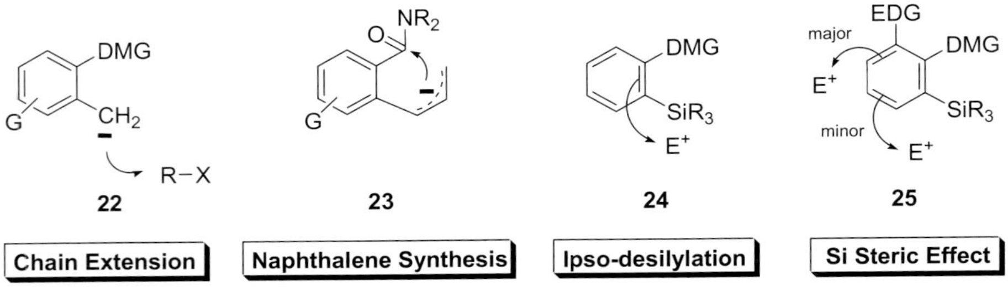

22	23	24	25
Chain Extension	Naphthalene Synthesis	Ipso-desilylation	Si Steric Effect

Scheme 6

- Silicon group introduction by DoM also provides opportunity for design of E^+-induced ipsodesilylation and sterically determined reactions. In the former, the operation of a β-cation effect induces rate enhancements in ipso-substitution in systems which are unreactive (EWG) or para-reactive (EDG) toward typical electrophilic reagents (**24**). The presence of strong EDG overrides the β-cation effect but the bulk of the silane increases the ortho/para ratio which enables development of substitution patterns and types which are not easily derived by classical electrophilic reaction means (**25**) [23].

1.2.1.2 Mechanism

The status, detailed in the excellent comprehensive review by Clayden [5a], and supplemented by coverage of the complex-induced proximity effect (CIPE) [15] may be summarized by the following, at times conflicting, points.

1.2.1.2.1 Kinetics

Large kinetic-isotope effects (KIE) for DMG = $CONR_2$ and CON^-R suggest tunneling during proton transfer and may be interpreted that DoM proceeds by fast reversible complexation (CIPE) followed by slow deprotonation. Kinetic and IR studies and direct observation of a complex in the α-deprotonation of N,N-dimethylbenzamide may be taken as indirect evidence for a CIPE in DoM. For DMG = OMe, KIE are much lower but show no evidence of prior BuLi–OMe coordination and may suggest that only the acidifying effect of OMe as a σ-EWG group drives the DoM reaction. The competitive efficiencies of deprotonation for 2-quinolone (C-8), N,N-diisopropylbenzamide, and 2-TMS-N,N-diisopropylbenzamide (C-6), 32,000:1800:1 respectively, are indicative of strong geometrical dependence of the DoM process. In contrast, a study on a series of benzyl alcohols suggests an out-of-plane relationship for favorable DoM.

1.2.1.2.2 NMR and Calculations

^{1}H and ^{13}C NMR studies show that for DMG = CH$_2$NMe$_2$, species occur as mixtures of three different dimers with strong intramolecular chelation which is maintained in the monomer on addition of TMEDA. ^{6}Li and ^{1}H NMR studies on DMG = OMe, NMe$_2$ show that, although a tetrameric anisole–n-BuLi complex is formed, it is unreactive and that only by addition of TMEDA does DoM proceed via an (n-BuLi)$_2$ TMEDA/anisole complex. Ab-initio calculations on DMG = OMe, OR systems support minimum RO–Li interactions and lead to the conclusion that inductive effects are of great significance and that precomplexation is not a factor in DoM (transition state **28**, Scheme 7) [24]. MNDO calculations for DMG = CON⁻R suggest that agostic Li–H interactions are important in prelithiation structures and provide indirect evidence for CIPE.

1.2.1.2.3 Solvent and Additive Effects

Direct observation of complex in the α-deprotonation of N,N-dimethylbenzamide and kinetic studies are interpreted in terms of a tetrameric cube-like transition state structure typical of solid-state RLi structures in which ligands bound to the Li centers facilitate the release of the α-carbanion species. The continuing existence of the RLi tetramer on addition of TMEDA contrasts with the dogma that this additive breaks up aggregates and is rationalized by TMEDA advancing the formation of the α-carbanion in the transition state analogous to the effect of the R carbanion character in the tetramer.

1.2.1.2.4 X-ray Structural Analysis

The solid-state structure of the species DMG = CH$_2$NMe$_2$ is a tetramer in which each Li atom is associated with three carbanion carbons and one nitrogen atom. Similar analysis of N,N-diisopropylbenzamide and 1-naphthamide shows that in the solid state the bulky amide substituent, being unable to bridge the amide carbonyl O with ortho-C via a single Li atom, escapes into bridged dimer structures in which one Li is planar with the amide and the other is planar with the aromatic ring.

1.2.1.2.5 Current Mechanistic Picture

Setting aside the lively debate regarding CIPE vs kinetically enhanced deprotonation [15], the emerging picture is that the DoM reaction involves two steps, complex formation and deprotonation. Which effect predominates is a function of the DMG – for strong DMG, e.g. CONR$_2$, CIPE predominates and geometry effects are significant; for weak DMG, e.g. OMe, coordination is electronically or geometrically impossible and inductive effects are responsible for the acidity of the ortho-hydrogen. For both, steric effects will, as expected, also play a role. The interconversion of organolithium homo- into hetero-aggregates also plays a role in observed structural and reactivity differences [25].

1.2.1.3 Scope and Limitations

1.2.1.3.1 Directed Metalation Groups (DMG)

For fast and efficacious DoM, the best DMG will combine strong Lewis-base character for coordination to Li and robust inductive effect for ortho-H acidification. The CIPE concept may be invoked as a controlling element of regioselectivity by altering the coordination/induction balance. Not to be denied is the consideration that the most powerful DMG also are also electrophilic groups, e.g. $OCONR_2$ and CO_2^-, which must be denied reactivity by low temperature, deactivation (e.g. CON^-R), and steric effect considerations (or combinations of these, e.g. N^-Boc, N^-Piv). DMG = halogen plays an important role in heteroaromatic DoM [26]. Use of DMG = Br, I, but not Cl for DoM of aromatic compounds is plagued by metal–halogen exchange or DoM followed by benzyne formation when RLi reagents are used. Regioselectivity can, however, be achieved by CIPE in dihalogen systems and $LiNR_2$ reagents do not participate in metal–halogen exchange processes. For the synthetic chemist the critical action is usually DMG manipulation into useful functional groups and target molecules.

The relative power of DMG (Table 1), established by experiments at low temperature and short reaction times and thus crudely representative of kinetic control conditions, may vary with inter- and intramolecular competition, conditions, and sometimes results are conflicting. Nevertheless, for synthetic practice this hierarchy follows a qualitative order consistent with CIPE and serves as a useful predictive chart. For thermodynamic control conditions, the pK_a chart of Fraser of 12 DMG [27], determined by equilibrium deprotonation using LiTMP ($pK_a = 37.8$), is a guide for lithium dialkylamide DoM reactions.

1.2.1.3.2 Analysis of Competing DMG

Coordination and inductive effects of DMG will also be affected by base, solvent, and additives. Illustrative of rationalizations for regiochemical consequence are **26** and **27** (Scheme 7). For both systems, **26**, under conditions of BuLi, coordination (CIPE) dominates with the consequence of deprotonation ortho to the stronger Lewis base site while under BuLi/TMEDA conditions the alkyllithium is complexed to the basic TMEDA (and THF) and, hence, coordination effects are mimized, acidifying effects are significant, and deprotonation occurs ortho to the more electronegative, therefore more Lewis-acidifying, OMe group. A similar rationale is suggested for the base-dependent difference in regioselectivity for **27**.

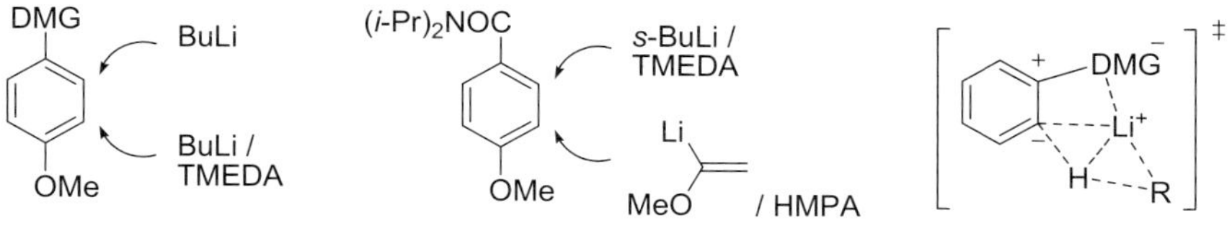

Scheme 7. Analysis of competing DMGs.

DMG = NMe$_2$, CH$_2$NMe$_2$

1.2.1.3.3 Compatible Functional Groups

As expected, functional group tolerance in the strongly basic alkyllithium or lithium dialkylamide DoM reaction is limited to non-electrophilic, nonacidic, and non-metal halogen exchange type groups (inter alia, CO$_2$R (at times), OR, SR, Cl, SiR$_3$, all DMG). Mixed Zn and Al bases may in the future overcome some of these drawbacks [28]. Once introduced, useful functionality requires either in situ (e.g. CHO → CH(OLi)NR$_2$) or separate protection. Retention of recalcitrant DMG until the end of a synthetic sequence may be advisable (e.g. CONEt$_2$ → CO$_2$H). Introduction of an ortho-bromo substituent by DoM provides a masked latent anion which may be revealed by metal–halogen exchange subsequent to manipulation of the molecule by unrelated chemistry [29].

1.2.1.3.4 Initial Test Conditions

Random but fairly extensive literature analysis enables production of a rough guide for the test DoM experimental conditions as a function of DMG (Table 1). In practice, variation of all commercially available RLi and LiNR$_2$ reagents, additives, solvents, and temperature is the norm.

1.2.1.4 DoM Methodology for Substituted Aromatics

Selected examples demonstrate the methodological value of DoM. The conversion 29 → 30 → 31 (Scheme 8) shows the merit of in-between metalation, OH$^+$ synthetic equivalent introduction, and metalation-mediated iodination to overcome the inability of electrophilic iodination owing to the presence of a new EDG [30].

The sequence 32 → 33 → 34 → 35 (Scheme 8) also illustrates in-between DoM but, in addition, the important conversion of amides into more useful FG and a method to override the normal Friedel–Crafts reactivity by use of a carbo-desilylation reaction (37) [31]. Amide DMG which have been tested in developing methods for conversion to corresponding acid derivatives and masked aldehyde DMG and the parent carboxylic acid are useful to consider (36).

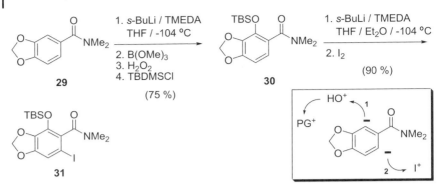

Scheme 8

DMG		
36	CON(Me)CH(TMS)$_2$	CON(Et)CH$_2$CH$_2$NEt$_2$
	CON(Me)*t*-Bu	CON(Et)CH(OMe)*t*-Bu
	CON(-C(Me)$_2$OCH$_2$C(Me)$_2$-)	CONLiC(Me)$_2$Ph
	CO$_2$Li	oxazoline
	CHO	CH(OLi)NR$_2$

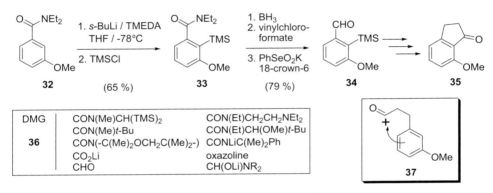

Scheme 9. Overriding Friedel–Crafts regiochemistry. Amides as precursors to more useful derivatives.

Chain extension methodology, **38** → **39** → **40** (Scheme 9) enables abbreviated preparation of systems which are difficult by classical, usually electrophilic substitution chemistry (**40**) [32] and heterannulation strategies **41**, **42** [19, 33].

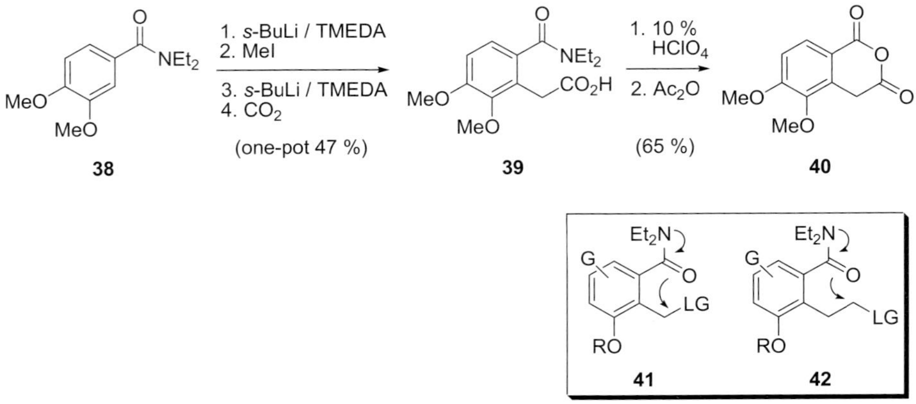

Scheme 10. DoM of benzamide followed by *o*-tolyl carboxylation.

Sulfonamide DoM in conjunction with cross coupling chemistry, **43** → **44** → **45** (Scheme 10) is instructive for double DoM sequences followed by further advantage of an introduced DMG [34].

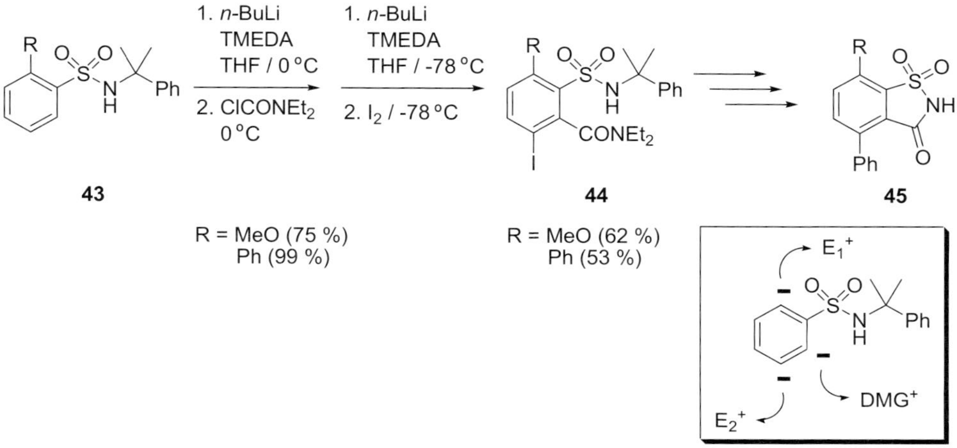

Scheme 11. Sequential Do of N-cumylsulfonamide and amide DMGs. Synthesis of 4,7-substituted saccharins.

The advantage of in-between DoM followed by sequential intramolecular trapping of the incipient benzyne and electrophile quench, **46** → **47** (Scheme 11) constitutes a powerful method in heteroaromatic synthesis [35].

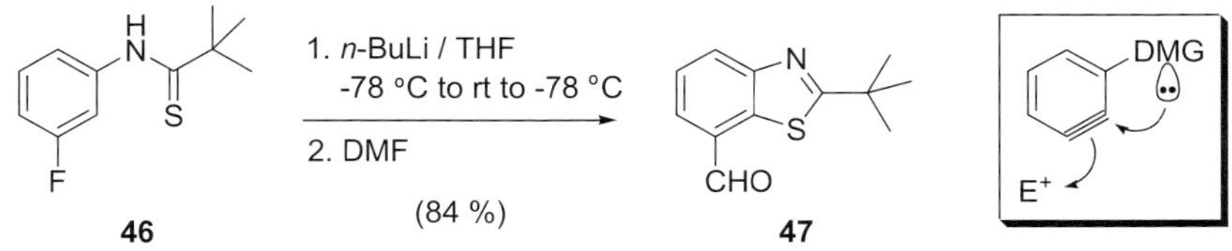

Scheme 12

1.2.1.5 DoM in Total Synthesis

Total synthesis has benefited from key DoM reactions. The sequence **48** → **49** → **50** → **51** (Scheme 12) en route to the natural product ochratoxin A (**52**) takes multiple advantage of anion chemistry (**53**) ortho-metalation of the powerful OCONEt$_2$ group (step 1), anionic Fries rearrangement (step 2), in-between DoM and chain-extension by Li-Mg transmetalation (step 3) [12].

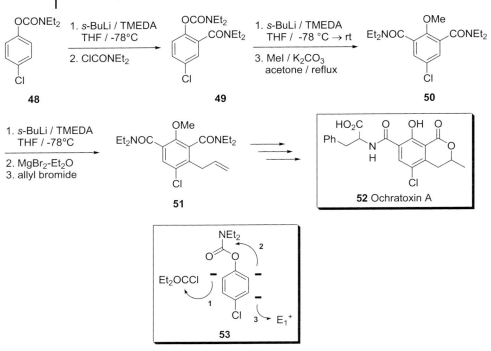

Scheme 13. ArOCONEt$_2$ DoM / ortho-Fries strategy towards the total synthesis of ochratoxin A.

2,6-Disubstitution via DoM, **54** → **55** (Scheme 14) constitutes an efficient starting point in the synthesis of the estrone precursor **56** [36].

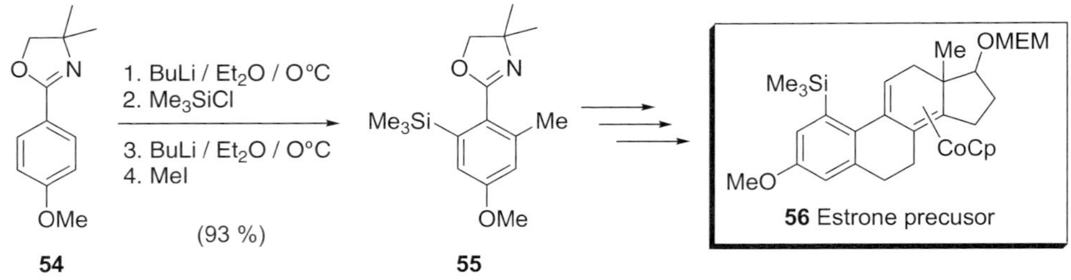

Scheme 14. Sequential DoM using oxazolinyl DMG towards the synthesis of co-steroidal complexes.

The route toward the purported natural product, plicadin [14] (**61**, Scheme 15) incorporates in-between DMG metalation-regioselective chromene construction (**57** → **58**), further DoM (→ **59**), a Sonogashira–Castro Stephens sequence (→ **60**), and an efficient key DreM-based concluding step.

The synthesis of pancratistatin [37] (**65**, Scheme 15), involving anionic ortho-Fries (**62** → **63**) and further DoM (→ **64**) is a pertinent showcase of anionic chemistry for preparation of a pentasubstituted aromatic using the conceptual framework **66**.

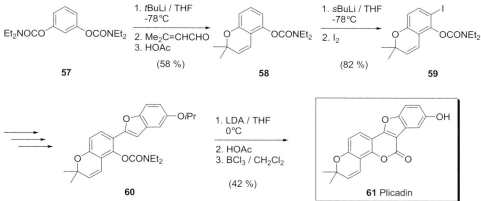

Scheme 15. total synthesis of plicadin.

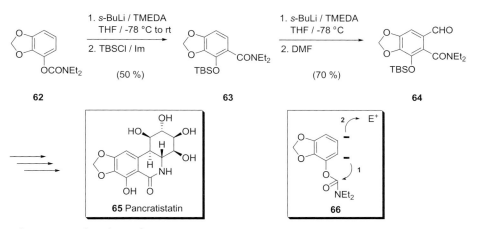

Scheme 16. Total synthesis of pancratistatin.

The synthesis of the angiotensin II inhibitor Losartan [38] (**69**, Scheme 16) fittingly illustrates the impact of DoM on current large-scale industrial practice. The powerful tetrazolyl DMG (**67**) is used in a 4-step sequence without isolation of intermediates to afford boronic acid **68** which, via a key Suzuki–Miyaura cross coupling, leads the commercial medicinal agent in multi-ton quantities.

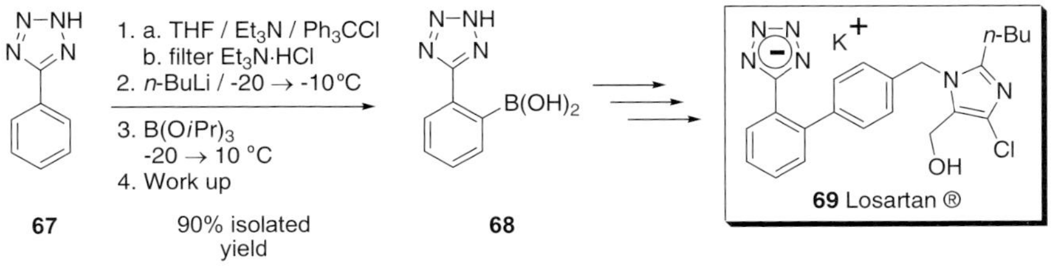

Scheme 17. DoM of N-tritylphenyltetrazole applied to industrial scale synthesis of losartan.

Experimental

2-Phenyl-4-phenanthrol [22]

MeLi (3.4 mL, 4.8 mmol, 1.4 M in Et_2O) was added to a stirred solution of o-(2-phenylallyl)-N,N-diisopropylnaphthamide (0.9 g, 2.4 mmol) in anhydrous THF (50 mL) at -78 °C under argon. The solution turned black immediately. The reaction mixture was left to warm to ambient temperature overnight, quenched with saturated aqueous NH_4Cl (50 mL) and the aqueous phase was extracted with Et_2O (3 × 50 mL). The combined organic layer was evaporated to dryness and the residue was purified by flash column chromatography on silica gel (4:1 hexanes–Et_2O) to afford 2-phenyl-4-phenanthrol (0.59 g, 90 %) as a light brown solid.

5-Chloro-N^1,N^1,N^3,N^3-tetraethyl-2-methoxyisophthalamide, 50 (Scheme 13) [12]

A solution of 4-chloro-2-(diethylcarbamoyl)phenyl diethylcarbamate 49 (2.06 g, 8.7 mmol) in anhydrous THF (10 mL) was added to a stirred solution of sec-BuLi (13.8 mL, 19.14 mmol, 1.39 M in cyclohexane) and TMEDA (2.9 mL, 19.14 mmol) in THF (170 mL) maintained at -78 °C under a nitrogen atmosphere. The stirred reaction mixture was left to warm to room temperature overnight and treated with satd aq NH_4Cl. The THF was removed under reduced pressure and the remaining solution was extracted with Et_2O (3 × 25 mL). The aqueous layer was acidified with 2 M HCl (pH 3), and the whole mixture was extracted with Et_2O–CH_2Cl_2 (1:1, 3 × 25 mL) and concentrated under reduced pressure to give 1.7 g of a solid that was heated under reflux with a mixture of MeI (10 mL, 22.8 mmol) and K_2CO_3 (3 g, 24.6 mmol) in acetone (30 mL) for 20 h. Standard workup followed by flash column chromatography on silica gel (1:1, EtOAc–hexane) afforded 50 (0.976 g, 42 %) as a colorless oil.

9-Isopropoxyplicadine (Scheme 15) [14]

To a stirred solution of freshly prepared LDA (4.367 mmol) in THF (5 mL) at 0 °C was added, via a cannula, a solution of 2,2-dimethyl-5-(N,N-diethyl-O-carbamoyl)-6-(2-(isopropoxy)benzo[b]furan)-2H-1-benzopyran 60 (398 mg, 0.89 mmol) in THF (20 mL). After 15 min, the reaction mixture was concentrated under reduced pressure and AcOH (15 mL) was added. The stirred mixture was heated to reflux for 10 min, cooled, CH_2Cl_2 (20 mL) and H_2O (10 mL) were added, and the organic layer was separated, dried (Na_2SO_4), and concentrated under reduced pressure. Flash column chromatography (EtOAc–hexanes 1.5:8.5) afforded 9-isopropoxyplicadine (376 mg, 84 %) as a yellow solid.

1.2.2
Electrophilic Metalation of Arenes
Vladimir V. Grushin

1.2.2.1 Introduction

Discovered over a century ago, electrophilic mercuration is probably the oldest known C–H bond activation reaction with a metal compound. The earliest examples of aromatic mercuration were reported by Volhard (mercuration of thiophene) [1], Pesci (mercuration of aromatic amines) [2], and Dimroth [3] who was the first to mercurate benzene and toluene, generalize the reaction, and assign the correct structures to the products originally observed by Pesci. Since the work of Dimroth, electrophilic aromatic metalation reactions with compounds of other metals, such as Tl(III), Pb(IV), Sn(IV), Pt(IV), Au(III), Rh(III), and Pd(II), have been discovered [4]. In this chapter, we will focus on intermolecular S_EAr reactions involving main-group metal electrophiles and resulting in the formation of isolable metal aryls which find numerous important applications in synthesis [5]. Well-known electrophilic cyclometalation reactions, such as cyclopalladation, can be found in other chapters of this book and will not be reviewed here.

In comparison with other methods to arylate metals, direct metalation of aromatic C–H bonds is particularly attractive due to its simplicity. Like any S_EAr reaction [6], electrophilic metalation may lead to a mixture of positional isomers, depending on the nature of the substrate, metal reagent, solvent, and reaction conditions. On the other hand, electrophilic metalation is a cost-efficient process which provides useful organometallic compounds directly from arenes and inexpensive inorganic precursors. Alternative methods employ more costly functionalized aromatic compounds (e.g. haloarenes), active organometallic reagents (e.g. aryl lithiums and Grignard reagents), and purified dry solvents. Unlike C–H activation via oxidative addition to electron-rich derivatives of metals in low oxidation states, electrophilic metalation always involves metals in high oxidation states.

1.2.2.2 Mercuration

Aromatic mercuration (Scheme 1), the oldest S_EAr metalation reaction known, has been the subject of several reviews and monographs [7–11].

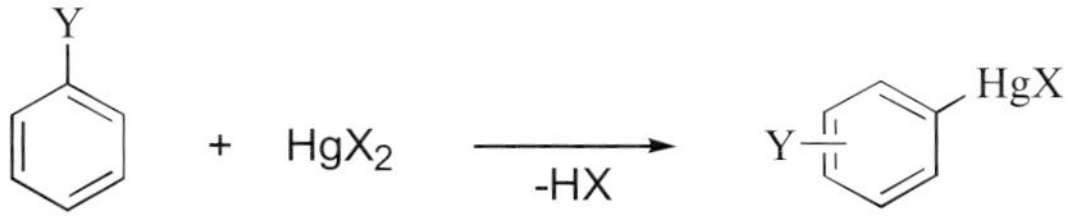

Scheme 1. Mercuration of a substituted benzene.

Numerous studies have revealed an electrophilic mechanism of aromatic mercuration. For instance, the reactivity of a mercurating agent HgX_2 strongly depends on the nature of the counter-ion X. Weaker anionic donors X result in enhanced electrophilicity of the mercury center. Thus, mercuric cyanide and halides have proven to be unreactive (or barely reactive), whereas Hg(II) acetate has greater electrophilic character and hence is more often used to mercurate aromatics. In the presence of ionizing $HClO_4$, aromatic mercuration with $Hg(OAc)_2$ occurs even faster [7]. Furthermore, mercury trifluoroacetate (TFA) in trifluoroacetic acid (TFAH) is a remarkably powerful reagent which mercurates benzene 6.9×10^5 times faster than $Hg(OAc)_2$ in acetic acid [12]. Up to five Hg atoms can be introduced into the benzene ring with $Hg(TFA)_2$ [13], whereas use of $Hg(OAc)_2$ leads to the mono- and dimercurated products only. Icosahedral closo-carboranes $C_2B_{10}H_{12}$ are unreactive toward $Hg(OAc)_2$ but undergo smooth mercuration with $Hg(TFA)_2$ in TFAH [14].

Kinetic experiments have demonstrated that the mercuration of benzene with mercury acetate or nitrate is second order, first order in each the mercury salt and the arene [7]. Electron-donating substituents on the ring increase mercuration rates, whereas electron-withdrawing groups slow down the reaction [7–11]. For example, mercuration of nitrobenzene requires 150 °C to occur, and the reaction of benzene with $Hg(OAc)_2$ in AcOH takes hours at 90–120 °C, whereas N,N-dimethylaniline readily reacts with $Hg(OAc)_2$ in aqueous EtOH at room temperature within a few minutes [7].

The mechanism of mercuration is shown in Scheme 2. In the first step, the mercury salt forms a π-complex with the aromatic substrate [15, 16]. In 1982, Lau, Huffman, and Kochi [17] reported the first isolation and full characterization (including X-ray molecular and crystal structure) of such an intermediate, a complex of hexamethylbenzene with $Hg(TFA)_2$. The X-ray structure revealed a $Hg_2(\mu\text{-TFA})_4$ framework with a molecule of C_6Me_6 η^2-coordinated to each of the Hg atoms. Analogous π-complexes have also been observed and studied by Dean and co-workers [18] and more recently by Barron's [19] and Gabbai's [20] groups. The π-complex intermediate can rearrange to the σ-complex (a Wheland intermediate) directly, or sometimes via electron transfer, to produce a radical ion pair which then collapses (Scheme 2) [21,22].

Hundreds of reports have been published on electrophilic mercuration of a variety of aromatic compounds and uses of the resulting organomercurials in synthesis [5, 7–11]. Although in this chapter, we will not discuss synthetic applica-

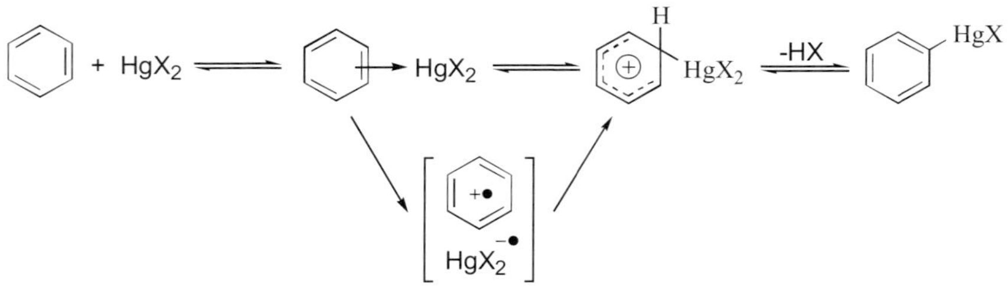

Scheme 2. Mechanism of electrophilic aromatic mercuration.

tions of aromatic mercuration due to space limitations, one interesting recent example is noteworthy. Buzhansky and Feit [23] have reported the synthesis of optoelectronic-relevant unsubstituted oligo(α-thiophenes) by one-pot mercuration of thiophene with Hg(TFA)$_2$, followed by homocoupling of the resulting organomercurial with PdCl$_2$/Et$_3$N. The procedure can then be repeated with the 2,2′-bithienyl produced to give a tetramer which, in turn, can be further dimerized in the same manner. The yields for the individual isolated oligomers were in the range of 70–92 % [23].

1.2.2.3 Thallation

The earliest examples of direct aromatic thallation, reported by Gilman and Abbott [24] and Glushkova and Kocheshkov [25], employed weak thallium electrophiles TlCl$_3$ and Tl(i-PrCO$_2$)$_3$, respectively. It was not until the late 1960s and early 1970s that Taylor, McKillop and co-workers [26] prepared thallium(III) trifluoroacetate, Tl(TFA)$_3$, and demonstrated its utility as a potent electrophile for aromatic thallation (Scheme 3). A number of reviews have been published on thallium(III) aryls [27–30]; the review article by Usyatinskii and Bregadze [29] provides particularly comprehensive coverage of the thallation reaction and its applications in synthesis.

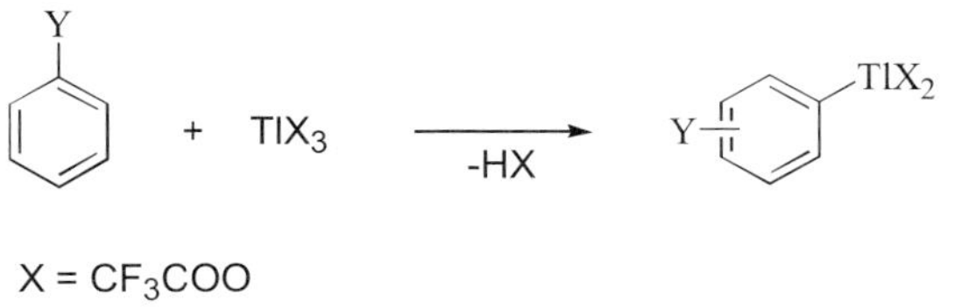

X = CF$_3$COO

Scheme 3. Thallation of a substituted benzene with Tl(TFA)$_3$.

In general, Tl^{3+} is a weaker electrophile than Hg^{2+}; as a result, the rate of thallation is approximately 200–400 times slower than the rate of mercuration under similar conditions [29]. Electron-enriched arenes such as alkylbenzenes undergo thallation with Tl(TFA)$_3$ in TFAH within minutes at room temperature. Slightly deactivated aromatic rings (Y = halide in Scheme 3) can be thallated in less than an hour in TFAH under reflux, but with less nucleophilic arenes (e.g. Y = CF$_3$) the reaction takes days to go to completion. Being bulkier and less electrophilic, Tl(TFA)$_3$ metalates substituted benzenes with higher positional selectivity than Hg(TFA)$_2$. For example, the thallation of monoalkylbenzenes is much more para-selective than the mercuration. Remarkably, however, some substituents Y containing a donor atom direct the thallation in the ortho position with uncommonly high efficiency [26]. For example, benzoic acid is thallated with Tl(TFA)$_3$ in TFAH to give the ortho derivative with 95 % selectivity, the other 5 % being the expected meta isomer [26]. Benzyl alcohol undergoes the thallation exclusively in the ortho position. This unusual ortho-selectivity is explained by the original coordination of the donating heteroatom of Y to Tl^{3+}, followed by intramolecular electrophilic substitution of the ortho-H. Such "chelate intermediates" are stable and therefore operative only for 5- or 6-membered rings, otherwise the ortho-selectivity is decreased considerably or lost altogether. Indeed, the ortho selectivity of thallation of a series of $Ph(CH_2)_nCOOH$ was found to be 95, 92, 29, 6, and 5 % for $n=0, 1, 2, 3$, and 4, respectively [26]. Also, being a reversible process, electrophilic aromatic thallation can lead to mixtures of positional isomers in different ratios depending on reaction temperature and time (kinetic vs. thermodynamic control) [26, 29].

The electrophilic thallation path involves $[Tl(TFA)_2]^+$ as the active electrophile which forms a π-complex with the substrate [16, 31]. Single electron transfer from electron-enriched polymethylarenes can be significant, leading to the formation of biaryls and side-chain thallated products [31]. The electron transfer path during electrophilic metalation is much more characteristic of $Tl(TFA)_3$ than $Hg(TFA)_2$ [16]. A review article [29] includes useful analysis of mechanistic features of aromatic thallation with $Tl(TFA)_3$ and some other Tl(III) reagents, including the even more electrophilic triflate.

Once introduced into the aromatic ring, the TlX_2 functionality can be displaced with a variety of groups such as halides, CN, SCN, SeCN, NO_2, and SO_2Ar. These reactions and their possible mechanisms have been extensively reviewed [27–30]. In this chapter, we will mention only one such reaction (Scheme 4) [32] furnishing valuable fluoroarenes and thus providing a rare alternative [33] to the Balz–Schiemann reaction – the synthesis of ArF via the thermal decomposition of arenediazonium tetrafluoroborates or hexafluorophosphates. The method (Scheme 4) is limited to electron-rich arenes only. Arylthallium fluorides containing no electron-donating groups on the ring produce only trace quantities of the desired aryl fluorides [32].

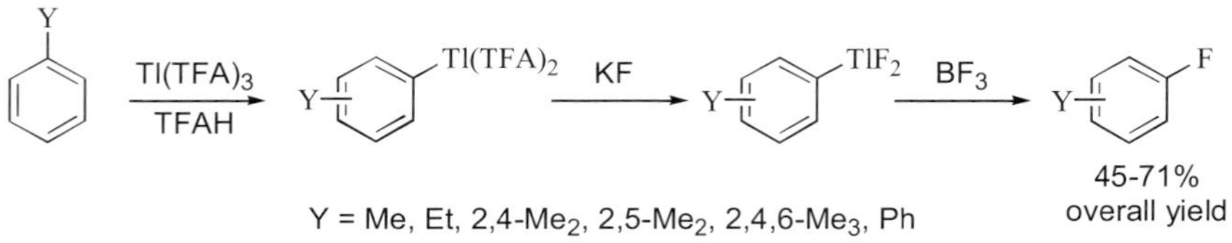

Y = Me, Et, 2,4-Me$_2$, 2,5-Me$_2$, 2,4,6-Me$_3$, Ph

45-71% overall yield

Scheme 4. Synthesis of fluoroarenes via electrophilic thallation.

1.2.2.4 Plumbylation (Plumbation)

In recent years, there has been much interest in the use of aryllead compounds in organic synthesis, due to the ability of compounds of the type $ArPb(O_2CR)_3$ to smoothly arylate a variety of nucleophiles [34–37]. Aryllead tricarboxylates can be prepared by direct plumbylation of aromatic compounds (Scheme 5), a reaction that has been known for a long time [38].

Lead(IV) acetate is most commonly employed to plumbylate arenes, often using chloroform as the solvent. Being a relatively weak electrophile, $Pb(OAc)_4$ can react only with electron-rich aromatic compounds, for example anisole and polyalkoxybenzenes. The electrophilicty of the lead reagent is, however, enhanced substan-

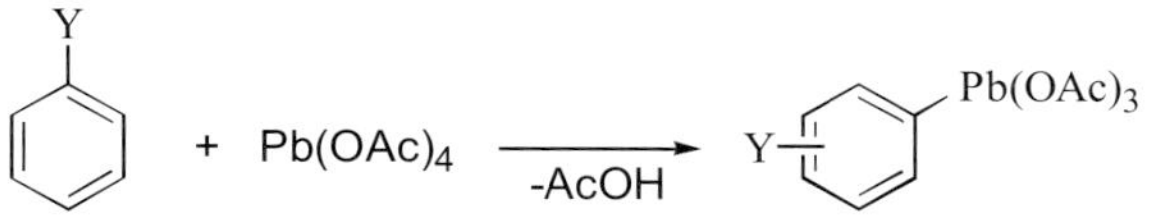

Scheme 5. Plumbylation of a substituted benzene with lead(IV) acetate.

tially in the presence of acetic acid and especially its polychlorinated derivatives [37]. Although anisole is poorly reactive toward Pb(OAc)$_4$ in benzene, the reaction is efficient in AcOH [39] and even faster in dichloroacetic acid [37]. At room temperature, toluene is readily plumbylated with Pb(OAc)$_4$ in Cl$_2$CHCOOH [40] or Cl$_3$CCOOH [37] but not in AcOH.

Lead(IV) trifluoroacetate in TFAH is a very reactive electrophile that is capable of plumbylating less electron-rich arenes. Nonetheless, the use of trifluoroacetate anions in the plumbylation reactions should be avoided, because aryllead(IV) trifluoroacetates are unstable compounds that readily decompose to the corresponding aryl trifluoroacetates and biaryls [34–37, 40, 41]. It has been reported [41] that 4-FC$_6$H$_4$Pb(TFA)$_3$ is reasonably stable and can be isolated from the reaction of Pb(TFA)$_4$/TFAH with fluorobenzene. A mechanistic study [41] indicated an electrophilic substitution path for the plumbylation reaction, which was considerably more para-selective than mercuration and thallation. For example, the plumbylation of toluene with Pb(OAc)$_4$ in dichloroacetic acid has been reported [41] to occur with >90% para-selectivity.

The products of direct aromatic plumbylation, ArPb(OAc)$_3$, are attractive reagents, because of their accessibility and ability to arylate various nucleophiles under mild conditions [36]. These arylation reactions might involve ligand exchange at the Pb(IV) center, followed by reductive elimination [35]. Among nu-

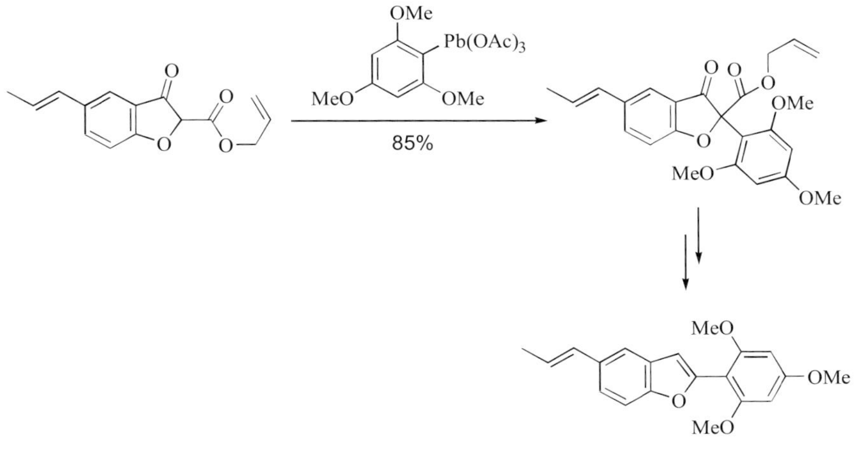

Scheme 6. Arylation with an organolead compound as a key step in the synthesis of nor-neolignan.

merous applications of aryllead(IV) carboxylates is the formation of a quaternary carbon center on C-arylation; this is particularly valuable for natural product synthesis [36]. This type of reaction can be exemplified by the synthesis of nor-neolignan, involving an aryllead reagent in a key step (Scheme 6) [42].

1.2.2.5 Stannylation

Discovered by Thorn and first communicated only a few years ago [43], electrophilic aromatic stannylation is the newest of all metalation reactions described in this chapter. The reactive Sn-electrophile can be prepared by dearylation of Ph_4Sn with TFAH or, preferably, by reaction of SnO with trifluoroperacetic acid generated from H_2O_2, TFAH, and trifluoroacetic anhydride. Because the structure of the Sn(IV) species (or mixtures of species) produced this way is unknown, the reactive electrophile is depicted as "$Sn(TFA)_4$", for simplicity. The stannylation reactions of benzene and p-xylene were shown to produce tetratin clusters containing two "inorganic" and two "organometallic" Sn centers (Scheme 7).

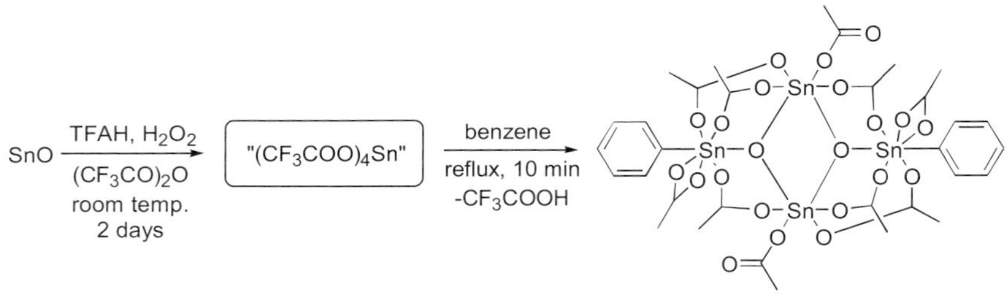

Scheme 7. Electrophilic stannylation of benzene.

The structures of both the phenyl and 2,5-dimethylphenyl tin clusters were established by single-crystal X-ray diffraction. Under similar conditions, toluene was stannylated to produce a mixture of organotin compounds. Bromination of this mixture yielded 3-bromotoluene and 4-bromotoluene in a 2:1 ratio, with trace amounts of 2-bromotoluene. Similarly, PhBr was produced upon treatment of the phenyltin product with aqueous Br_2. No applications of the stannylation reactions have yet been reported.

Experimental

Mercuration of N,N-dimethylaniline [2, 7, 44, 45]

Solutions of $Hg(OAc)_2$ (60 g) in 50% EtOH (400 mL) and of N,N-dimethylaniline (50 g) in 96% EtOH are poured together. The initially clear mixture thickens after a few minutes into a mash of fine needles of p-dimethylaminophenylmercury acetate, m.p. 165 °C (from EtOH). The product was isolated in 94% yield when half the amount of the mercuric acetate solution was used under similar conditions [44]. The compound has been characterized by X-ray diffraction, analysis, and spectroscopy

(NMR, IR, ES-MS) data [45]: Anal. Calcd. for $C_{10}H_{13}NO_2Hg$, %: C, 31.62; H, 3.45; N, 3.67. Found, %: C, 31.76; H, 3.38; N 3.58. NMR (CDCl$_3$): ^{1}H δ = 2.08 (s, CCH$_3$), 2.95 (s, NCH$_3$), 7.12, 6.69 (two d, J = 8.8 Hz, H$_{aryl}$); ^{13}C δ = 23.5 (CCH$_3$), 40.3 (NCH$_3$), 113.0 (C3, $^3J_{C-Hg}$ = 212 Hz), 128.5 (C1), 136.8 (C2, $^2J_{C-Hg}$ = 142 Hz), 151.0 (C4), 177.5 (C=O); ^{199}Hg δ = 930. ES-MS (MeOH, HPF$_6$) m/z 382 [M + H]$^+$. IR (KBr, cm^{-1}) 945 m, 926 m, 803 vs, 794 sh, 752 w, 690 s, 650 w, 614 w, 570 w, 509 m, 476 w.

o-Iodophenylacetic acid via thallation of phenylacetic acid [26]
A vigorously stirred suspension of 50 g of thallium(III) oxide in 200 mL TFAH containing 25 mL water was heated under reflux for 12 h in a flask wrapped with aluminum foil. Filtration of the cooled reaction mixture to remove a small amount of residual solid then gave an approximately 0.88 M solution of Tl(TFA)$_3$ in TFAH. Evaporation of the TFAH from the colorless solution under reduced pressure gave Tl(TFA)$_3$ as a granular solid in 90–100 % yield. A solution of 2.72 g (0.02 mol) phenylacetic acid in 20 mL of a 1 M solution of Tl(TFA)$_3$ in TFAH was stirred for 48 hr at room temperature. The reaction mixture was suspended in 100 mL water and KI (8 g, 0.05 mol) was added. The suspension was heated under reflux for 5 hr, and sodium metabisulfite (1 g) was then added to reduce iodine that had been formed during the reaction. Heating was continued for another 30 min, and the precipitated thallium(I) iodide was filtered off (without prior cooling). The collected inorganic material was thoroughly washed with acetone. The combined filtrate and acetone washings were extracted with three 40-mL portions of ether, and the ether extracts were dried over anhydrous sodium sulfate and evaporated to give crude o-iodophenylacetic acid (5.10 g, 97.5 %), m.p. 80–90 °C. Recrystallization from petroleum ether gave 3.80 g of the purified acid with, m.p. 108–110 °C. Methylation of a small sample with diazomethane and subsequent inspection of the resulting methyl ester by GC showed the compound to be pure methyl o-iodophenylacetate.

Plumbylation of 1,3,5-trimethoxybenzene [37]
A solution of 1,3,5-trimethoxybenzene (10.03 g, 0.06 mol) in dry CHCl$_3$ (20 mL) was added dropwise over 15 min to a well-stirred solution of Pb(OAc)$_3$ (10.00 g, 0.022 mol) in dry CHCl$_3$ (75 mL) at room temperature and the mixture was stirred for 8 h. The mixture was washed with H$_2$O (3 × 100 mL), the organic phase dried, concentrated to a volume of 20 mL, and added dropwise to light petroleum ether (500 mL), whereupon a bright yellow solid precipitated out. This solid was collected by suction filtration and dried to give tris(acetoxy)(2,4,6-trimethoxyphenyl)plumbane as yellow plates; yield 8.25 g (68 %); m.p. 175–179 °C (dec).

Stannylation of Benzene [43]
Aqueous H$_2$O$_2$ (30 %; 0.40 mL) was slowly added (*Caution: exothermic reaction*) dropwise to a vigorously stirred mixture of SnO (0.50 g), TFAH (3 mL), and trifluoroacetic anhydride (4 mL). This mixture was stirred at room temperature for 2 days until the dark solids (SnO) had all dissolved. The resulting white opaque solution was evaporated under N$_2$ and the residue was treated first with trifluoro-

acetic anhydride (6 mL) and then with a solution of trifluoroperacetic acid freshly prepared by slowly adding 0.63 mL of 30% H_2O_2 to $(CF_3CO)_2O$ (10 mL; *Caution: exothermic reaction*) at 0–5 °C. The mixture was stirred at room temperature for 30 h, filtered through glass-wool under N_2, and reduced in volume to ca. 2–4 mL. Dry CH_2Cl_2 (30 mL), followed by hexane (5 mL) were added, and the mixture was stirred until the originally precipitated oil solidified. The solid was separated, washed with hexane, and dried under vacuum. The yield of "$Sn(TFA)_4$" (see above) was 0.69 g. To this solid, was added dry benzene (6 mL) and the mixture was stirred under reflux (N_2) for 10 min. Most of the solids dissolved to give a pale pinkish–tan solution which was separated warm from a small amount of viscous oil and left at room temperature. After 2 days, white crystalline material precipitated which was washed with hexane and dried under vacuum. The yield was 0.39 g. X-Ray analysis of one of the crystals found in the solid product indicated the structure shown in Scheme 7. ^1H NMR (CD_2Cl_2, 20 °C), $\delta = 7.7$ (m, 3H, *m,p*-C_6H_5); 7.9 (m, 2H, $J_{Sn-H} = 177$ Hz, *o*-C_6H_5). A weak singlet at 7.4 ppm from benzene was also observed. ^{19}F NMR (CD_2Cl_2, 20 °C), $\delta = 75.7$ (s). To determine the amount of phenyltin species in this product, 301 mg of it was treated, in $CDCl_3$ (2 mL), with excess aqueous bromine (4 mL). After 1 h of vigorous stirring at room temperature all the solids had dissolved. ^1H NMR analysis of the organic phase with 1,2-dichloroethane as an internal standard indicated the presence of bromobenzene (48 mg), benzene (8 mg) and no other products. The amount of bromobenzene produced translated into 91% purity of $[(Ph)_2Sn_4O_2(TFA)_{10}]$ under the cautious assumption that this was the only Ph-Sn species formed in the stannylation reaction.

1.2.3
Iridium-Catalyzed Borylation of Arenes

Tatsuo Ishiyama and Norio Miyaura

1.2.3.1 Introduction and Fundamental Examples

Aromatic boron derivatives are an important class of compound the utility of which has been amply demonstrated in various fields of chemistry. Traditional methods for their synthesis are based on the reactions of trialkylborates with aromatic lithium or magnesium reagents derived from aromatic halides. Pd-catalyzed cross-coupling of aromatic halides with tetra(alkoxo)diborons or di(alkoxo)boranes is a milder variant in which the preparation of magnesium and lithium reagents is avoided. Alternatively, transition metal-catalyzed aromatic C–H borylation of aromatic compounds **1** by pinacolborane (HBpin, pin= $O_2C_2Me_4$) **2** or bis(pinacolato)diboron **3** [1, 2], which was first reported in 1999 [3], is highly attractive as a convenient, economical, and environmentally benign process for synthesis of aromatic boron compounds **4** without any halogenated reactant. Among the catalysts developed to date, an Ir system has excellent activity and selectivity for the aromatic C–H borylation (Scheme 1).

Aromatic C–H borylation is catalyzed by an Ir complex with a small and electron-donating ligand, for example PMe_3, 1,2-bis(dimethylphosphino)ethane (dmpe), 2,2′-bipyridine (bpy), and 4,4′-di-*tert*-butyl-2,2′-bipyridine (dtbpy) (Table 1)

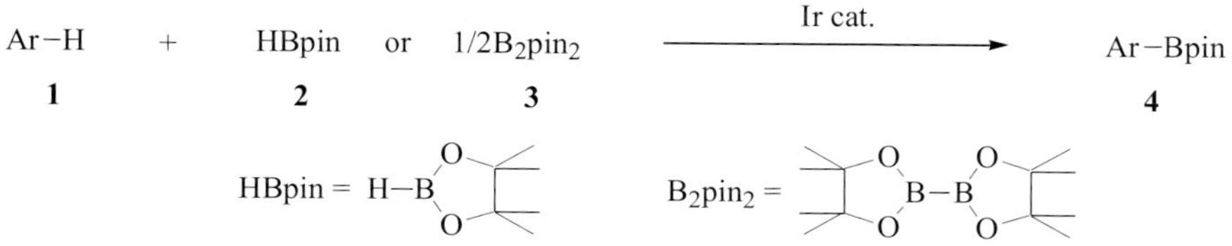

Ar—H + HBpin or 1/2B$_2$pin$_2$ $\xrightarrow{\text{Ir cat.}}$ Ar—Bpin
 1   **2**   **3**   **4**

Scheme 1. Iridium-catalyzed aromatic C–H borylation.

[3–10]. The maximum turnover number achieved in the reaction of benzene with **2** is 4500 when using Ir(η^5-C$_9$H$_7$)(cod)-dmpe at 150 °C; for reaction with **3** the TON is 8000 when using 1/2[IrCl(coe)$_2$]$_2$-dtbpy at 100 °C. The bpy-based Ir catalysts are effective for borylation by either **2** or **3** at lower temperature. 1/2[IrCl(coe)$_2$]$_2$-dtbpy catalyzes the reaction even at room temperature.

Table 1. Representative catalysts and their activity.

C$_6$H$_6$—H (excess) + **2** or 1/2**3** $\xrightarrow{\text{Ir cat.}}$ C$_6$H$_5$—Bpin

Reagent	Ir cat.	Temp. (°C)	Time (h)	Yield (%)	TON
2	Ir(η^5-C$_9$H$_7$)(cod)-2PMe$_3$ (2.0 mol%)	150	18	88	–
2	Ir(η^5-C$_9$H$_7$)(cod)-dmpe (0.02 mol%)	150	61	90	4500
2	1/2[IrCl(cod)]$_2$-bpy (3.0 mol%)	80	16	80	–
3	1/2[IrCl(cod)]$_2$-bpy (1.5 mol%)	80	16	95	–
3	1/2[IrCl(coe)$_2$]$_2$-dtbpy (0.01 mol%)	100	16	80	8000
3	1/2[IrCl(coe)$_2$]$_2$-dtbpy (2.5 mol%)	25	4.5	83	–

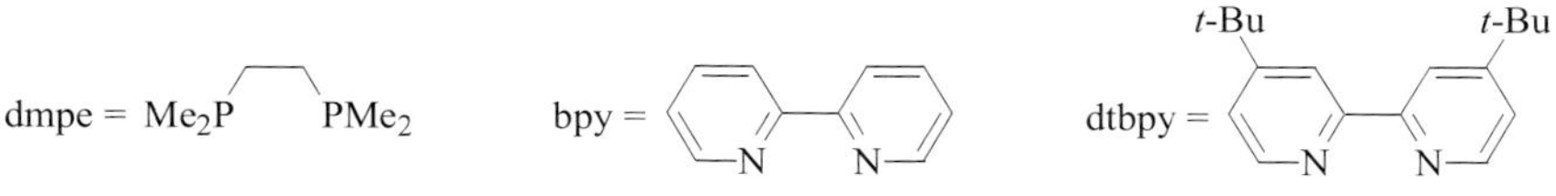

dmpe = Me$_2$P⌒PMe$_2$ bpy = 2,2'-bipyridine   dtbpy = 4,4'-di-*t*-Bu-2,2'-bipyridine

Functional group tolerance of the borylation is quite high. The reaction occurs selectively at the C–H bond for **1** bearing MeO, Me, I, Br, Cl, F$_3$C, MeO$_2$C, and NC groups (Scheme 2) [4–10]. In particular, it is interesting to note that aromatic C–H bonds are selectively borylated even in the presence of weaker benzylic C–H bonds or C-halogen bonds.

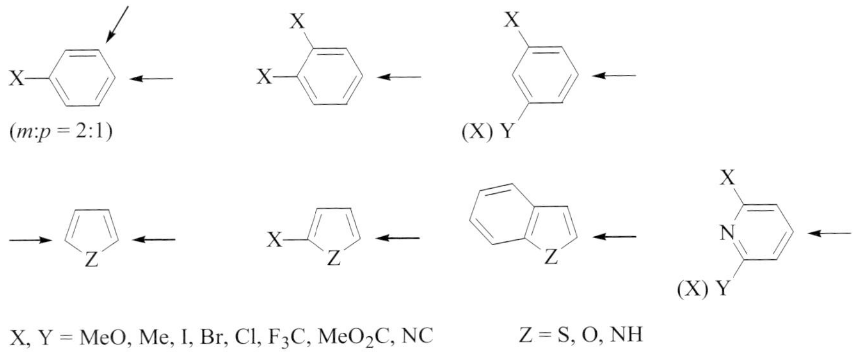

X, Y = MeO, Me, I, Br, Cl, F₃C, MeO₂C, NC   Z = S, O, NH

Scheme 2. Functional group tolerance and regiochemistry.

Regiochemistry of the borylation is summarized in Scheme 2. For arenes, regiochemistry is primarily controlled by the steric effects of substituents [4–6, 8, 10]. Both electron-rich and electron-poor monosubstituted arenes produce regioisomeric mixtures of meta and para borylation products in statistical ratios (ca. 2:1) indicating no significant electronic influence of substituents on regioselectivity. Thus, 1,2-disubstituted arenes bearing identical substituents yield **4** as single isomers. The borylation of 1,3-disubstituted arenes proceeds at the common meta position; therefore, regioisomerically pure **4** are obtained even for two distinct substituents on the arenes. In the case of five-membered heteroarenes, the electronegative heteroatom causes the α-C–H bonds to be active for the borylation [7–10]. Unsubstituted substrates yield mixtures of 2-borylated and 2,5-diborylated products, but both products are selectively obtained by reactions with the appropriate ratio of substrate and reagent. On the other hand, monoborylation selectively occurs for 2-substituted or benzo-fused substrates. For pyridines 2,6-disubstituted substrates undergo smooth borylation at the γ-position [5], whereas pyridine itself is significantly less reactive, because of the strong coordinating ability of the basic nitrogen for the catalyst [7].

1.2.3.2 Mechanism

The mechanism proposed for aromatic C–H borylation of aromatic compounds **1** by B₂pin₂ **3** catalyzed by the Ir-bpy complex is depicted in Scheme 3 [6–9]. A tris(boryl)Ir(III) species [5, 6, 11] **6** generated by reaction of an Ir(I) complex **5** with **3** is chemically and kinetically suitable to be an intermediate in the catalytic process. Oxidative addition of **1** to **6** yields an Ir(V) species **7** that reductively eliminates an aromatic boron compound **4** to give a bis(boryl)Ir(III) hydride complex **8**. Oxidative addition of **3** to **8** can be followed by reductive elimination of HBpin **2** from **9** to regenerate **6**. **2** also participates in the catalytic cycle via a sequence of oxidative addition to **8** and reductive elimination of H₂ from an 18-electron Ir(V) intermediate **10**. Borylation of **1** by **2** may occur after consumption of **3**, because the catalytic reaction is a two-step process – fast borylation by **3** then slow borylation by **2** [6].

Catalytic borylation by **2** may proceed through intermediates **6**, **7**, **8**, and **10**, in which **6** can be generated by the reaction of **5** with **2** [10]. A similar process is also postulated for the borylation of **1** by **2** catalyzed by phosphine-based Ir complexes [5]. Although the possibility of an Ir(I)–Ir(III) cycle cannot be completely eliminated, the synthetic [4–10], mechanistic [5, 6], and theoretical studies [12] support the Ir(III)–Ir(V) cycle.

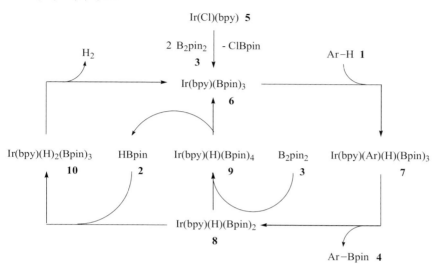

Scheme 3. Proposed mechanism for borylation by B_2pin_2 **3**.

1.2.3.3 Scope and Limitations

Although C–H borylation of aromatic compounds **1** by HBpin **2** or B_2pin_2 **3** is usually performed using an excess of substrate or reagent at high temperature, reactions at room temperature with stoichiometric amounts of **1** toward **2** or **3** in an inert solvent is desirable for thermally unstable, expensive, or solid substrates. It has recently been reported that an Ir complex comprised of 1/2[Ir(OMe)(cod)]$_2$ and dtbpy is a highly active catalyst for the aromatic C–H borylation by **2** or **3**. This high activity enables room-temperature borylation of **1** by **2** or **3** with a stoichiometric amount of **1** toward **2** or **3** in hexane to produce the corresponding aromatic boron compounds **4** in high yields with high regioselectivity. With this advance, the aromatic C–H borylation provides a practical tool for preparing **4**. Representative examples are shown in Table 2 [8–10].

Table 2. Stoichiometric borylation of aromatic compounds **1** by HBpin **2** or B$_2$pin$_2$ **3** catalyzed by 1/2[Ir(OMe)(cod)]$_2$-dtbpy in hexane at room temperature; a: 5 equiv. of **1** to **2**, b: 10 equiv. of **1** to **3**, c: 0.45 equiv. of **1** to **2**, d: 1 equiv. of **1** to **3**.

Product	Yield (%) 2	Yield (%) 3	Product	Yield (%) 2	Yield (%) 3
Cl-C$_6$H$_3$(Cl)-Bpin	73 (8 h)	82 (8 h)	2-Bpin-thiophene	75[a] (0.5 h)	91[b] (1 h)
I-C$_6$H$_3$(Cl)-Bpin	67 (8 h)	82 (4 h)	2,5-bis(Bpin)-thiophene	86[c] (2 h)	83[d] (0.5 h)
F$_3$C-C$_6$H$_3$(OMe)-Bpin	73 (24 h)	81 (8 h)	2-Me-5-Bpin-furan	83 (2 h)	85 (2 h)
MeO$_2$C-C$_6$H$_3$(Cl)-Bpin	70 (24 h)	80 (8 h)	indole-N-H-Bpin	99 (0.5 h)	88 (0.5 h)
NC-C$_6$H$_3$(Br)-Bpin	74 (2 h)	83 (2 h)	2-Cl-6-Me-pyridine-Bpin	75 (2 h)	84 (4 h)

Experimental

3-Bromo-5-(4,4,5,5-tetramethyl-1,3,2-dioxaborolan-2-yl)benzonitrile

A 25-mL flask containing a magnetic stirring bar and assembled with a septum inlet, a condenser, and a bubbler was charged with [Ir(OMe)(cod)]$_2$[13] (0.015 mmol) and 4,4'-di-*tert*-butyl-2,2'-bipyridine (0.03 mmol), and then flushed with nitrogen. Dry hexane (6 mL), pinacolborane (1.1 mmol), and 3-bromobenzonitrile (1.0 mmol) were added, and the mixture was stirred at 25 °C for 2 h. The reaction mixture was treated with water at room temperature, extracted with benzene, washed with brine, and dried over MgSO$_4$. Kugelrohr distillation in vacuo gave an analytically pure sample. The purity, determined by NMR and GC analysis was, >95%; ^{1}H NMR (CDCl$_3$, 400 MHz) δ 1.35 (s, 12 H), 7.85 (s, 1 H), 8.01 (s, 1 H), 8.13 (s, 1 H); ^{13}C NMR (CDCl$_3$, 100 MHz) δ 24.81, 84.85, 113.77, 117.32, 122.60, 136.70, 136.74, 141.73; exact mass calcd for C$_{13}$H$_{15}$BBrNO$_2$ 307.0379, found 307.0387.

2-(4,4,5,5-Tetramethyl-1,3,2-dioxaborolan-2-yl)indole

A 25-mL flask containing a magnetic stirring bar and assembled with a septum inlet, a condenser, and a bubbler was charged with [Ir(OMe)(cod)]$_2$[13] (0.015 mmol), 4,4'-di-*tert*-butyl-2,2'-bipyridine (0.03 mmol), and bis(pinacolato)diboron (1.0 mmol), and then flushed with nitrogen. Dry hexane (6 mL) and indole (2.0 mmol) were added, and the mixture was stirred at 25 °C for 0.5 h. The reaction mixture was treated with water at room temperature, extracted with benzene, washed with brine, and dried over MgSO$_4$. Kugelrohr distillation in vacuo gave an analytically pure sample. The purity determined, by NMR and GC analysis, was >95%; ^{1}H NMR (CDCl$_3$, 400 MHz) δ 1.36 (s, 12 H), 7.09 (t, 1 H, J = 7.7 Hz), 7.11 (s, 1 H), 7.23 (t, 1 H, J = 8.3 Hz), 7.38 (d, 1 H, J = 8.3 Hz), 7.67 (d, 1 H, J = 7.8 Hz), 8.56 (br s, 1 H); ^{13}C NMR (CDCl$_3$, 100 MHz) δ 24.78, 84.13, 111.25, 113.83, 119.75, 121.57, 123.59, 128.24, 138.18; exact mass calcd for C$_{14}$H$_{18}$BNO$_2$ 243.1431, found 243.1433.

1.2.4
Transition-metal Catalyzed Silylation of Arenes
Fumitoshi Kakiuchi

1.2.4.1 Introduction and Fundamentals

Catalytic C–H bond transformation is a very attractive research subject in organic and organometallic chemistry [1]. There are several procedures for C–H bond transformation. One is conversion of C–H bonds to C–Si bonds, i.e. the direct silylation of C–H bonds. Three procedures have been developed for silylation of C–H bonds – use of hydrosilanes (reaction 1 in Scheme 1), disilanes (reaction 2 in Scheme 1), and vinylsilanes (reaction 3 in Scheme 1) as silylating reagents.

In Scheme 1, reaction 1, two hydrogen atoms remain during the reaction and these should be removed from the metal center to regenerate the low-valent metal species. Generation of molecular hydrogen by reductive elimination is most likely, but this process is usually thermally unfavorable, so the co-presence of an efficient hydrogen acceptor, or use of photoirradiation, is usually required to accomplish the reaction catalytically. In reaction 2, H–M–SiR$_3$ species should be converted to the low-valent metal by formal dissociation of H–SiR$_3$ from the metal. In reaction 3 a vinylsilane functions as a silylating reagent and an acceptor of hydrogen. Thus, during the course of the reaction ethylene should be generated. Among these three reactions reaction 1 is the most frequently used in the catalytic silylation of C–H bonds.

For catalytic C–H bond transformation, there are two types of reaction. One involves non-chelation-assisted C–H bond cleavage, the other is chelation-assisted.

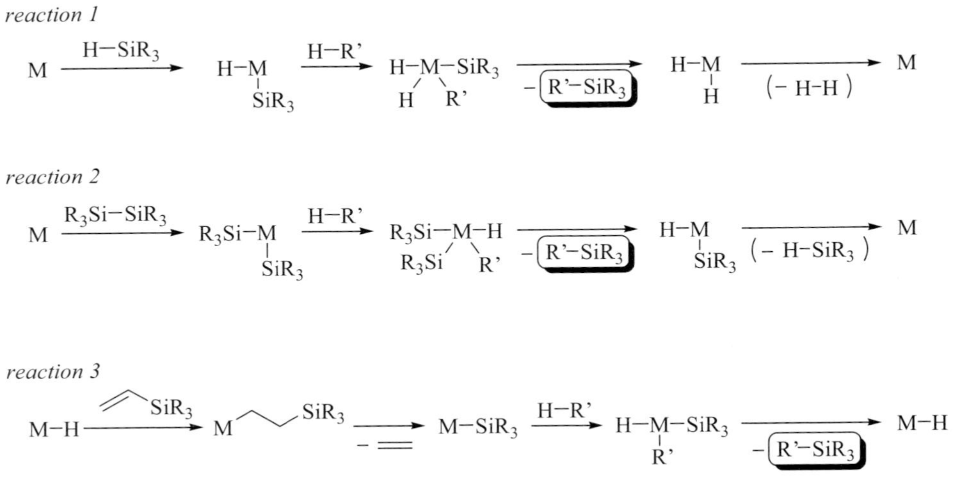
reaction 1
M
H–SiR3
H–M
SiR3
H–R'
H–M–SiR3
H R'
– R'–SiR3
H–M
H
(– H–H)
M
reaction 2
M
R3Si–SiR3
R3Si–M
SiR3
H–R'
R3Si–M–H
R3Si R'
– R'–SiR3
H–M
SiR3
(– H–SiR3)
M
reaction 3
M–H
SiR3
M
SiR3
M–SiR3
H–R'
H–M–SiR3
R'
– R'–SiR3
M–H

Scheme 1. Outlines of reactions for silylation of C–H bonds.

In the former reaction regioisomers are possibly formed, so regioselectivity cannot be controlled. Chelation-assisted reactions, on the other hand, occur regioselectively, usually at the position ortho to the directing group.

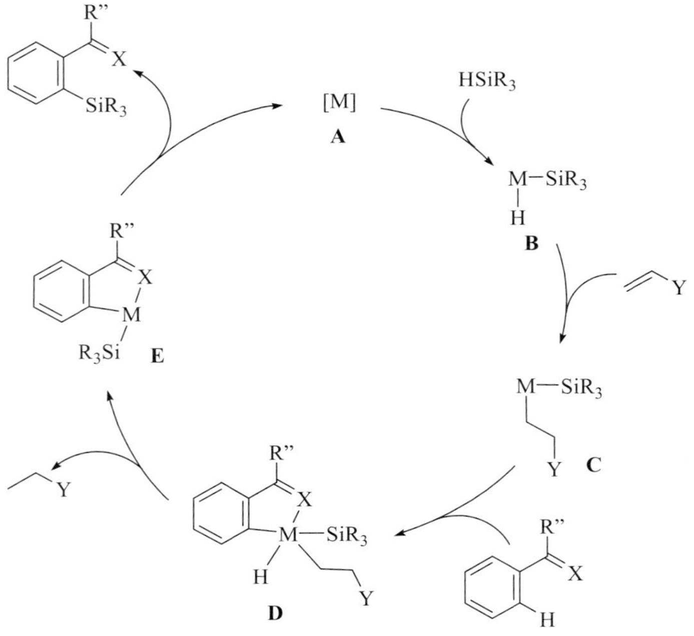
R''
X
SiR3
[M]
A
HSiR3
M–SiR3
H
B
Y
M–SiR3
Y
C
R''
X
H
R''
X
M–SiR3
H
Y
D
Y
R''
X
M
R3Si
E

Scheme 2. One of the possible reaction pathways for the silylation of C–H bonds with hydrosilanes.

1.2.4.2 Mechanism

As mentioned above, three types of reaction are used for silylation of C–H bonds. In this section, we describe the reaction pathway of chelation-assisted silylation of C–H bonds with hydrosilanes (reaction 1).

Several reaction pathways for reaction 1 are possible. A clear reaction mechanism has not been elucidated. Although it is premature to discuss the details of the reaction pathway for this silylation reaction, one possible pathway for the chelation-assisted silylation of C–H bonds is shown in Scheme 2. The catalytic reaction is initiated by oxidative addition of hydrosilane to **A**. Intermediate **B** reacts with an olefin to give **C**. Then, addition of a C–H bond to **C** leads to intermediate **D**. Dissociation of alkane from **D** provides Ru(silyl)(aryl) intermediate **E**. Reductive elimination making a C–Si bond gives the silylation product and the active catalyst species **A** is regenerated. Another pathway, addition of a C–H bond to **A** before addition of hydrosilane to **A** is also possible. At present, these two pathways cannot be distinguished.

1.2.4.3 Scope and Limitations

1.2.4.3.1 Silylation with Hydrosilanes

The first example of silylation of C–H bonds in arenes with hydrosilanes was reported by Curtis [2]. Later, silylation of C–H bonds with triethylsilane using a rhodium catalyst was reported (Scheme 3) [3, 4]. The reaction of arenes with bis(hydrosilane) using a platinum catalyst involves a bis(silyl)platinum species in the coupling reaction (Scheme 3) [5]. In these non-chelation-assisted reactions possible regioisomers should be formed.

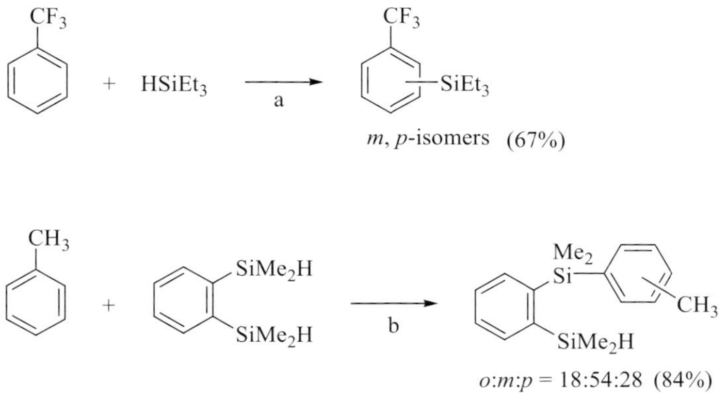

Scheme 3. Silylation of C–H bonds via non-chelation-assistance: (a) 8 mol% (η^5-C$_5$Me$_5$)Rh(H)$_2$(SiEt$_3$)$_2$, *tert*-butylethylene, 150 °C, 250 min; (b) 2 mol% Pt$_2$(dba)$_3$, 110 °C, 84 h.

For high regioselectivity, one of the most reliable procedures is chelation-assistance, which involves coordination of a heteroatom to a metal. A representative example of regioselective silylation is the Ru$_3$(CO)$_{12}$-catalyzed reaction of aryloxazolines with hydrosilanes in the presence of *tert*-butylethylene as a scavenger of the hydrogen [6]. In this reaction the nitrogen atom in the oxazolyl group functions as a directing group and C–SiR$_3$ bond formation occurs exclusively at the position ortho to the oxazolyl group (Table 1). For this silylation, triethyl-, dimethylphenyl, *tert*-butyldimethyl-, and triphenylsilanes have high activity, but hydrosilanes with a heteroatom on the silicon atom have no reactivity. This type of silylation of C–H bonds can be applied to a variety of aromatic compounds with an ester, amide, imino, azo, amino, pyridyl, imidazolyl, pyrazolyl, triazolyl, or tetrazolyl group (Scheme 4) [7, 8]. The directing groups remain intact in the coupling products. The functional group compatibility of this reaction is good. The reaction is tolerant of both electron-donating (Me, OMe, and NMe$_2$) and electron-withdrawing (CF$_3$, F, and Cl) groups [8]. Nearly always the corresponding ortho-silylation products are obtained in good to excellent yields. In these reactions the silyl group can be introduced at aromatic C–H bonds (sp^2 C–H bonds) selectively.

Table 1. Silylations of C–H bonds in aromatic and heteroaromatic compounds and of benzyl C–H bonds; 6 mol% Ru$_3$(CO)$_{12}$, HSiEt$_3$, *tert*-butylethylene or norbornene, toluene, reflux, 20 h.

Starting material	Products	Yield (%)	Ref.
naphthyl-oxazoline	naphthyl-oxazoline with SiEt$_3$ ortho	100	5,7
MeO-C$_6$H$_4$-CH=N-Bu^t	MeO-C$_6$H$_3$(SiEt$_3$)-CH=N-Bu^t	96	7
furan-2-C(O)NPr^i$_2$	3-SiEt$_3$-furan-2-C(O)NPr^i$_2$	65	7
PhCH$_2$NMe$_2$	o-SiEt$_3$-C$_6$H$_4$-CH$_2$NMe$_2$	58	6
PhCH$_2$-(2-pyridyl)	ortho-SiEt$_3$ isomer : benzyl-SiEt$_3$ isomer 6 : 1	90	6

Table 1. Continued

Starting material	Products		Yield (%)	Ref.
Cl-C6H4-O-quinoline	Cl-C6H3(SiEt3)-O-quinoline : Cl-C6H3(SiEt3)-O-quinoline(SiEt3) (1 : 1.6)		81	6
Ph-tetrazole-Me	C6H4(SiEt3)-tetrazole-Me : C6H3(SiEt3)(Et3Si)-tetrazole-Me (1 : 3)		93	7
8-methylquinoline	8-(CH2SiEt3)quinoline		78	8
1-(3-methylpyridin-2-yl)-2-methylnaphthalene	silylated product (CH2SiEt3)		72	8

Silylation of benzyl C–H bonds using hydrosilanes can also be performed with the aid of $Ru_3(CO)_{12}$-catalyst (Table 1) [9]. This silylation occurs only at benzylic CH_3 groups. Pyridine, pyrazole, and hydrazones function as good directing groups. Benzylamines, oxime ethers, dimethylanilides, and aryl pyridyl ethers have no activity in this silylation.

1.2.4.3.2 Silylation with Disilanes

Disilanes are also effective as silylation reagents. Silylation of arenes with strained cyclic disilane **1** occurs with the aid of $Ni(PEt_3)_4$ as catalyst [10]. This silylation preferentially occurs at the benzene ring. In the reaction of toluene with disilane **1**, the silylation products **2** and **3** (meta and para isomers) are obtained. A bis(silyl)-nickel intermediate is involved as a key intermediate in this reaction. Platinum(0) complexes have catalytic activity in the silylation of arenes with disilane. Thus, for silylation with disilane, chelation-assistance is also effective for achieving regioselectivity. The silylation with hexaorganodisilanes of aromatic imines **4** with an electron-withdrawing group on the aromatic ring proceeds in the presence of $Pt_2(dba)_3/P(OCH_2)_3CEt$ catalyst to give the silylation products **5** and **6** [11].

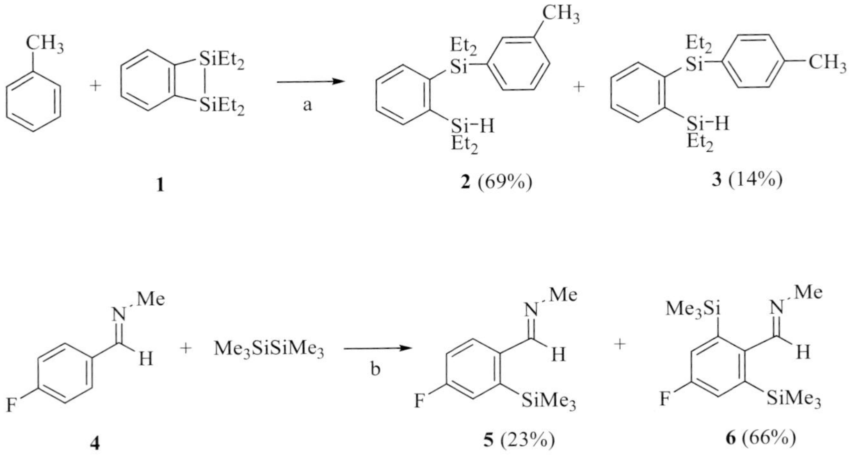

Scheme 4. Silylation of C–H bonds with disilanes: (a) 5 mol% Ni(PPh$_3$)$_4$, disilane 1, toluene, reflux 20 h; (b) 1 mol% Pt$_2$(dba)$_3$-P(OCH$_2$)$_3$CEt, toluene, 160 °C, 20 h.

1.2.4.3.3 Silylation with Vinylsilanes

Vinylsilanes can function as a silylating reagents for C–H bonds in heteroaromatic compounds (Scheme 5) [12]. The reaction of furan **7** with vinylsilane **8** using Ru$_3$(CO)$_{12}$ as a catalyst affords the β-silylation product in 98 % yield. This silylation can be applied only to heteroaromatic esters, ketones, and amides. For this reaction, RuHCl(CO)(PPh$_3$)$_3$ also has catalytic activity.

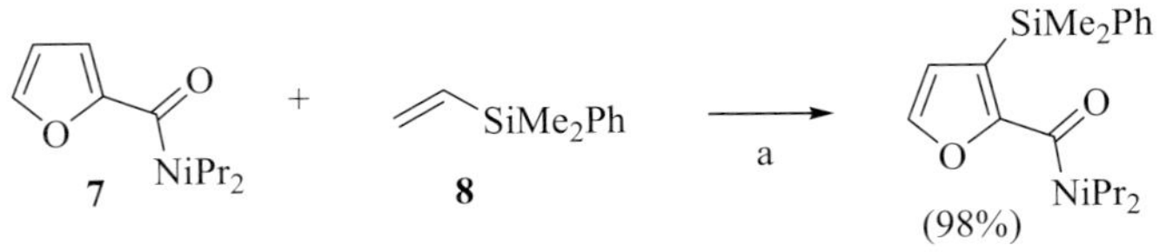

Scheme 5. Silylation of C–H bonds with vinylsilanes: (a) 6 mol% Ru$_3$(CO)$_{12}$, toluene, 115 °C, 20 h.

Experimental

4,4-Dimethyl-2-(2-methyl-6-triethylsilanylphenyl)-4,5-dihydrooxazole

Ru$_3$(CO)$_{12}$, (38.4 mg, 0.06 mmol), toluene (0.5 mL), 4,4-dimethyl-2-(2-methylphenyl)-4,5-dihydrooxazole (189 mg, 1 mmol), triethylsilane (581 mg, 5 mmol), and 3,3-dimethyl-1-butene (421 mg, 5 mmol) are placed in a 10-mL two-necked flask equipped with a reflux condenser connected to a nitrogen line, a rubber septum, and a magnetic stirring bar, all previously flame-dried under a flow of nitrogen. The resulting solution is heated under reflux under a nitrogen atmosphere for

20 h. The reaction mixture is evaporated to remove volatile materials. TLC (silica gel, hexane–ethyl acetate, 20:1) $R_F = 0.30$. The product ($R_F = 0.30$) is isolated by silica gel column chromatography (Merck silica gel 60, 230–400 mesh; column 50 mm i.d. × 80 mm length; eluent, hexane–ethyl acetate, 10:1): 210 mg (73 % isolated yield; 93 % GC yield). ^{1}H NMR (CDCl$_3$) 0.81–0.97 (m, 15 H, CH$_2$CH$_3$), 1.43 (s, 6 H, CH$_3$), 2.35 (s, 3 H, ArCH$_3$), 4.11 (s, 2 H, CH$_2$), 7.19 (d, $J = 7.16$ Hz, 1 H, ArH), 7.26 (dd, $J = 7.29$ Hz, 7.16 Hz, 1 H, ArH), 7.35 (dd, $J = 7.29$ Hz, 1.35 Hz, 1 H, ArH); ^{13}C NMR (CDCl$_3$) 3.86 (SiCH$_2$), 7.55 (CH$_2$CH$_3$), 19.92 (ArCH$_3$), 28.68 (CH$_3$), 68.20 (C(CH$_3$)$_2$), 78.73 (CH$_2$), 128.32, 130.66, 132.85, 134.82, 135.72, 136.74 (Ar), 163.07 (C=N).

1.3
Alkylation and Vinylation of Arenes

1.3.1
Friedel–Crafts-type Reactions

1.3.1.1 Comparison of Classical and Fancy Catalysts in Friedel–Crafts-type Reactions
Gerald Dyker

1.3.1.1.1 Introduction and Fundamental Examples
Friedel–Crafts alkylations and acylations are still among the most important C–C-coupling reactions (also see next Chapter, 1.3.1.2). Besides the classical Friedel–Crafts catalysts – Bronstedt and Lewis acids – related transition metal-catalyzed methods have emerged in recent years featuring outstanding regio- or even stereoselectivity. Because some of these new methods are assumed to proceed via an initial metalation step at the arene rather than by an activation of the electrophile, they are not usually published under the label "Friedel-Crafts reaction", although with in respect of the product formed they could not be distinguished. For example, for the AuCl$_3$–catalyzed reaction of 2-methylfuran (**1**) with methyl vinyl ketone (**2**, MVK) at least two mechanistic pathways must be discussed [1], a Michael-type process with a metalation of the furan **1** as the first step and a conjugate addition reaction as a second step. According to NMR studies AuCl$_3$ indeed interacts with furan **1**. Alternatively a Friedel–Crafts-type process might have occurred, either directly with AuCl$_3$ as Lewis-acidic catalyst or with Bronstedt-acidic catalysts H[AuCl$_3$OH], H[AuCl$_4$] or even hydrochloric acid as hydrolysis product.

The uncertainty about the active catalyst prompted us to test a variety of classical acidic catalysts in comparison with gold(III)chloride in coupling reactions of arenes with MVK **2**. As a second model reaction we chose the propargylation of arenes with propargylic alcohols. Of course these studies might not answer the mechanistic questions, but they are certainly of practical value.

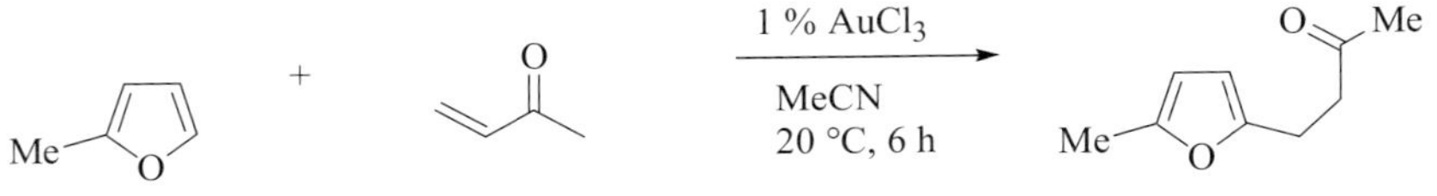

Scheme 1. Gold-catalyzed C–C-coupling reaction: Friedel–Crafts-type or Michael-type reaction?

1.3.1.1.2 Scope and Limitations of the Alkylation of Arenes with MVK

On the basis of mechanistic considerations $HAuCl_4$ and HCl were tested as alternative catalysts for the reaction depicted in Scheme 1, but both acids proved much less effective than $AuCl_3$. HCl, especially, is uneconomical, because of the competing addition reaction forming 4-chlorobutan-2-one as byproduct. Nevertheless,

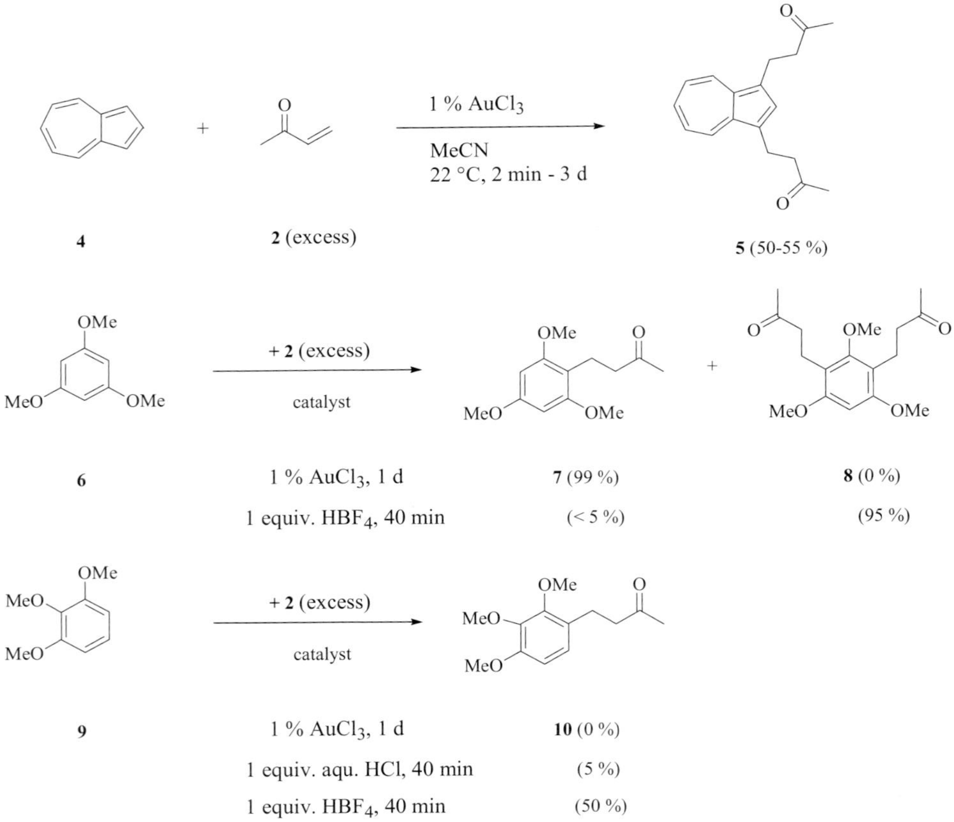

Scheme 2. Reactivity and selectivity of $AuCl_3$, HCl, and HBF_4 as catalysts in Friedel–Crafts-type reactions.

para-toluenesulfonic acid – a less nucleophilic Bronstedt acid – gave a similar result to AuCl$_3$.

The advantage of AuCl$_3$ becomes obvious in the reaction with the acid sensitive azulene (**4**) – with protic catalysts of higher acidity such as hydrochloric acid it is recommended to begin working up after about two minutes reaction time, because of competing and product-consuming oligomerization processes, whereas with AuCl$_3$ a rather constant 50–55 % yield of the bis-product **5** is obtained both after 2 min and after 3 days reaction time.

For the 1,3,5-substituted trimethoxybenzene **6**, AuCl$_3$ is highly selective for the mono-adduct **7**, whereas the more reactive HBF$_4$ is appropriate for obtaining the bis-adduct **8**, again with excellent yield. Surprisingly AuCl$_3$ fails in the reaction with the 1,2,3-substituted trimethoxybenzene **9**. Obviously, this substitution pattern already induces significant steric hindrance at the 4-position, thus HBF$_4$ is needed as catalyst to achieve an acceptable yield. For the alkylation of indole (**11**) in the 3-position just 1 % AuCl$_3$ is sufficient to achieve an excellent yield. BF$_3$-etherate – one equivalent is used – even catalyzes the alkylation of 3-methylindole (**13**) at the 2-position, but is somewhat inferior to bismuth nitrate [2, 3], which is reported to give a 46 % yield in this reaction. It should be emphasized that Friedel–Crafts reaction of MVK is limited to especially activated arenes. The catalysts tested failed, for example, with mesitylene, pyrene, 2,7-dimethoxynaphthalene, and even 1,4-dimethoxybenzene.

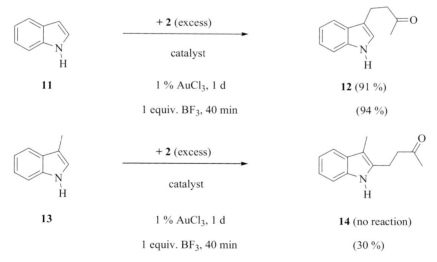

Scheme 3. Friedel–Crafts-type reactions with indoles.

1.3.1.1.3 Alkylation of Arenes with Propargylic Alcohols

Because of its potential in synthesis, the propargylation of arenes is a focus of current interest. According to Yamamoto et al. the dinuclear cationic ruthenium complex **15** is exceptionally active in this type of reaction when applied at in 5 %

amounts at 60 °C [4]. For example, with propargylic alcohol **16** as the standard model reactant good to excellent yields of the mono-substitution products **17** and **19** are obtained from reaction with excess 1,3,5-trimethoxybenzene **6** and excess azulene **4**, respectively. Electrophilic (η-propargyl)ruthenium species are proposed as intermediates.

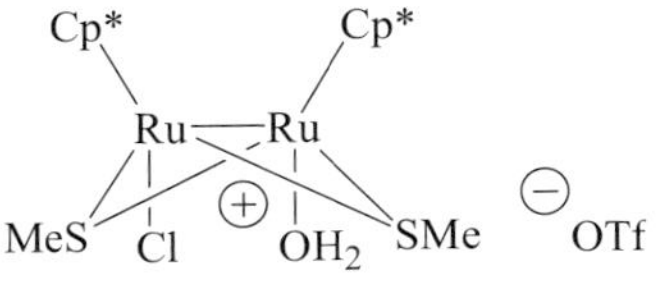

15

We have tested AuCl$_3$ and BF$_3$ etherate in comparison. Merely 0.3 % of the gold catalyst is sufficient to obtain 96 % isolated yield of the mono-substitution product **17** [5]. The reaction is performed at room temperature with two equivalents of propargylic alcohol **16**, thus demonstrating both outstanding reactivity and remarkable selectivity for the mono-substitution. Application of the classical Friedel–Crafts catalyst BF$_3$ etherate in stoichiometric amounts or in excess necessitates cooling, and results in rather clean formation of the sterically crowded bis-substitution product **18** as a mixture of the two diastereoisomers in the ratio 1:1.

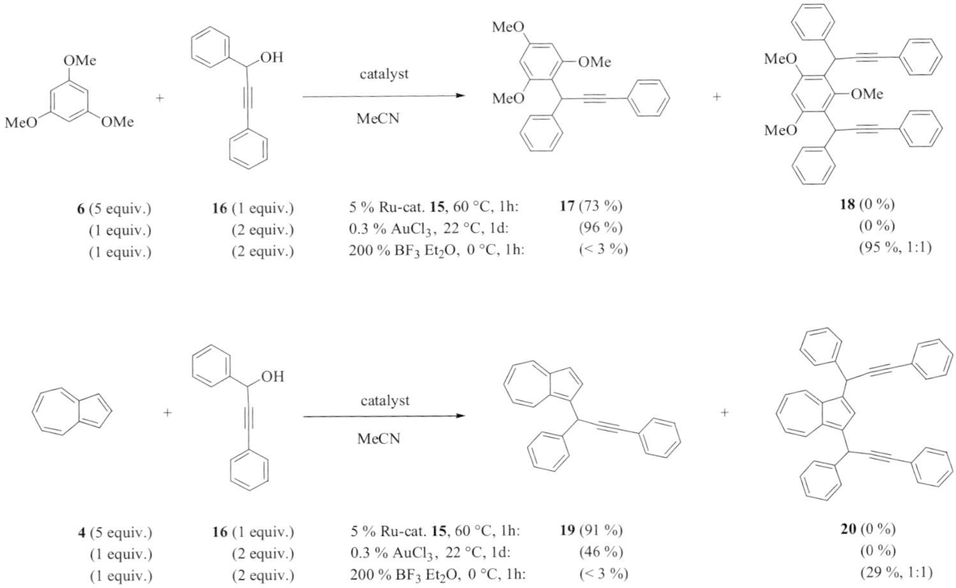

Scheme 4. Propargylation of electron-rich arenes [4, 5].

Recently Toste et al. [6] introduced the air- and moisture-tolerant Re-complex (dppm)Re(O)Cl₃ as an alternative. Applied as catalyst in 5% amounts (5 h at 65 °C) the propargylation of 1,2,3-trimethoxybenzene (**9**) with model reactant **15** is achieved with 92% yield of the corresponding mono-substitution product. The potential of this method is highlighted by the successful synthesis of natural products such as β-apopicropodophyllin (**24**) (Scheme 5).

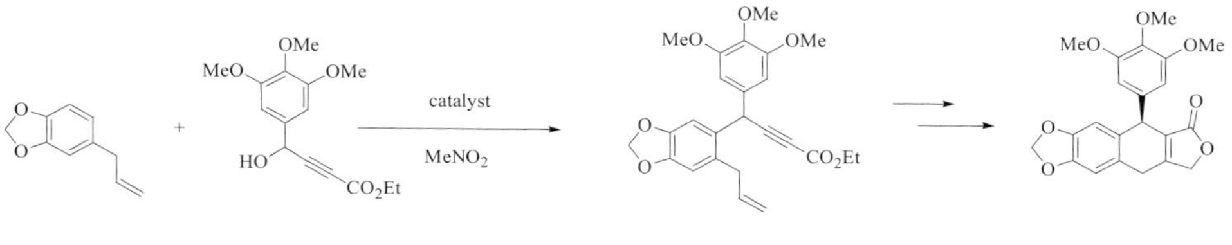

Scheme 5. Synthesis of β-apopicropodophyllin (**24**).

1.3.1.1.4 Conclusion

Classical Friedel–Crafts catalysts such as BF₃ etherate or Bronstedt acids, which are applied in stoichiometric amounts or even in excess, still have their value, because they are relatively inexpensive yet powerful catalysts for C–H transformation at arenes and, most importantly, they are often successful when modern catalysts fail. Gold(III) chloride, at less than 1%, has impressive reactivity under very moderate reaction conditions and its selectivity is exceptionally good. Friedel–Crafts type reactions with this catalyst are, however, restricted to very electron-rich arenes. Further studies should concentrate on increasing the electrophilicity of gold catalysts. New catalysts have emerged, for instance based on ruthenium and rhenium, which promise broad applicability based on alternative mechanisms. Catalysts based on rare earth metals are discussed in the next chapter.

Experimental

1,3-Bis-(3-oxobutyl)-azulene (5)

A mixture of 211 mg (3.00 mmol) MVK, 950 mg acetonitrile, and 300 mg (10 µmol, 0.2 mol%) AuCl₃ was slowly added under argon to a vigorously stirred solution of 193 mg (1.50 mmol) azulene (**4**) in 2 mL acetonitrile. After 1 day at room temperature the reaction mixture was filtered through a small pad of silica (2 g) with an additional 5 mL acetonitrile as eluent. The solvent was removed in vacuo at 50 °C (from 15 mbar to 0.2 mbar) and the residue purified by flash chromatography on silica gel (petroleum ether *tert*-butyl methyl ether 2:1; $R_F = 0.10$). Recrystallization from CH₂Cl₂–diethyl ether yielded 201 mg (50%) **15** as dark blue–green crystals (m.p. 63 °C). ¹H NMR (CDCl₃, 500 MHz): $\delta = 2.12$ ppm (s, 6H), 2.87 (m, 4H), 3.31 (m, 4H), 7.00 (t, $J = 9.9$ Hz), 7.49 (t, $J = 9.9$ Hz), 7.60 (s, 1H), 8.17 (d, $J = 9.9$ Hz). ¹³C NMR (CDCl₃, 125 MHz): $\delta = 21.17$ ppm (t), 30.11 (q), 45.16 (t), 121.12 (d), 127.12 (s), 133.07 (d), 135.92 (s), 136.47 (d), 137.73 (d), 208.42 (s).

1.3.1.2 Lanthanoid Triflates in Catalytic Amounts for Friedel–Crafts-type Reactions
Shu Kobayashi

1.3.1.2.1 Introduction and Fundamental Examples

Friedel–Crafts acylation and alkylation are among the most fundamental and useful reactions for introducing functional substituents to aromatic rings [1], and are often used for industrial production of pharmaceuticals and agricultural chemicals, plastics, and liquid crystals, etc. These reactions were originally performed by using more than stoichiometric amounts of Lewis acids, for example aluminum trichloride ($AlCl_3$), because the Lewis acids are consumed by coordination with the aromatic ketones produced. As a result, a large amount of $AlCl_3$ and its waste after aqueous work-up procedures often causes serious environmental problems. To address this issue, several much effort was devoted to performing reactions using catalytic amounts of acidic promoters [2], or to recycling metal oxide catalysts, etc. [3]. Usually, however, substrates are limited and examples of truly efficient catalysts are few.

On the other hand, rare-earth trifluoromethanesulfonates (rare earth triflate, $RE(OTf)_3$) have been found to work efficiently as Lewis acids even in aqueous media or in the presence of amines [4]. A catalytic amount of $RE(OTf)_3$ enables several synthetically useful reactions, for example aldol, Michael, allylation, Mannich, Diels–Alder reactions, etc., to proceed. It has also been demonstrated that a small amount of $RE(OTf)_3$ is enough to complete the reactions and that $RE(OTf)_3$ can easily be recovered from the reaction mixture and can be reused. A key to accomplishing the catalytic processes was assumed to be the equilibrium between Lewis acids and Lewis bases, for example water, carbonyl compounds, and amines, etc. A similar equilibrium was expected between Lewis acids and aromatic ketones, and, thus, $RE(OTf)_3$-catalyzed Friedel–Crafts acylation was investigated [5].

Reaction of anisole (**1**) with acetic anhydride was chosen as a model, and ytterbium trifluoromethanesulfonate (ytterbium triflate, $Yb(OTf)_3$) was the first $RE(OTf)_3$ representative used. Several reaction conditions were examined; the results are summarized in Table 1. When acetic anhydride, acetonitrile, or nitromethane was used as a solvent (entries 4–10), the reaction mixture became homogeneous and the acylation reaction proceeded smoothly. Nitromethane gave the highest yield of 4-methoxyacetophenone (**2**) (entries 7–10). On the other hand, in carbon disulfide, dichloroethane, or nitrobenzene (entries 1–3), the reaction mixture was heterogeneous and the yield of **2** was low. It was noted that the acylation proceeded quantitatively when a catalytic amount of $Yb(OTf)_3$ was used (0.2 equiv., entry 9). Even when 0.05 equiv. of the catalyst was employed, **2** was obtained in 79 % yield (entry 10).

Other $RE(OTf)_3$ were also examined as catalysts in the reaction of **1** with acetic anhydride (Table 2). Catalytic amounts of all the $RE(OTf)_3$ listed effectively mediated the acylation of **1**. Among these, scandium trifluoromethanesulfonate (scandium triflate, $Sc(OTf)_3$)[6] or $Yb(OTf)_3$ was superior to other $RE(OTf)_3$ and afforded the acylation product **2** quantitatively. When, on the other hand, lanthanum trifluoromethanesulfonate (lanthanum triflate, $La(OTf)_3$) was used, the yield of **2** was relatively low. The yields shown in Table 2 may reflect the catalytic activity of respective $RE(OTf)_3$.

Table 1. Yb(OTf)$_3$-catalyzed Friedel–Crafts acylation of **1**.

MeO–C$_6$H$_5$ (**1**) + Ac$_2$O (1.0 equiv.) →[Yb(OTf)$_3$, 50 °C] MeO–C$_6$H$_4$–COMe (**2**)

Entry	Yb(OTf)$_3$ equiv.	Time h	Solvent	Yield %
1	1.0	24	CS$_2$	ND[a]
2	1.0	24	ClCH$_2$CH$_2$Cl	8
3	1.0	20	PhNO$_2$	32
4	1.0	3	Ac$_2$O	53
5	0.2	18	Ac$_2$O	48
6	1.0	24	MeCN	52
7	1.0	20	MeNO$_2$	78
8	0.2	48	MeNO$_2$	60
9[b]	0.2	18	MeNO$_2$	99
10[b]	0.05	48	MeNO$_2$	79

[a] Not detected. [b] Two equiv. of Ac$_2$O were used.

Table 2. Effect of rare earth metal triflates (RE(OTf)$_3$).

MeO–C$_6$H$_5$ (**1**) + Ac$_2$O (2.0 equiv.) →[RE(OTf)$_3$ (0.2 equiv.), 50 °C, 18 h] MeO–C$_6$H$_4$–COMe (**2**)

RE	Yield/%	RE	Yield/%
Sc	99	Gd	77
Y	69	Dy	85
La	41	Ho	79
Pr	87	Er	78
Nd	79	Tm	84
Sm	80	Yb	99
Eu	78	Lu	81

The difference between the catalytic activity of Yb(OTf)$_3$ and Sc(OTf)$_3$ is shown in Table 3. When the acylation of **1** was performed in a shorter reaction time (4 h), Sc(OTf)$_3$ gave a higher yield of **2** than Yb(OTf)$_3$. In the acylation of mesitylene (**3**), Sc(OTf)$_3$ also gave a much higher yield of aromatic ketone **4** in a shorter reaction time (73 %, 1 h) than Yb(OTf)$_3$ (16 %, 18 h). These results show that Sc(OTf)$_3$ is the most active Friedel–Crafts catalyst among RE(OTf)$_3$. In the periodic table, scandium lies above yttrium and the lanthanides and its chemical behavior is known to be intermediate between aluminum and lanthanides [7]. Scandium oxide is more basic than aluminum oxide and more acidic than lanthanide oxides. In addition, its ionic radius is the smallest among rare earths, and thus it would be reasonable to conclude that Sc(OTf)$_3$ is the strongest Lewis acid among the RE(OTf)$_3$. The effect of using different amounts of Sc(OTf)$_3$ in the acetylation of **1** is summarized in Table 4. The reaction proceeded smoothly in a few hours if a catalytic amount of Sc(OTf)$_3$ was used, and 6200 % yield of acetylation product **2** (based on the catalyst) was obtained when 0.01 equiv. Sc(OTf)$_3$ was employed (entry 3).

Table 3. Catalytic activity of Yb(OTf)$_3$ and Sc(OTf)$_3$.

$$\text{Ar-H} + \text{Ac}_2\text{O (2 equiv.)} \xrightarrow[\text{MeNO}_2,\ 50\ ^\circ\text{C}]{\text{Catalyst (0.2 equiv.)}} \text{Ar-COMe}$$

Ar-H	Catalyst	Product	Time/h	Yield/%
MeO–C$_6$H$_5$ (**1**)	Yb(OTf)$_3$	MeO–C$_6$H$_4$–COMe (**2**)	4	55
	Sc(OTf)$_3$		4	89
mesitylene (**3**)	Yb(OTf)$_3$	2,4,6-trimethylaryl–COMe (**4**)	18	55
	Sc(OTf)$_3$		1	89

Table 4. Effect of the amount of Sc(OTf)$_3$.

$$\text{MeO-C}_6\text{H}_5\ (\mathbf{1}) + \text{Ac}_2\text{O (1.0 equiv.)} \xrightarrow[\text{MeNO}_2,\ 50\ ^\circ\text{C, 4 h}]{\text{Sc(OTf)}_3} \text{MeO-C}_6\text{H}_4\text{-COMe}\ (\mathbf{2})$$

Entry	Sc(OTf)$_3$/equiv.	Yield/%[a]
1	0.2	89 (445)
2	0.05	76 (1520)
3	0.01	62 (6200)

[a] The number in parentheses is the yield based on Sc(OTf)$_3$.

In the RE(OTf)$_3$-catalyzed Friedel–Crafts acylation, the acylation of aromatic compounds with electron-donating substituents, for example anisole and mesitylene, proceeded smoothly whereas the reactivity of benzene, toluene, and xylenes was low under the same conditions. On the other hand, it was revealed that the catalyst activity of RE(OTf)$_3$ was increased when combined with LiClO$_4$, and that the acceleration effect was strongly dependent on the amount of LiClO$_4$.

The reaction of *m*-xylene (**5**) with acetic anhydride was studied in the presence of Sc(OTf)$_3$ or Yb(OTf)$_3$ (a representative RE(OTf)$_3$). The yields were increased, depending on the amount of LiClO$_4$ in both cases (Table 5). In the absence of LiClO$_4$, Sc(OTf)$_3$ catalyzed the acylation of **5** to afford the desired aromatic ketone (**6**) in only 12% yield. When Sc(OTf)$_3$ was combined with 10 equiv. LiClO$_4$, however, the yield was improved to 89%. With Yb(OTf)$_3$ no product was obtained in the absence of LiClO$_4$ and 83% yield of the desired product was obtained in the presence of 6 equiv. LiClO$_4$.

Table 5. Acceleration of acylation by LiClO$_4$.

RE = Sc[a]		RE = Yb[b]	
LiClO$_4$ equiv.	Yield %	LiClO$_4$ equiv.	Yield %
None	12	None	ND[c]
0.4	12	0.4	8
0.8	28	0.8	17
1.0	36	1.0	22
2.0	51	2.0	38
4.0	61	4.0	67
6.0	82	6.0	83
8.0	88		
10.0	89		

[a] 1 h. [b] 4 h. [c] Not detected.

1.3.1.2.2 Mechanism

Although the precise mechanism of RE(OTf)$_3$-catalyzed Friedel–Crafts acylation has not been reported, a key step is believed to be ready release of RE(OTf)$_3$ from RE(OTf)$_3$–aromatic ketone complexes. Whereas LiClO$_4$ worked well in combination with RE(OTf)$_3$, other metal salts such as NaClO$_4$, Mg(ClO$_4$)$_2$, LiOTf, and LiBF$_4$, etc., were not effective. It was reported that ionic species such as CH$_3$CO$^+$SbF$_6^-$, CH$_3$CO$^+$PF$_6^-$, CH$_3$CO$^+$BF$_4^-$, and CH$_3$CO$^+$AsF$_6^-$, were active acylating agents in which the acylium cation was stabilized by the weakly basic counter anion [8]. It was also known that acetyl perchlorate, CH$_3$CO$^+$ClO$_4^-$, generated in situ by reaction of carboxylic acid or carboxylic anhydride with perchloric acid in aqueous media, had high reactivity in the acylation of aromatic compounds [9]. Furthermore, it was reported that lithium perchlorate (LiClO$_4$) formed a stable acylium cation when mixed with an acylating agent in the presence of an antimony or hafnium compound. [10] It was therefore expected that an acylium cationic species such as acyl perchlorate generated in the presence of RE(OTf)$_3$ and LiClO$_4$ would react with aromatic compounds to afford the corresponding aromatic ketones.

As an trial to generate an active acylium cation species in the presence of RE(OTf)$_3$, a combination of Sc(OTf)$_3$ (0.2 equiv.) and an alkali or alkaline earth metal salt (2.0 equiv.) was investigated in the reaction of *m*-xylene (**5**) with acetic anhydride. Without the metal salt, acylation product **6** was obtained only in 12% yield. When, however, the acylation was accelerated by adding LiClO$_4$, the yield was improved to 50%. Interestingly, sodium or magnesium perchlorate was not effective. In the absence of Sc(OTf)$_3$, LiClO$_4$ was not soluble in the reaction mixture and no acylation product was obtained. When Sc(OTf)$_3$ was added to a suspension of LiClO$_4$, acetic anhydride, and *m*-xylene (**5**) in nitromethane, the heterogeneous mixture changed to a dark-red homogeneous solution and the acylation reaction began to proceed.

1.3.1.2.3 Scope and Limitation

Several substituted benzenes were subjected to Yb(OTf)$_3$ or Sc(OTf)$_3$-catalyzed Friedel–Crafts acylation using acetic anhydride, acetyl chloride, benzoic anhydride, or benzoyl chloride as acylating agent (Table 6). Acetylation of *o*- and *m*-dimethoxybenzene proceeded smoothly to afford the desired acetylated products in excellent yields. Although xylenes gave the acetylated adducts in lower yields, benzene and toluene reacted sluggishly. Acetyl chloride was effective in the reaction, and yields were 5–10% lower than those obtained using acetic anhydride. Both benzoic anhydride and benzoyl chloride enabled successful benzoylation. Anisole (**1**) reacted with benzoic anhydride and benzoyl chloride smoothly in the presence of a catalytic amount of Sc(OTf)$_3$ to give 4-methoxybenzophenone in high yields (entries 1 and 2). In each reaction formation of other isomers was not observed.

1.3 Alkylation and Vinylation of Arenes

Table 6. RE(OTf)$_3$-catalyzed Friedel–Crafts acylation.

$$\text{Ar-H} + \text{Acylating Agent} \xrightarrow[\text{MeNO}_2,\ 50\ °C]{\text{RE(OTf)}_3\ (0.2\ \text{equiv.})} \text{Ar-COMe}$$

Entry	Ar-H	RE(OTf)$_3$	Acylating agent	Time/h	Product	Yield/%
1	MeO-C$_6$H$_5$ (**1**)	Sc(OTf)$_3$	PhCOCl	18	MeO-C$_6$H$_4$-COPh	79
2		Sc(OTf)$_3$	(PhCO)$_2$O	18		90
3	1,3-(MeO)$_2$C$_6$H$_4$	Sc(OTf)$_3$	Ac$_2$O	1	(MeO)$_2$C$_6$H$_3$-COMe	99
4	1,2-(MeO)$_2$C$_6$H$_4$	Sc(OTf)$_3$	Ac$_2$O	1	(MeO)$_2$C$_6$H$_3$-COMe	89
5	o-xylene	Yb(OTf)$_3$	Ac$_2$O	18	Me$_2$C$_6$H$_3$-COMe	22
6		Sc(OTf)$_3$	Ac$_2$O	1		14
7	m-xylene (**5**)	Yb(OTf)$_3$	Ac$_2$O	18	Me$_2$C$_6$H$_3$-COMe (**6**)	25
8		Sc(OTf)$_3$	Ac$_2$O	1		11

The reusability of the catalyst is one of the major advantages of using RE(OTf)$_3$ as a Friedel–Crafts catalyst. RE(OTf)$_3$ can be easily recovered from the reaction mixture by simple extraction. The catalyst is soluble in the aqueous layer rather than in organic layer, and is recovered by removing water to give a crystalline residue, which can be re-used without further purification. The efficiency of recovery and the catalytic activity of the reused RE(OTf)$_3$ were examined in the reaction of **1** with acetic anhydride using Yb(OTf)$_3$ and Sc(OTf)$_3$. As shown in Table 7, more than 90% of Yb(OTf)$_3$ and Sc(OTf)$_3$ were easily recovered, and the yields of acylation product **2** in the second and the third uses were almost the same as in the first.

In the Sc(OTf)$_3$–LiClO$_4$-system, wider substrate scope was observed as is shown in Table 8. Each acylation reaction in the Table gave a single acylation product and formation of other isomers was not observed. Acetylation of anisole (**1**) resulted in excellent yield of the product (entry 1). Mesitylene (**3**) and xylenes were transformed to 2,4,6-trimethylacetophenone and dimethylacetophenones, respectively, in moderate yields (entries 2–5). It is noteworthy that toluene was acylated by the Sc(OTf)$_3$–LiClO$_4$ system to give 4-methylacetophenone in 48% yield (entry 6) but the acylation did not proceed in the absence of LiClO$_4$. Furthermore, recovery and reuse of the RE(OTf)$_3$–LiClO$_4$ system were performed successfully. As shown in Table 9, the yields of **6** in the second and third uses of the catalyst system were almost the same as that in the first use.

Table 7. Reuse of RE(OTf)$_3$ in the acetylation of **1**.

MeO–C$_6$H$_5$ (**1**) + Ac$_2$O $\xrightarrow[\text{MeNO}_2, \ 50\,°\text{C},\ 4\ \text{h}]{\text{RE(OTf)}_3\ (0.2\ \text{equiv.})}$ MeO–C$_6$H$_4$–COMe (**2**)

RE(OTf)$_3$	No. of times used	Yield/%	Recovery of RE(OTf)$_3$/%
Sc(OTf)$_3$[a]	1	89	94
	2	93	90
	3	91	92
Yb(OTf)$_3$[b]	1	70	93
	2	69	90
	3	71	92

[a] One equiv. of Ac$_2$O was used. [b] Two equiv. of Ac$_2$O were used.

Table 8. Sc(OTf)$_3$-catalyzed acylation in LiClO$_4$–MeNO$_2$.

Ar–H + Ac$_2$O (1.0 equiv.) $\xrightarrow[\text{MeNO}_2,\ 50\,°\text{C},\ 1\ \text{h}]{\text{Sc(OTf)}_3\ (0.2\ \text{equiv.)},\ \text{LiClO}_4\ (4.0\ \text{equiv.)}}$ Ar–COMe

Entry	Ar-H	Product	Yield/%
1	MeO–C$_6$H$_5$ (**1**)	MeO–C$_6$H$_4$–COMe (**2**)	96
2	3,5-dimethylbenzene (**3**)	3,4,5-trimethyl COMe product (**4**)	93
3	o-xylene	Me$_2$C$_6$H$_3$–COMe	55
4	m-xylene (**5**)	Me$_2$C$_6$H$_3$–COMe (**6**)	61
5	toluene	Me-C$_6$H$_3$(COMe)-	16
6[a]	benzene	Me-C$_6$H$_4$–COMe	47

[a] 20 h.

Table 9. Catalytic activity of recovered $RE(OTf)_3$– $LiClO_4$.

$$5 + Ac_2O \text{ (1.0 equiv.)} \xrightarrow[\text{MeNO}_2, 50\,°C, 1\,h]{RE(OTf)_3 \text{ (0.2 equiv.)} \atop LiClO_4 \text{ (4.0 equiv.)}} 6$$

RE	Number of times used	Yield/%	Recover of catalyst/%
Sc	1	61	96
	2	55	94
	3	53	87
Yb[a]	1	68	95
	2	73	99
	3	70	91

[a] $LiClO_4$ (10.0 equiv.) was used and the reaction time was 4 h.

For catalytic Friedel–Crafts acylation, several excellent catalysts other than RE have recently been reported [11]. Although some are more active than RE, most are water-sensitive and cannot be recovered and reused after the reactions.

Experimental

Typical Procedure for RE(OTf)$_3$-catalyzed Friedel–Crafts Acylation

A mixture of Yb(OTf)$_3$ (620 mg, 1 mmol), anisole (1, 540 µL, 5 mmol), and acetic anhydride (940 µL, 10 mmol) in nitromethane (5 mL) was stirred at 50 °C for 4 h. After dilution with water (10 mL), the mixture was extracted with dichloromethane. The organic layers were combined and dried over NaSO$_4$. After filtration and evaporation of the solvents, the crude mixture was purified by column chromatography on silica gel to afford 4-methoxyacetophenone (**2**). The aqueous layer was concentrated in vacuo to give a crystalline residue, which was heated at 190 °C for 4 h in vacuo to afford 576.6 mg (93 %) Yb(OTf)$_3$ as colorless crystals. The recovered Yb(OTf)$_3$ was reused in the next acylation reaction. All products of the acylation of aromatic compounds shown in this chapter are known compounds and are commercially available.

Typical Procedure for RE(OTf)$_3$-LiClO$_4$ System-catalyzed Friedel–Crafts Acylation

Acetic anhydride (470 µL, 5 mmol) was added to a suspension of Sc(OTf)$_3$ (490 mg, 1 mmol), *m*-xylene (610 µL, 5 mmol), and LiClO$_4$ (2130 mg, 20 mmol) in nitromethane (5 mL) and the suspension changed to a dark-red homogeneous solution. After stirring at 50 °C for 1 h, the reaction mixture was diluted with water (10 mL), and the aqueous layer was extracted with dichloromethane (3 × 10 mL). The organic layers were combined and dried over NaSO$_4$. After filtration and evaporation of the solvents, the crude mixture was purified by column chromatography on silica gel to afford 2,4-dimethylacetophenone (**6**). The aqueous layer was concentrated to give crystalline residues which were heated at 190 °C for 4 h in vacuo to afford a mixture of LiClO$_4$ and Sc(OTf)$_3$ (2510 mg, 96 %). The recovered catalyst was reused in the next acylation reaction.

1.3.1.3 Enantioselective Friedel–Crafts Type Alkylation Reactions
Marco Bandini, Alfonso Melloni, and Fabio Piccinelli

1.3.1.3.1 Introduction and Fundamental Examples

Although Friedel–Crafts (F–C) alkylation reactions are among the longest known organic transformations requiring Lewis acids (LA) as promoters [1], only in recent decades significant improvements in the design and development of new catalytic stereoselective procedures have been reported [2]. In this scenario, two different, and to some extent complementary, approaches have been recognized as being highly effective in enantioselective electrophilic alkylations of electron-rich aromatic compounds: (1) the use of small chiral organic molecules mimicking the catalytic activity of enzymes, and (2) the employment of chiral organometallic complexes, made by use of enantiomerically pure organic ligands (*chiraphor unit*) with metal salts (*catalaphor unit*), acting as chiral Lewis acids (Scheme 1).

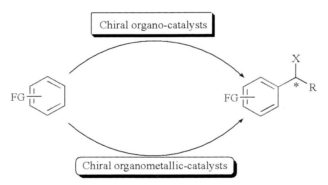

Scheme 1. Organic and organometallic chiral catalysts: different approaches for stereocontrolled alkylation of arenes.

Both strategies involve the key interaction between the chiral catalyst and the electrophilic reagent (usually a carbonyl compound) but their complementary behavior lies in the nature of organic molecules that can be coupled with the aromatic ring. As a matter of fact, although organic catalysts such as chiral imidazolidinone salts (**1**) have proven successful in the enantioselective alkylation of arenes with aromatic and aliphatic α,β-unsaturated aldehydes (1,4-addition), enantiomerically pure organometallic catalysts (e.g. bisoxazoline, binaphthyl based ligand, and Schiff base–metal complexes **2–4**, Scheme 2) afforded remarkable results in the presence of ketones and carboxylic substrates (1,4-addition) and carbonyl moieties C=X (1,2-addition). It is important to note that whereas organometallic catalysts usually require strictly anhydrous reaction conditions to guarantee high levels of chemoselectivity and stereoselectivity, organic catalyst-based approaches are usually performed in aqueous media (iPrOH, H_2O) or in reagent-grade solvents (CH_2Cl_2, THF), to properly dissolve the imidazolidinone salt and achieve satisfying catalytic turnover numbers (TON) and turnover frequencies (TOF).

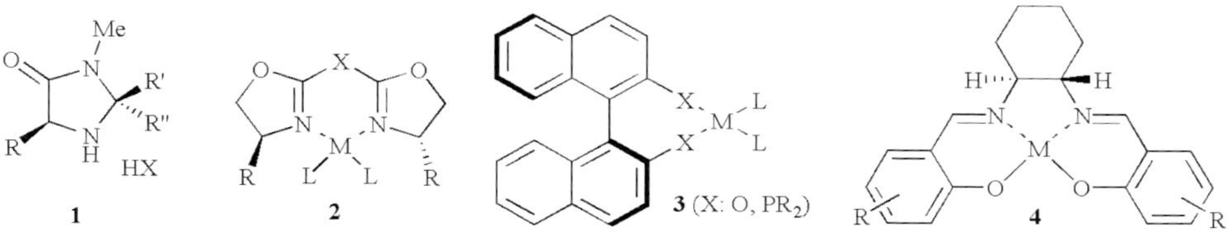

Scheme 2. Chiral catalysts for enantioselective F–C alkylations of aromatic compounds.

Two examples of such applications are outlined in Scheme 3. In the first one imidazolidinone **1a** 10 mol% catalyzed the conjugated addition of a plethora of variously substituted N,N-disubstituted anilines to α,β-unsaturated aldehydes to afford highly enantiomerically enriched β-aryl aldehydes **5** (ee: 89–97%) in good yields (>73%) [3]. On the other hand, the use of cationic BINAP-Cu complexes **6** (1–5 mol%) was found to be effective in the stereocontrolled 1,2-addition of indoles to the N-tosyl-α-imino ester **7** in THF at -78 °C [4].

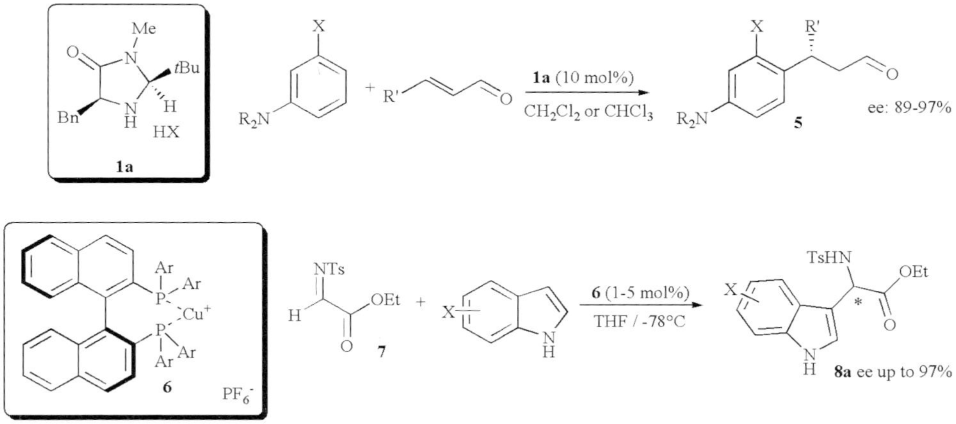

Scheme 3. Strategies for LUMO-lowering activation of electrophiles in stereoselective F–C alkylation.

1.3.1.3.2 Mechanism

Detailed mechanistic studies have been carried out for both the strategies of activation of electrophiles (E). In particular, the former stoichiometric Lewis acid mediated F–C version has been mechanistically well established and the formation of transient [LA-E] species is well known to be involved in the course of the reaction [5]. More recently, the combined use of Lewis acids (In^{III}, Ga^{III}, Bi^{III}, RE^{III}), guaranteeing kinetically labile formation of the catalyst–product adduct, with effective scavengers enabled widespread substitution of the pioneering stoichiometric technologies with cleaner catalytic alternatives. Analogously, in organo-catalyst-mediated F–C alkylations the intermediate formation of the highly reactive chiral iminium ion **9** enables remarkable LUMO-lowering activation of the α,β-unsaturated aldehyde with consequent regioselective conjugate addition of the aromatic system (i.e. pyrrole) [6]. The final hydrolysis of the chiral quaternary ammonium salt **10** yields the desired β-pyrrolyl aldehyde and releases the chiral auxiliary for the ongoing process. A pictorial representation of the proposed catalytic cycle is depicted in Scheme 4. In this context, the mineral acid used as co-catalyst (20 mol%) revealed to be very important in the stereochemical outcome of the process and in the alkylation of pyrroles TFA provides the optimum reaction conditions. Finally, on the basis of theoretical computations the authors suggest a hypothetical incoming trajectory of the aromatic rationalizing the role of the chiral catalyst in determining the absolute configuration of the product.

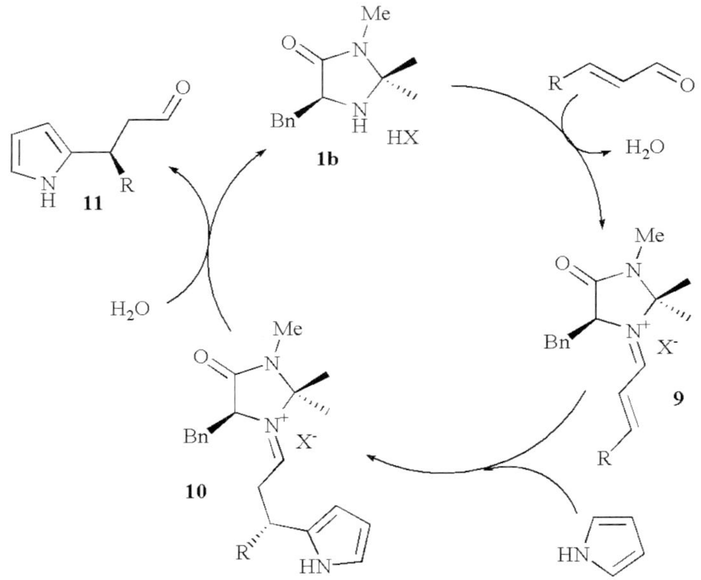

Scheme 4. Proposed mechanistic pathway for the stereoselective organo-catalyzed F–C alkylation of pyrroles.

1.3.1.3.3 Scope and Limitations

Until now, three main F–C transformations have been used for catalytic stereoselective formation of benzylic carbon stereocenters – 1,2-addition of arenes to carbonyl (C=X, X: O, NR) moieties, 1,4-addition of arenes to electron-deficient C–C double bonds, and ring-opening reaction of epoxides.

In the first strategy chiral organometallic complexes are used to differentiate the enantiotopic faces of the carbonyl moiety, providing a powerful route to the preparation of enantiomerically enriched α-aryl/heteroaryl esters. The presence of a two-site binding (chelating) interaction between the Lewis acid metal center and the electrophilic substrate is generally required to achieve high stereoselectivity. Therefore, 1,2- and 1,3-dicarbonyls have been found to be suitable candidates for such transformations. A brief summary of the most recent, straightforward enantioselective 1,2-type alkylation procedures is presented in Table 1. Highly enantiomerically pure α-hydroxy esters bearing tertiary and quaternary benzylic stereocenters (**14b–d**) are synthesized in good yields by using chiral Cu(I) and Ti(IV)-based Lewis acids. Moreover, synthetic useful optically active aromatic and heteroaromatic α-amino esters **8a-b** are isolated in excellent yields by alkylation of electron-rich aromatic C–H bonds with α-imino esters (**7, 12e**).

Table 1. Asymmetric catalyzed aromatic C–H addition of aromatics to C=X.

Carbonyl	Arene	Catalyst	F-C product	Yield/ee
12a[7a]	13a	3a	14a	Y: 66-97% ee: 68-90%
12b[7b]	13b	2a	14b	Y: 19-84% ee: 77-95%
12b[7c]	13b	3a	14b	Y: 85-99% ee: 80-97%
12c[7d]	13c	2a	14c	Y: 61-94% ee: 83-94%
12d[7e]	3d	2b	14d	Y: 72% ee: 92%
12e[7f]	13b,c/furanes	6	8b	Y: 24-82% ee: 59-96%

154 | 1 C–H Transformation at Arenes

The second approach is particularly suitable for the synthesis of carbonyl and carboxylic compounds bearing β-aryl/heteroaryl benzylic stereocenters. Both organic [9] and organometallic chiral catalysts provide high levels of chemical and optical yield in these C–H transformations (Table 2). The high regioselectivity normally

Table 2. Asymmetric F–C alkylations by conjugated C–H addition to electron-poor C=C double bonds.

Carbonyl	Arene	Catalyst	F-C product	Yield/ee
15a[8a]	13c	2a	16a	Y: 50-99% ee: 46-69%
15a[8b]	13c	2b	16a	Y: 73-99% ee: 60-92%
15b[6]	13e	1b	11	Y: 68-90% ee: 87-99%
15c[8c]	6/furanes/13d	2a	16b	Y: 65-99% ee: 60-99%
15d[8d]	13c	4a	16c	Y: 70-98% ee: 49-89%
15b[8e]	13c	1a	16d	Y: 80-98% ee: 89-97%
15e[8f]	13c	2c	16e	Y: 51-88% ee: 80-99%
15f[8g]	13c	3b 2 SbF$_6$⁻	16f	Y: 20-80% ee: 56-86%

encountered in these 1,4-additions (enones and α,β-unsaturated carboxylic compounds) is ascribed to the preferential attack of the arenes (soft nucleophiles) on the softer electrophilic β-position of the conjugate system. On the other hand, it should be stressed that the attempt to carry out a Lewis acid mediated Michael-type F–C alkylation with α,β-unsaturated aldehydes usually affords a complex mixture of compounds which results from competitive oligomerization of the electron-rich arene with the desired conjugate C–H addition. Then, use of a suitable metal-free chiral organic activator prevents this possibility providing the β-aryl aldehydes with high enantiomeric excesses.

The third route to catalytic stereoselective activation of C–H (sp^2) bonds has been investigated only recently by our group. It involves stereoselective kinetic resolution of racemic aromatic epoxides by indoles in the presence of catalytic amounts of [SalenCrSbF$_6$] complex (**4c**) [10]. Peculiarly, kinetic resolution processes enable the synthesis of both unconverted epoxides and the F–C adduct in high ee simply by a careful tuning of the initial ratio of reagents. In particular, by using excess epoxide, the ring-opening product is isolated in high yield and excellent enantioselectivity (up to 91%, Scheme 5a). *meso*-stilbene oxide can, moreover, also be effectively desymmetrized (ee: 90–98%) by use of a range of substituted indoles in the presence of commercially available chiral [SalenCrCl] **4b** (Scheme 5b).

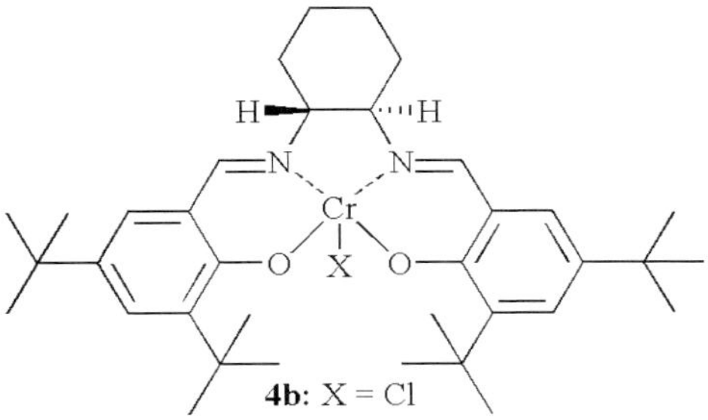

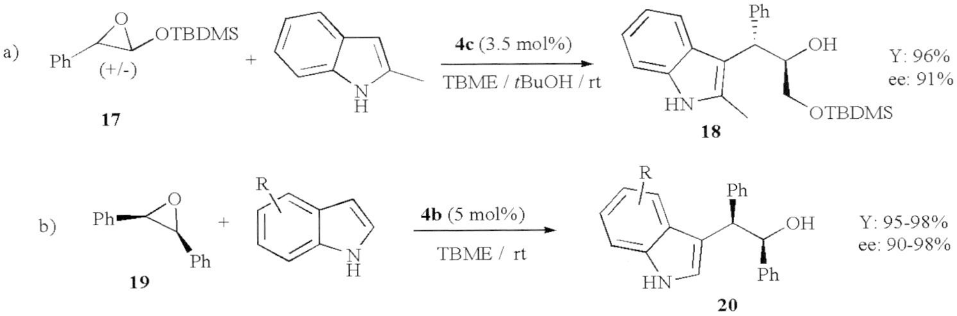

Scheme 5. Enantioselective kinetic resolution of 1,2-disubstituted epoxides (a) and desymmetrization of *meso* stilbeneoxide (b) with indoles catalyzed by **4b–4c**.

Experimental

(R)-3-(1-Methyl-1H-indol-3-yl)-butanal (16d)

A solution of (2S,5S)-5-benzyl-2-*tert*-butyl-3-methyl-imidazolidin-4-one (**1a**, 24.6 mg, 0.10 mmol) in CH_2Cl_2 (0.85 mL) and isopropanol (0.15 mL), was treated with TFA (7.7 µL, 0.10 mmol) and stirred for 5 min before addition of crotonaldehyde (125 µL, 1.50 mmol). After stirring for an additional 10 min the 1-methyl-1H-indole (**13c**, 64 µL, 0.50 mmol) was added in one portion. The resulting suspension was stirred at constant temperature (-83 °C) until complete consumption of the indole was observed by TLC (19 h). The reaction mixture was then passed cold through a silica gel plug with Et_2O and then concentrated. The resulting residue was purified by silica gel chromatography (benzene) to afford the title compound as a colorless oil (82 % yield, 92 % ee).

IR (film): 3054, 2960, 2824, 2722, 1720, 1616, 1550, 1474, 1374, 1329, 1241, 740 cm^{-1}. ^{1}H NMR (300 MHz, CDCl$_3$) 9.75 (dd, J = 2.1, 2.1 Hz, 1H), 7.63 (d, J = 7.8 Hz, 1H), 7.32–7.21 (m, 2H), 7.12 (ddd, J = 1.5, 7.4, 8.1 Hz, 1H), 6.84 (s, 1H), 3.75 (s, 3H), 3.68 (dt, J = 6.9, 13.8 Hz, 1H), 2.88 (ddd, J = 2.7, 6.9, 16.2 Hz, 1H); 2.71 (ddd, J = 2.7, 6.9, 16.2 Hz, 1H); 1.44 (d, J = 7.2 Hz, 3H); ^{13}C NMR (75 MHz, CDCl$_3$) 202.8, 137.2, 126.6, 125.2, 121.8, 119.1, 118.9, 118.8, 109.5, 51.2, 32.8, 26.0, 21.9; HRMS (CI) exact mass calcd. for ($C_{13}H_{15}NO$) requires m/z 201.1154, found m/z 201.1152. $[\alpha]_D$ = -4.2 (c 1.0, CHCl$_3$). The enantioselectivity was determined by subjecting approximately 10 mg of the title compound to excess $NaBH_4$ and 1 mL ethanol. After 15 min, the solution was treated with saturated aqueous $NaHCO_3$, and the mixture was extracted with CH_2Cl_2. The organic layer was separated, filtered through a silica gel plug and subjected to chiral HPLC analysis. Chiracel AD column (EtOH–*n*-hexanes 2:98, 1 mL min^{-1}); *S* isomer t_R = 25.2 min, *R* isomer t_R = 27.8 min.

(S,R-2-(2-Methyl-3-indolyl)-1-dimethyl-*t*-butylsilyloxy-2-phenylethanol (18)

A flamed, two-necked flask equipped with a magnetic stirring bar was charged with [SalenCrCl] **4b** (0.01 mmol) and activated molecular sieves 4 Å (100 mg). TBME (0.2 mL) was then added and the resulting solution was stirred under nitrogen at room temperature for 5 min. The solution was cooled to 0 °C and then **17** (0.3 mmol) and 2-methylindole (0.1 mmol) were added to the reaction mixture. Finally *t*-BuOH (0.1 mmol) was added and the reaction mixture stirred at 0 °C until GC analysis indicated complete conversion of 2-methylindole. The crude reaction mixture was diluted with Et_2O and filtered through celite. After evaporation of the solvent under reduced pressure the resulting residue was purified by silica gel chromatography (cyclohexane–Et_2O, 7:3) to afford the title compound as a white solid (96 % yield, 91 % ee).

Mp 109–111 °C. ^{1}H NMR (200 MHz, CDCl$_3$) -0.08 (s, 3H); 0.03 (s, 3H); 0.91 (s, 9H); 2.45 (s, 3H); 2.77 (d, J = 4.5 Hz, 1H); 3.45 (dd, J = 6.3, 10.2 Hz, 1H); 3.65 (dd, J = 3.6, 10.2 Hz, 1H); 4.32 (d, J = 10.2 Hz, 1H); 4.78–4.84 (m, 1H); 7.06–7.18 (m, 5H); 7.26–7.32 (m, 2H); 7.53 (dd, J = 0.9, 9.0 Hz, 1H); 7.63 (d, J = 7.8, 1H); 7.80 (br, 1H). ^{13}C NMR (50 MHz, CDCl$_3$) -5.35 (2), 12.42, 18.25, 25.90 (3), 45.91, 65.62, 72.18, 110.24, 111.92, 119.32 (2), 120.82, 125.89, 127.45, 128.08 (2), 128.44 (2),

131.37, 135.23, 142.59. GC–MS m/z (relative intensity) 59 (8), 75 (20), 115 (10), 130 (15), 177 (15), 207 (25), 220 (100), 247 (15), 395 (10). Elem. Anal. Calcd. for $C_{24}H_{33}NO_2Si$: C, 72.86 %; H, 8.41 %; N, 3.54 %. Found: C, 72.96 %; H, 8.30 %; N, 3.49 %. Enantiomeric excess was evaluated by HPLC analysis: Chiracel OF, isocratic (n-hexane–i-PrOH, 88:12), flow 0.5 mL min^{-1}; R,S isomer t_R = 10.7 min; S,R isomer t_R = 15.7 min. $[\alpha]^D$ = +34.6 (c 0.53, CHCl$_3$).

1.3.1.4 Gold-catalyzed Hydroarylation of Alkynes
Manfred T. Reetz and Knut Sommer

1.3.1.4.1 Introduction and Fundamental Examples

The conventional way of forming a C–C bond between an aromatic moiety and an alkyne with the formation of hydroarylation products involves the use of an arylmetal reagent in a Cu- or Zr-catalyzed process or Ar$_2$CuLi in a stoichiometric reaction, followed by protonation of the vinylmetal intermediate [1]. Although these and other carbometalation reactions are reliable, the overall process requires several steps beginning with halogenation of an arene followed by transformation into the arylmetal reagent (Scheme 1). Atom and step economy would be greatly improved if a direct reaction between the arene and the alkyne could be developed (Scheme 1). Formally, this would involve C–H-activation of the arene.

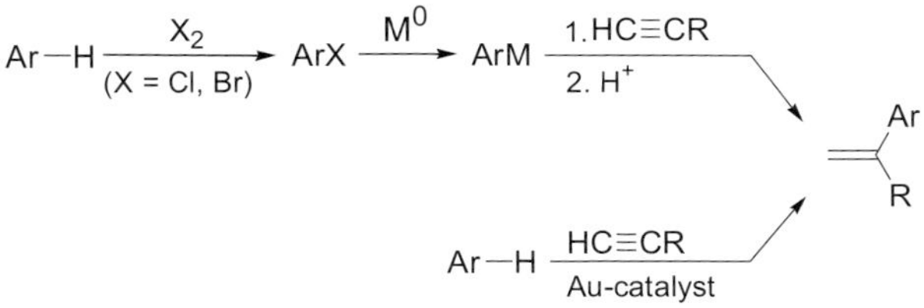

Scheme 1. Conventional (top) and direct (bottom) hydroarylation of alkynes.

If the R-group in the alkyne is a strongly electron-withdrawing moiety such as a carbonyl function, regioselectivity may be reversed, leading to the problem of E/Z-selectivity (Scheme 2). Irrespective of the type of alkyne, the one-step atom-economical process would need to be catalyzed because a simple thermal process does not provide the products.

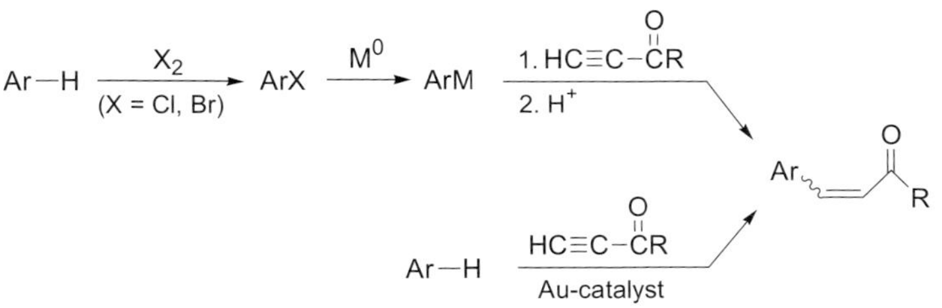

Scheme 2. Conventional (top) and direct (bottom) hydroarylation of electron-poor alkynes.

Recent work has shown that appropriate gold-catalysts are particularly suitable for direct hydroarylation of both kinds of alkyne [2], adding to previous catalytic systems based on other metals [3]. Depending upon the type of alkyne used, different gold catalysts must be employed. A typical reaction involving alkynes of the type noted in Scheme 1 is the hydroarylation of phenylacetylene **2** by mesitylene **1** (excess) with formation of olefin **3**. Whereas $AuCl_3$ affords only 20% of the desired product **3** with rather long reaction times (60 °C/16 h), "cationization" with concomitant activation of $AuCl_3$ (5 mol%) by $AgSbF_6$ affords an excellent catalyst which mediates the smooth transformation within one hour (Scheme 3). The reaction can be optimized so that only 1.5 mol% of $AuCl_3/2AgSbF_6$ needs to be used (50 °C/4 h; 86 % **3**). This type of Au-catalyst is active for a fairly wide range of substrates, if the arene is electron-rich [2].

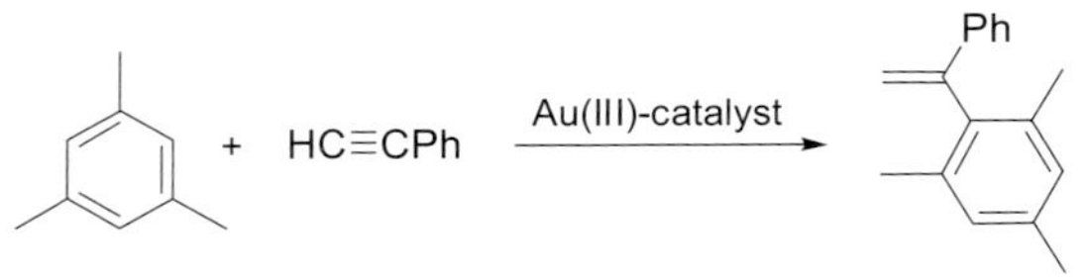

Scheme 3. Typical Au(III)-catalyzed hydroarylation of alkyne; 5 mol% [$AuCl_3$/3$AgSbF_6$], 60 °C, 1 h, 3 equiv. **1**, yield determined by GC using the internal standard method.

For electron-poor alkynes (Scheme 2) such as acetylene carboxylic acid ethyl ester **4**, the activated Au(III) catalysts lead to about 60% of the product **5** as a 3.6:1 Z/E mixture. In contrast, Au(I) complexes of phosphines activated by $AgSbF_6$ or Lewis acids are much better suited, delivering nearly quantitative yields of **5** with complete Z-selectivity (Scheme 4) [2]. Only traces of the doubly vinylated compound **6** are observed as a side product. A particularly effective Lewis acid is $BF_3 \cdot OEt_2$, which also makes the catalytic system cheaper than those using $AgSbF_6$ as the activating agent.

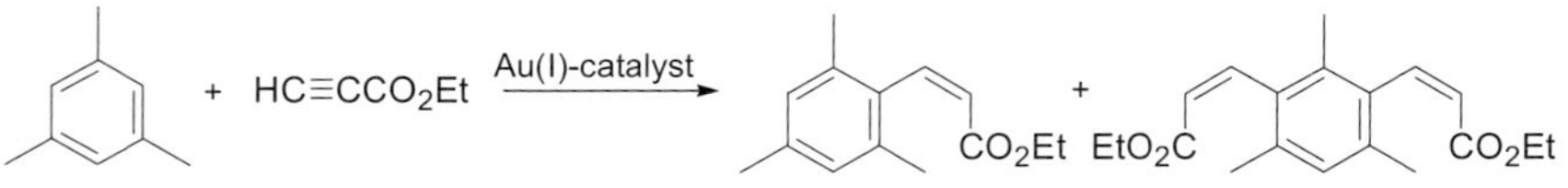

Scheme 4. Typical Au(I)-catalyzed hydroarylation of electron-poor alkyne; 1 mol% [Ph_3PAuCl/5 $BF_3.OEt_2$], 50 °C, 14 h, 30 equiv. **1**, yields determined by GC using the internal standard method.

1.3.1.4.2 Mechanism

So far a detailed mechanistic study has not been completed, although initial observations are strongly suggestive [2, 4]. The synthetic results must be viewed within the context of other gold-catalyzed nucleophilic addition reactions of alkynes [5]. Cationization of $AuCl_3$ or $AuCl$ complexes by $AgSbF_6$ or Lewis acids is required, but electrophilic auration of the arene as part of the hydroarylation seems unlikely in view of the known stoichiometric reaction of arylgold compounds with alkynes, which yields diarylacetylenes and not olefins [6d]. Thus, activation of the alkyne by Au-π-complexation is likely to be involved, as has been postulated for other Au-catalyzed nucleophilic addition reactions [5]. For phenylacetylene **2**, the π-complex undergoes a type of electrophilic aromatic substitution with the electron-rich arene to form an intermediate vinyl-Au species [2, 4]. The concomitantly released H^+ protonates this intermediate, setting free the product **3** and regenerating the catalyst (Scheme 5). Regioselectivity is determined by electronic factors. Scheme 5 may be a simplification, because complete cationization with formation of true ion pairs may not occur at all steps. Strongly polarized neutral species could also be involved [7].

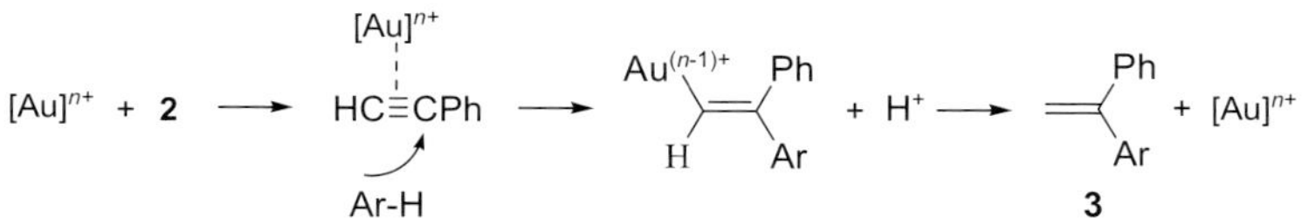

$$[Au]^{n+} + \mathbf{2} \longrightarrow HC\equiv CPh \cdot [Au]^{n+} \xrightarrow{Ar-H} \underset{H}{\overset{Au^{(n-1)+}}{>}}=\underset{Ar}{\overset{Ph}{<}} + H^+ \longrightarrow \underset{Ar}{\overset{Ph}{>}}=< + [Au]^{n+}$$

Scheme 5. Proposed mechanism for the hydroarylation of phenylacetylene **2**.

The mechanism of the Au(III) catalysis proposed in Scheme 5 implies the stereoselective formation of the new C–C bond which, of course, cannot be observed in the final product when terminal alkynes are used (the aryl group and the former alkyne hydrogen are situated at the same side of the double bond in the vinyl-Au intermediate). For the reaction of 1-phenyl-1-propyne and mesitylene **1** (see below, Table 1) the proposed mechanism should lead to preferential formation of the *Z* isomer which is, in fact, observed [2]. The formation of a small amount of *E* isomers can be explained by isomerization of the initially formed *Z* compound. Such isomerization was, in fact, observed directly in the case of related electron-poor alkynes [4].

In the Au(I) catalysis of electron-poor alkynes such as **4**, the catalytically active species is likely to be a cationic ligand-stabilized gold(I) π-complex, as in previously reported additions of oxygen nucleophiles to alkynes [5]. Gold catalysts are very soft and thus *carbophilic* rather than *oxophilic*. On the basis of this assumption a plausible mechanism can be formulated as shown in Scheme 6. The cationic or strongly polarized neutral Au(I)-catalyst coordinates to the alkyne, and nucleophilic attack of the electron-rich arene from the opposite side leads to the formation of a vinyl-gold intermediate **7** which is stereospecifically protonated with final formation of the Z-olefin **8** [2, 4]. Regioselectivity is dominated by electronic factors in a type of gold-catalyzed Michael addition. Preliminary NMR experiments are in line with this mechanism [2, 4].

Table 1. Hydroarylation of internal aryl-substituted alkynes with mesitylene **1**.[a]

Entry	Alkyne	Time (h)	Product	Yield (%)[b] (Z:E)
1	PH-C≡C-CH$_3$	16	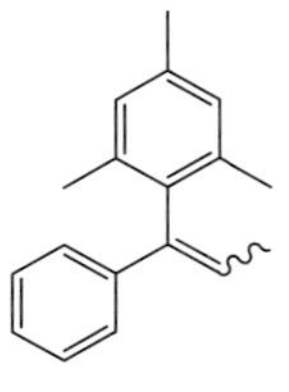	(81) (70:30)[d]
2[c]	PH-C≡C-CH$_3$	4		79 (79:21)[d]
3[c]	PH-C≡C-CH$_3$	8		(94) (76:24)[d]
4	PH-C≡C-Ph	4	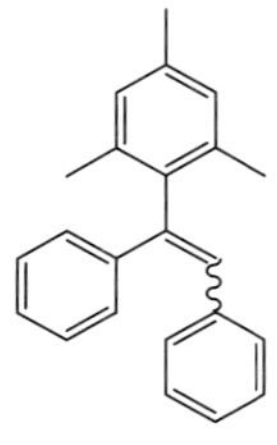	(5) (one isomer, double bond structure not determined)

[a] Conditions (unless otherwise stated): 1.5 mol% AuCl$_3$, 3.0 mol% AgSbF$_6$, 50 °C, CH$_3$NO$_2$, 10 equiv. mesitylene **1**
[b] Isolated yield. Values in parentheses refer to GC yields, determined with n-hexadecane as an internal standard
[c] 4.5 mol% AgSbF$_6$ employed
[d] Determined by GC and/or NMR spectroscopy.

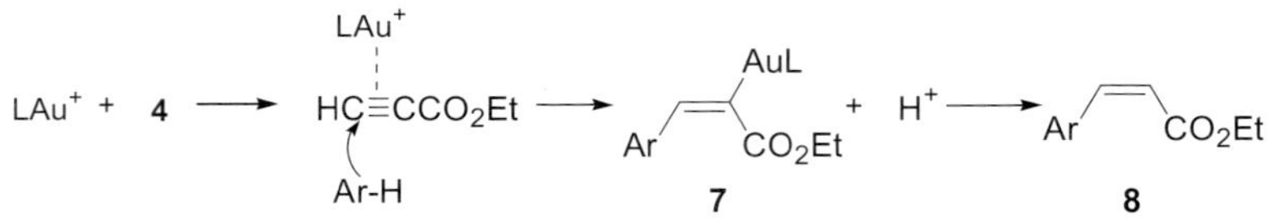

Scheme 6. Proposed mechanism for the hydroarylation of **4**.

1.3.1.4.3 Scope and Limitations

The hydroarylation of aryl-substituted alkynes using activated AuCl$_3$ is fairly general, if electron-rich arenes are used as arylating compounds [2, 4]. Benzene or toluene do not react. Scheme 7 summarizes most of the results published on this subject. Cyclization reactions using AuCl$_3$ have been reported [8], but activated forms as shown above should lead to improved yields [4].

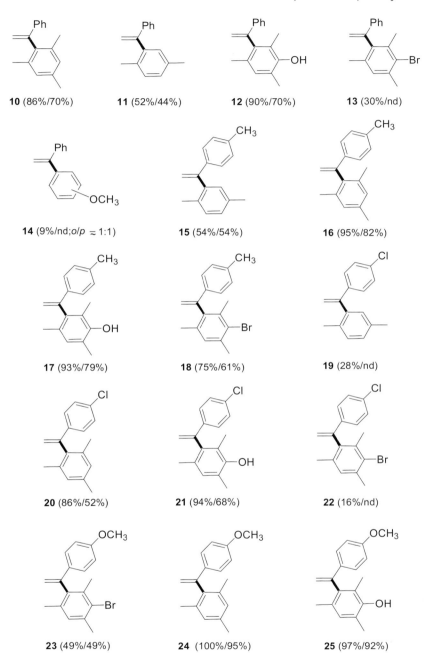

Scheme 7. Scope of the hydroarylation reaction with aryl-substituted alkynes. Conditions: 1.5 mol% AuCl$_3$, 3.0 mol% AgSbF$_6$, 10 equivalents arene, CH$_3$NO$_2$, 50 °C, 4 h. (GC yield/isolated yield); nd = not determined.

Table 2. Hydroarylation of acetylene carboxylic acid ester **4** by various arenes.[a]

Entry	Arene (equiv.)	Co-catalyst	Time (h)	Temp (°C)	Product	Yield (%)[b] (Z:E)[c]
1	Pentamethyl-benzene (3)	$BF_3 \cdot OEt_2$	14	50	pentamethylphenyl-CH=CH-CO_2Et	98 (100:0)
2	Mesitylene (30)	$BF_3 \cdot OEt_2$	14	50	mesityl-CH=CH-CO_2Et	90 (100:0)
3	p-Xylene (15)	$BF_3 \cdot OEt_2$	14	50	xylyl-CH=CH-CO_2Et	55 (95:5)
4	Toluene (15)	$BF_3 \cdot OEt_2$	14	50	H_3C-C$_6H_4$-CH=CH-CO_2Et	(15) (100:0) o:p = 60:40
5	2,4,6-Trimethyl-phenol (10)	$BF_3 \cdot OEt_2$	14	50	HO-trimethylphenyl-CH=CH-CO_2Et	83 (100:0)
6	Phenol (3)	$BF_3 \cdot OEt_2$	14	50	HO-C$_6H_4$-CH=CH-CO_2Et	(0)
7	Furan (20)	$AgSbF_6$	3	40	furan-2-yl-CH=CH-CO_2Et	82 (>99:1)[d]
8	2-Methylfuran (20)	$AgSbF_6$	4	r.t.	5-methylfuran-2-yl-CH=CH-CO_2Et	(80) (Z, traces of the E-2-isomer are formed) 3 isomers (82:15:3)[e]

[a] Conditions: 1 mol% Ph_3PAuCl, 1 mol% $AgSbF_6$ or 5 mol% $BF_3 \cdot OEt_2$, CH_3NO_2, excess arene
[b] Isolated yields of purified products. GC Yields (determined with n-hexadecane as an internal standard) in parentheses
[c] Determined by GC or NMR
[d] GC of the crude product shows minor amounts of two other isomers and side products (according to GC–MS analysis hydroarylation products containing two furyl moieties or two alkenyl moieties, respectively)
[e] The product consists of 82 % Z-2 isomer, 15 % Z-4 isomer, and 3 % of a third isomer, probably the Z-3 isomer (this isomer was characterized only by GC–MS). Traces of a side product which, according to GC–MS analysis contains two methylfuryl moieties are formed

Internal aryl-substituted alkynes also undergo Au(III)-catalyzed hydroarylation with electron-rich alkynes (Table 1) [2, 4]. The reactivity of diphenylactylene, however, is very low.

As delineated above, hydroarylation of electron-poor alkynes such as acetylene carboxylic acid ethyl ester **4** is best catalyzed by Au(I) complexes. In addition to the previously mentioned $Ph_3PAuCl/BF_3 \cdot OEt_2$, several other Au(I)-complexes have been shown to be active catalysts, if activation by silver salts or Lewis acids is ensured [2, 4]. Remarkably, the Au(I)-precursor LAuCl can be prepared in situ by reacting AuCl with a phosphorus ligand.

The scope and limitation of hydroarylation of acetylene carboxylic acid ester **4** using Au(I)-catalysis is revealed in Table 2. Electron-rich arenes participate well, but also furan and 2-methylfuran. *p*-Xylene leads to a 55 % yield, whereas an unacceptably low conversion results in the case of toluene.

The hydroarylation of such acetylenic ketones as 3-butyn-2-one **26** poses no problems. In these cases nearly complete *E*-selectivity was observed, because of isomerization of the kinetically formed *Z* compounds [2, 4]. Four different electron-rich arenes, ArH, were shown to react smoothly, affording products **27** in yields of 77–96 %.

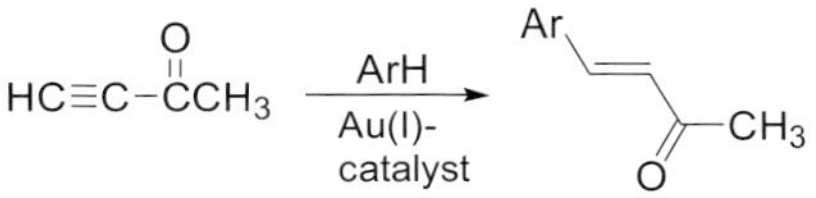

$$HC \equiv C-\overset{O}{\underset{}{C}}CH_3 \quad \xrightarrow[\text{catalyst}]{\text{ArH}\atop \text{Au(I)-}} \quad \underset{\underset{O}{}}{\overset{Ar}{\diagdown}}=\diagup CH_3$$

    26   **27**

Scheme 8. Hydroarylation of 3-butyn-2-one **26**.

In conclusion, Au(I) and Au(III) catalysts are well suited to the step- and atom-economical hydroarylation of alkynes. In general cationization is essential [2,4], which is now used in many other Au-catalyzed reactions. Previous work has shown that other metal catalysts such as Pd, In, and Sc salts are also active [3]. In some cases these catalysts are more effective (e.g. when benzene is used); in other cases less attractive (Pd-catalysis requires trifluoroacetic acid as solvent) [3a, b]. With gold catalysis, aqueous workup and neutralization are not necessary which is a substantial advantage. Recently, a similar gold-catalyzed process for the hydroarylation of electron-poor alkynes using $AuCl_3/3AgOTf$ (0.5–5 mol%) was published [9]. In contrast with our results, the authors favor a mechanism via aryl–gold intermediates.

Experimental

General Procedure for Hydroarylation of Aryl-substituted Alkynes (Small-scale Experiments)

Liquid Arenes

Stock solutions of $AuCl_3$ ($c = 0.015$ M) and $AgSbF_6$ ($c = 0.030$ M) in nitromethane were prepared. Each solution (0.90 mL) was added to a reaction flask under argon. An orange precipitate was observed. Arene (9.0 mmol) and alkyne

(0.90 mmol) (solid alkyne as a concentrated solution in nitromethane) were then added using syringes. The reaction mixture was stirred for 4 h at 50 °C. After cooling, 100 µL n-hexadecane and 2 mL ethyl acetate were added. An aliquot (ca. 0.5 mL) was taken and filtered through a small plug of silica gel (elution with ethyl acetate). The yield of hydroarylation product was determined by GC. The double bond structure was determined by ^{1}H NMR of the crude product after all volatiles had been removed under vacuum.

Solid Arenes
Solid arenes (9.0 mmol) were weighed into the reaction flask in air. The flask was then evacuated for several minutes and charged with argon (3×). Nitromethane (2 mL) was added. The Au- and Ag-solutions and the alkyne (0.90 mmol) were added rapidly, and the reaction mixture was stirred at 50 °C for 4 h. Workup was conducted as described above.

General Procedure for Hydroarylation of Aryl-substituted Alkynes (Large-scale Experiments)

Liquid Arenes
AuCl$_3$ (45.5 mg, 0.150 mmol) and AgSbF$_6$ (103 mg, 0.300 mmol) were weighed into a 100-mL Schlenk-flask under argon. Nitromethane (20 mL) was added. The arene (100 mmol) and the alkyne (10 mmol; solid alkyne as a concentrated solution in nitromethane) were added, and the reaction mixture was stirred for 4 h at 50 °C. After cooling, the solvent and excess arene were removed under vacuum. The residue was purified by column chromatography (silica gel, hexanes–ethyl acetate). After removal of all the volatiles, an oil remained which was further purified by distillation under diffusion pump vacuum (<10^{-3} mbar).

Solid Arenes (2,4,6-Trimethylphenol)
2,4,6-Trimethylphenol (13.62 g, 100.0 mmol) was weighed into a 100-mL Schlenk-flask in air. The flask was then evacuated for several minutes and charged with argon (3×). Nitromethane (20 mL) was added. Solutions of 45.5 mg AuCl$_3$ (0.150 mmol) and 103 mg AgSbF$_6$ (0.300 mmol) in 8.0 mL nitromethane and the alkyne (10 mmol, solid alkyne as a concentrated solution in nitromethane) were added rapidly. The reaction mixture was stirred at 50 °C for 4 h. After cooling, the solvent was removed under vacuum. Excess 2,4,6-trimethylphenol was removed by sublimation under vacuum (ca. 10^{-2} mbar) at 80 °C. The further workup procedure was the same as described above.

General Procedure for Hydroarylation of 4 by 1 with *in-situ* Prepared Gold Complexes
AuCl (6.28 mg, 0.027 mmol) and a (solid) phosphorus ligand (0.027 mmol) were weighed into a reaction flask in air. The flask was evacuated for several minutes and charged with argon (3×). Nitromethane (2 mL) was added. Liquid phosphorus ligands (0.027 mmol) were now added (either with microliter syringes or as stock solutions in nitromethane). Yellow suspensions were obtained which became

clear colorless solutions within a few minutes of stirring. Stock solution of $BF_3 \cdot OEt_2$ in nitromethane ($c = 0.027$ M, 1.0 mL) was added. Finally, mesitylene 1 (0.38 mL, 2.7 mmol) and acetylene carboxylic acid ester 4 (92 µL, 0.90 mmol) were added rapidly. The reaction mixture was stirred for 14 h at 50 °C. After cooling, 100 µL n-hexadecane and 2 mL ethyl acetate were added. An aliquot (ca. 0.5 mL) was taken and filtered through a small plug of silica gel (elution with ethyl acetate). The yields of 5 and 6 were determined by GC.

General Procedure for Hydroarylation of Alkynes 4 and 3-Butyn-2-one 26 by Different Arenes with Ph₃PAuCl under Optimized Conditions (Small-scale Experiments)

Liquid Arenes

Ph₃PAuCl (4.45 mg, 0.009 mmol) was weighed into a reaction flask in air. The flask was then evacuated for several minutes and charged with argon (3×). Nitromethane (2.0 mL) was added, and then the co-catalyst (either 1.5 mL of a 0.030 M solution of $BF_3 \cdot OEt_2$ in nitromethane or 0.90 mL of a 0.010 M solution of $AgSbF_6$ in the same solvent) was added. Finally, the flask was charged with the arene (excess) and the alkyne (0.90 mmol). The reaction mixture was stirred at the temperature and for the time given in Table 2. Workup was conducted as described above.

Solid Arenes

Solid arenes and the gold catalyst were weighed into the flask in air. The flask was then evacuated for several minutes and charged with argon (3×). Nitromethane (2.0 mL) was added. A solution of the co-catalyst and the alkyne was then added. The further procedure was as described above for liquid arenes.

General Procedure for Hydroarylation of Alkynes 4 and 3-Butyn-2-one 26 by Different Arenes with Ph₃PAuCl under Optimized Conditions (Large-scale Experiments)

Liquid Arenes

Ph₃PAuCl (24.7 mg, 0.0500 mmol) and (where applicable, see Table 2) $AgSbF_6$ (17.2 mg, 0.0500 mmol) were weighed in a 100-mL Schlenk-flask under argon. Nitromethane (5.0 mL) and (where applicable, see Table 2) (32 µL, 0.25 mmol) $BF_3 \cdot OEt_2$ were added and the arene (excess) and alkyne (5.0 mmol) were then added rapidly. The reaction mixture was stirred for the time and at the temperature given in the Table 2. After cooling, the solvent was removed under reduced pressure, and the remaining residue was purified by column chromatography (silica gel, hexanes–ethyl acetate).

Solid Arenes

Solid arenes were weighed into a 100-mL Schlenk-flask in air with Ph₃PAuCl (24.7 mg, 0.0500 mmol). The flask was then evacuated for several minutes and charged with argon (3×). Nitromethane (20–35 mL) was added. Finally, $BF_3 \cdot OEt_2$ (32 µL, 0.25 mmol) and the alkyne (5.0 mmol) were added. The reaction mixture was stirred for the time and at the temperature given in the Table 2. The workup procedure was as described above for liquid arenes.

1.3.2
Alkylation and Vinylation via Intermediary Transition Metal σ-Complexes of Arenes

1.3.2.1 Ruthenium-catalyzed ortho-Activation of Carbonyl-substituted Arenes
Fumitoshi Kakiuchi and Shinji Murai

1.3.2.1.1 Introduction and Fundamental Examples

Since the early 1960s, carbon–hydrogen bond cleavage has been an attractive research subject for organometallic and organic chemists [1, 2] and many studies have been reported. These studies have usually focused on isolation and characterization of highly reactive H-M-R species formed by an oxidative addition of C–H bonds to transition metals, and on theoretical and kinetic considerations of the C–H bond-cleavage step. Catalytic C–H bond transformation such as conversion of C–H bonds to C–C bonds is one of the most challenging research subjects. Until the early 1990s much effort had been devoted to this subject but only a limited number of examples of catalytic C–H bond transformations had been reported. In 1993 the first example of highly efficient, selective catalytic addition of C–H bonds in aromatic ketones to olefins was reported [3]. This finding was epochal in this area. The situation in catalytic C–H bond transformation changed dramatically. Use of C–H bonds has since become a highly powerful, reliable synthetic tool in organic chemistry.

Among C–H bond transformations, one of the most important goals has been to achieve the one-step addition of a C–H bond across the double bond of an olefin (C–H/olefin coupling) and across the triple bond of an acetylene (C–H/acetylene coupling), because carbon–carbon bond formation is a highly important reaction in organic synthesis. This transformation must be highly efficient, selective, and widely applicable for the purpose of an organic synthesis. To achieve high regioselectivity chelation-assistance is often used in catalytic C–H bond transformation.

The pioneering work in catalytic additions of C–H bonds to olefins by means of chelation-assistance was the ethylation of phenols catalyzed by a ruthenium phosphite complex [4]. Later, the zirconium-catalyzed α-selective alkylation of α-picolines with olefins was reported [5]. In these cases, coordination of a heteroatom is the key to achieving high regioselectivity. Although these results were quite promising for catalytic C–H bond transformation, generality and efficiency were insufficient for use as a synthetic tool in organic synthesis.

In 1993, Murai found that the ortho-selective alkylation of aromatic ketones with olefins with a ruthenium catalyst such as $Ru(H)_2(CO)(PPh_3)_3$ or $Ru(CO)_2(PPh_3)_3$ proceeded with high efficiency and selectivity (Scheme 1) [3]. One of the most important factors in the success of this coupling reaction involves chelation-assistance by an oxygen of the ketone carbonyl group. Thus, coordination of the oxygen atom to the ruthenium brings the ruthenium center closer to the C–H bond and stabilizes the C–Ru–H species formed by oxidative addition to the ruthenium of an ortho C–H bond in aromatic ketones. In addition, use of the chelation-assistance leads to high regioselectivity, an essential factor in organic synthesis.

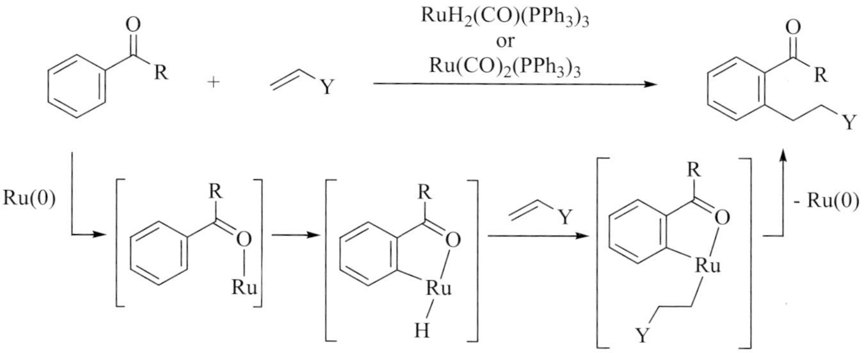

Scheme 1. The proposed reaction pathway of the ruthenium-catalyzed C–H/olefin coupling by means of the chelation-assistance.

This C–H/olefin coupling can be extended to coupling with acetylenes [6]. The reaction of aromatic ketones with internal acetylenes gives the ortho alkenylated product in high yield (Scheme 2), but reaction with terminal acetylenes does not afford the coupling product. With terminal acetylenes, dimerization of acetylenes occurs as a predominant reaction.

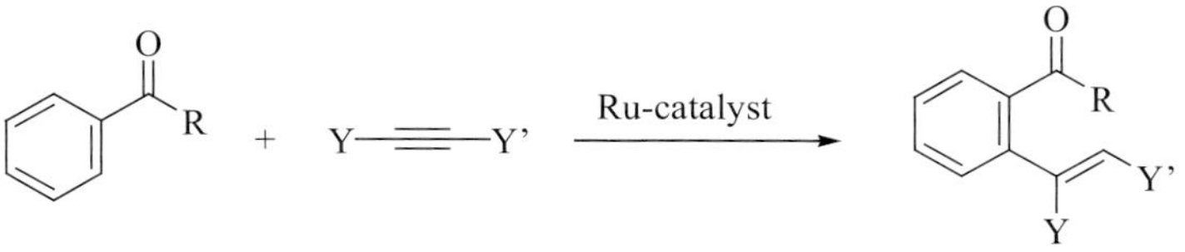

Scheme 2. Ruthenium-catalyzed C–H/acetylene coupling.

This chelation-assisted C–H/olefin and C–H/acetylene coupling can be applied to a variety of aromatic compounds with a directing group such as ester, aldehyde, imine, azo, oxazolyl, pyridyl, and nitrile [7]. In this section, we describe the coupling reactions of aromatic carbonyl compounds with olefins using a transition metal catalyst.

1.3.2.1.2 Mechanism

In general, it has been suggested that oxidative addition of a C–H bond to a transition metal giving a C–M–H species and reductive elimination leading to C–C bond formation proceed via a concerted pathway. Ruthenium-catalyzed chelation-assisted coupling of aromatic compounds with olefins, C–H bond cleavage, and C–C bond formation are, however, believed to proceed stepwise through the pathway shown in Scheme 3. Therefore, the C–H bond cleavage occurs as a result of nucleophilic attack of low-valent ruthenium on the ortho carbon of aromatic compounds followed by migration of the ortho hydrogen to the ruthenium center; C–C bond formation proceeds as a result of attack of the alkyl group on the ruthenium atom [8]. The most important feature of this coupling reaction is that each step before the reductive elimination is in rapid equilibrium – thus, C–H bond cleavage is not rate-determining and C–C bond formation (reductive elimination) is rate-determining [8]. Murai proposed this unusual pathway for ruthenium-catalyzed C–H/olefin coupling on the basis of experimental evidence [8]. Later, Morokuma and Koga revealed the validity of Murai's proposed pathway on the basis of ab initio theoretical calculation [9].

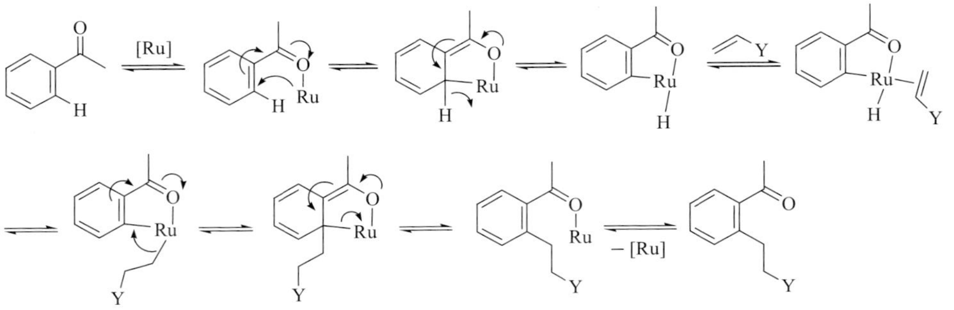

Scheme 3. Proposed stepwise pathway for the ruthenium-catalyzed C–H/olefin coupling.

In $(C_5Me_5)Rh(C_2H_3SiMe_3)_2$-catalyzed C–H/olefin coupling the effect of the coordination of the ketone carbonyl is different from that in the ruthenium-catalyzed reaction [10]. In the rhodium-catalyzed reaction all C–H bonds on the aromatic ring are cleaved by the rhodium complex without coordination of the ketone carbonyl. Thus, C–H bond cleavage and addition of Rh-H to olefins proceed without coordination of the ketone carbonyl. After addition of the Rh–H species to the olefin, a coordinatively unsaturated Rh(aryl)(alkyl) species should be formed. Coordination of the ketone carbonyl group to the vacant site on the rhodium atom leads

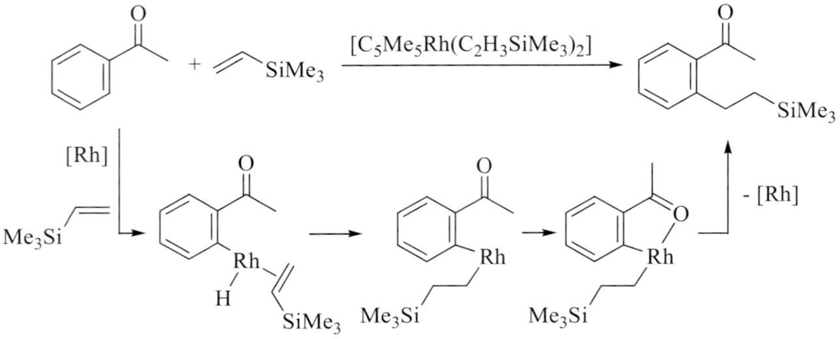

Scheme 4. The reaction pathway of the $(C_5Me_5)Rh(C_2H_3SiMe_3)_2$-catalyzed reaction of aromatic ketones with olefins.

to the 18-electron rhodium species. Coordination of the ketone carbonyl facilitates the reductive elimination.

Although there are several examples of transition metal-catalyzed addition of C–H bonds to acetylenes, there is neither experimental evidence nor theoretical consideration in respect of a reaction mechanism for C–H/acetylene coupling. This coupling reaction is believed to proceed through a pathway similar to that proposed for C–H/olefin coupling.

1.3.2.1.3 Scope and Limitations

The ruthenium-catalyzed addition of C–H bonds in aromatic ketones to olefins can be applied to a variety of ketones, for example acetophenones, naphthyl ketones, and heteroaromatic ketones. Representative examples are shown in the Table 1. Terminal olefins such as vinylsilanes, allylsilanes, styrenes, *tert*-butylethylene, and 1-hexene are applicable to this C–H/olefin coupling reaction. Some internal olefins, for example cyclopentene and norbornene are effective in this alkylation. The reaction of 2-acetonaphthone **1** provides the 1-alkylation product **2** selectively. Alkylations of heteroaromatic ketones such as acyl thiophenes **3**, acyl furans, and acyl pyrroles proceed with high yields. In the reaction of di- and tri-substitued aromatic ketones such as **4**, which have two different ortho positions, C–C bond formation occurs at the less congested ortho position. Interestingly, in the reaction of *m*-methoxy- and *m*-fluoroacetophenones C–C bond formation occurs at the congested ortho position (2'-position).

Table 1. Examples of the coupling of aromatic ketones with olefins.

Aromatic ketone	Olefin	Catalyst	Product	Yield (%)	Ref.
	SiMe₃	A	SiMe₃	97	3,7,8
		A		89	3,7,8
1	Si(OEt)₃	A	(EtO)₃Si · 2	88	3,7,8

Table 1. Continued

Aromatic ketone	Olefin	Catalyst	Product	Yield (%)	Ref.
2-acetylthiophene (**3**)	$CH_2=CHSi(OEt)_3$	A	3-[2-(triethoxysilyl)ethyl]-2-acetylthiophene	93	3,7,8
3'-(N-ethylacetamido)acetophenone (**4**)	$CH_2=CHSi(OEt)_3$	A	alkylated product (**5**)	92	7,8

note: **A**: $Ru(H)_2(CO)(PPh_3)_3$

Several related examples of transition metal-catalyzed addition of C–H bonds in ketones to olefins have been reported (Table 2) [11–14]. The alkylation of diterpenoid **6** with olefins giving **7** proceeds with the aid of $Ru(H)_2(CO)(PPh_3)_3$ (**A**) or $Ru(CO)_2(PPh_3)_3$ (**B**) as catalyst [11]. Ruthenium complex **C**, $Ru(H)_2(H_2)(CO)(PCy_3)_2$, has catalytic activity in the reaction of benzophenone with ethylene at room temperature [12]. The alkylation of phenyl 3-pyridyl ketone **8** proceeds with **A** as catalyst [13]. Alkylation occurs selectively at the pyridine ring. Application of this C–H/olefin coupling to polymer chemistry using α,ω-dienes such as 1,1,3,3-tetramethyl-1,3-divinyldisiloxane **11** has been reported [14].

Table 2. Examples of couplings of aromatic ketones with olefins using transition metal catalysts.

Aromatic ketone	Olefin	Catalyst	Product	Yield (%)	Ref.
diterpenoid **6** (OMe, MeO₂C substituents)	$CH_2=CHSi(OEt)_3$	B	alkylated diterpenoid **7**	100	11
benzophenone	$CH_2=CH_2$	C	2,2'-diethylbenzophenone	96	12

Table 2. Continued

Aromatic ketone	Olefin	Catalyst	Product	Yield (%)	Ref.
benzophenone	$\diagup$SiMe$_3$	D	ortho-CH$_2$CH$_2$SiMe$_3$ benzophenone	99	10
phenyl 3-pyridyl ketone **8**	$\diagup$Si(OEt)$_3$	A	alkylated product **9**	48	13
4-morpholinoacetophenone **10**	divinyldisiloxane **11**	A	polymer **12**	85	14

note: **A**: Ru(H)$_2$(CO)(PPh$_3$)$_3$; **B**: Ru(CO)$_2$(PPh$_3$)$_3$; **C**: Ru(H)$_2$(H$_2$)$_2$(PCy$_3$)$_2$; **D**: (C$_5$Me$_5$)Rh(C$_2$H$_3$SiMe$_3$)$_2$

Chelation-assisted C–H/olefin coupling is also applicable to other aromatic carbonyl compounds (Table 3) [7]. For aromatic esters, electron-withdrawing groups such as CF$_3$ (**13**) and CN facilitate the C–H/olefin coupling reaction. In this alkylation, trimethylvinylsilane is more reactive (100% yield in 1 h) than triethoxyvinylsilane (97% yield in 24 h) [15]. In general, aldehydes are more reactive substrates towards nucleophilic reactions than ketones and esters. When aromatic aldehydes are exposed to low-valent transition metal complexes, a formyl group is attacked by the metal. Thus, several reactions such as decarbonylation of aldehydes and hydroacylations of olefins and acetylenes occur. To prevent nucleophilic attack on an aldehyde carbonyl group by a low-valent transition metal complex, two proposals have been made. One is steric effect (hypothesis A in Scheme 5), and the other is electronic effect (hypothesis B in Scheme 5). Reaction of benzaldehyde **14**, with a bulky substituent ortho to the formyl group, with an olefin provides the alkylation product **15** in good yield. Thiophenecarboxaldehyde **16** also reacts with an olefin to give alkylation product **17**. These results indicate that the hypotheses shown in Scheme 5 appears to operate in the alkylation of aromatic aldehydes. An amide carbonyl group can also function as a directing group. In this case, an iridium complex, Ir(C$_5$H$_5$)((R)-biphemp) (biphemp = 2,2′-bis(diphenylphosphino)-6,6′-dimethyl-1,1′-biphenyl), has catalytic activity [17]. This reaction gives the optically active alkylation products in moderate yields with high enantiomeric excess.

Table 3. Examples of coupling reactions of aromatic compounds with olefins.

Aromatic ketone	Olefin	Catalyst	Product	Yield (%)	Ref.
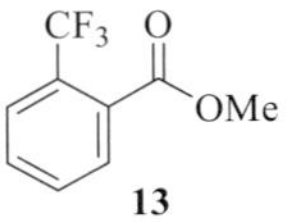 **13** (2-CF₃-C₆H₄-C(O)OMe)	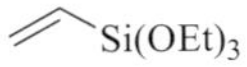 CH₂=CH-Si(OEt)₃	A	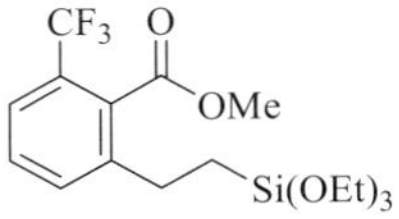 CF₃-C₆H₃(C(O)OMe)(CH₂CH₂Si(OEt)₃)	97	15
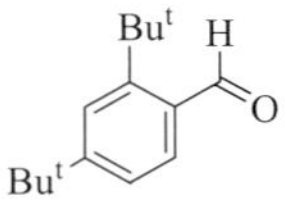 **14** (3,5-di-Buᵗ-benzaldehyde)	 CH₂=CH-Si(OEt)₃	A	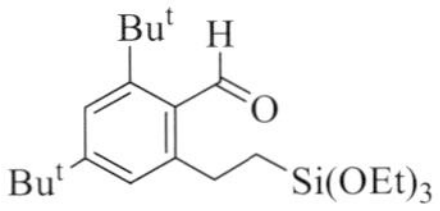 **15**	69	16
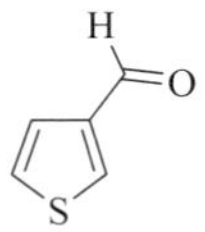 **16** (thiophene-3-carbaldehyde)	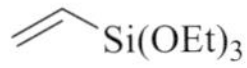 CH₂=CH-Si(OEt)₃	A	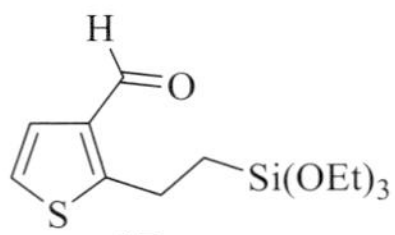 **17**	52	16
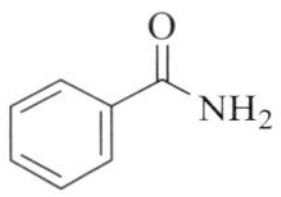 benzamide (C₆H₅-C(O)NH₂)	 norbornene	E	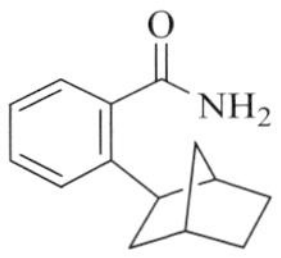 ortho-norbornyl benzamide	35, 87ee	17

note **E**: [Ir(C₅H₅)((R)-biphemp)]

hypothesis A

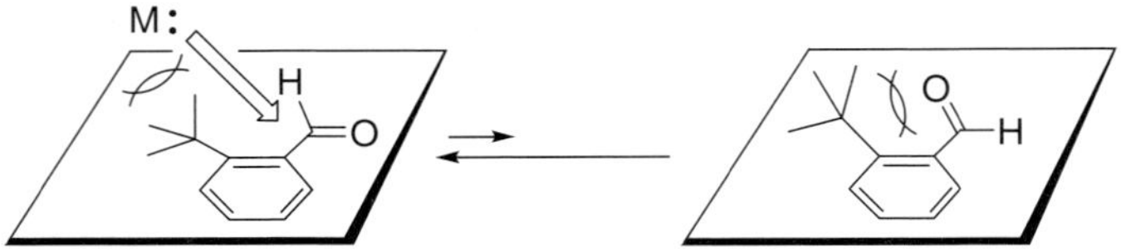

hypothesis B

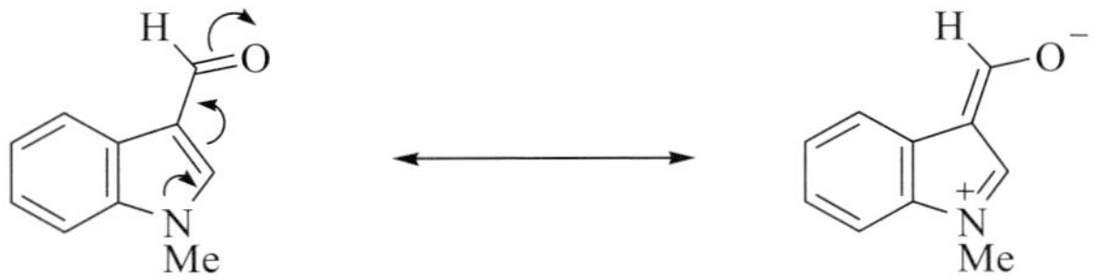

Scheme 5. Hypothesis A: suppression of the reactivity of the formyl group by steric effects; hypothesis B: suppression of the reactivity of the formyl group by electronic effects.

Acetylenes can also be used in ruthenium-catalyzed C–H bond transformations. Selected results are listed in Table 4. The C–H bond of α-tetralone **18**, one of the most reactive ketones in the couplings with olefins, undergoes addition to the triple bond of 1-trimethylsilyl-1-propyne **19** with complete regioselectivity and stereoselectivity [6]. This indicates that the addition of a Ru–H (or, less likely, a Ru–C) bond to the acetylene proceeds in syn fashion. This C–H/acetylene coupling provides highly congested trisubstituted olefins in high yields. Alkenylation of **6** also occurs. Self-polymerization of **20** leads to polymer **21** in high yield. Hypothesis B in Scheme 5 is also effective for the alkenylation of C–H bonds with acetylenes.

Table 4. Examples of the coupling of aromatic ketones with acetylenes.

Aromatic ketone	Acetylene	Catalyst	Product	Yield (%)	Ref.
18 (α-tetralone)	Me—≡—SiMe₃ **19**	A	trisubstituted alkenyl tetralone with SiMe₃	83 (*E* only)	6
6 (tricyclic OMe, MeO₂C ketone)	Ph—≡—SiMe₃	A	alkenylated **6** with Ph, SiMe₃	94 (*E:Z* = 50:50)	18
20 (4-ethynyl-SiMe₃ acetophenone)	A + styrene	polymer **21**	79	19
N-Me indole-3-carbaldehyde	Me—≡—SiMe₃	A	2-alkenylated indole with SiMe₃, Me	42 (*E:Z* = 85:15)	16

Experimental

8-(2-Triethoxylsilanylethyl)-3,4-dihydro-2H-naphthalen-1-one [8, 20]

Apparatus consisting of a 100-mL two-necked round-bottomed flask, a reflux condenser connected to a vacuum/N_2 line, an inlet tube sealed with a rubber septum, and a magnetic stirring bar was evacuated then flushed with nitrogen. This cycle was repeated four times. The apparatus was flame-dried under a flow of nitrogen, and then cooled to room temperature under a nitrogen atmosphere. Carbonyldihydridotris(triphenylphosphine)ruthenium(II), $Ru(H)_2(CO)(PPh_3)_3$, (0.918 g, 1.00 mmol) was placed in the flask under a slow flow of nitrogen. Addition of 20 mL toluene to the flask gave a suspension of white solid. To the suspension were added triethoxyvinylsilane (20.93 g, 110 mmol) and α-tetralone **18** (14.62 g, 100 mmol). The resulting mixture containing the white solids was heated under reflux. After heating for 30 min, the reaction mixture was cooled to room temperature. Volatile materials (toluene and triethoxyvinylsilane) were removed under reduced pressure. Distillation of the residue under reduced pressure gave 31–32.5 g (92–96 % yields) 8-(2-trimethylsilanylethyl)-3,4-dihydro-2H-naphthalen-1-one as a pale yellow liquid, b.p. 133–135 °C/0.2 mmHg. ^{1}H NMR (270 MHz, $CDCl_3$) 0.96–1.02 (c, 2 H, $SiCH_2$), 1.26 (t, $J = 7.02$ Hz, 9 H, CH_3), 2.07 (quint, $J = 6.48$ Hz, 2 H, $CH_2CH_2CH_2$), 2.64 (t, $J = 6.48$ Hz, 2 H, $ArCH_2$), 2.95 (t, $J = 6.48$ Hz, 2 H, $C(O)CH_2$), 3.10–3.16 (c, 2 H, $ArCH_2CH_2Si$), 3.88 (q, $J = 7.02$ Hz, 6 H, OCH_2), 7.09 (d, $J = 7.56$ Hz, 1 H, ArH), 7.14 (d, $J = 7.56$ Hz, 1 H, ArH), 7.33 (t, $J = 7.56$ Hz, 1 H, ArH); ^{13}C NMR (67.5 Hz, $CDCl_3$) 12.11, 18.06, 22.68, 28.41, 30.82, 40.79, 58.01, 126.56, 128.99, 130.17, 132.15, 145.48, 147.78, 199.28.

8-(1-Methyl-2-trimethylsilanylvinyl)-3,4-dihydro-2H-naphthalen-1-one [6]

Apparatus consisting of a 10-mL two-necked flask equipped with a reflux condenser connected to a nitrogen line, a rubber septum, and a magnetic stirring bar was flame dried under a flow of nitrogen. In the flask were placed the ruthenium complex (0.110 g, 0.12 mmol), 3 mL toluene, α-tetralone **18** (0.292 g, 2 mmol), and 1-trimethylsilylpropyne **19** (0.448 g, 4 mmol). The resulting mixture was heated under reflux under a nitrogen atmosphere. After heating for 3 h the reaction mixture was cooled to room temperature. Volatile materials (toluene and 1-trimethylsilylpropyne) were removed under reduced pressure. Bulb-to-bulb distillation under reduced pressure gave 0.428 g (83 % yield) 8-(1-methyl-2-trimethylsilanylvinyl)-3,4-dihydro-2H-naphthalen-1-one as a pale yellow liquid, 110 °C/2 mmHg. ^{1}H NMR ($CDCl_3$) 0.02 (s, 9H, CH_3), 2.04 (s, 3H, CH_3), 2.10 (quint, $J = 6.3$ Hz, 2H, CH_2), 2.63 (t, $J = 6.3$ Hz, 2H, $C(O)CH_2$), 2.96 (t, $J = 6.3$ Hz, 2H, $ArCH_2$), 5.24 (s, 1H, CH=), 7.01 (d, $J = 7.6$ Hz, 1H, ArH), 7.14 (d, $J = 7.6$ Hz, 1H, ArH), 7.35 (t, $J = 7.6$ Hz, 1H, ArH); ^{13}C NMR ($CDCl_3$) –0.29, 22.81, 23.02, 30.51, 40.20, 124.42, 127.33, 128.01, 129.26, 132.26, 144.87, 149.76, 157.65, 197.68.

1.3.2.2 Ruthenium-Catalyzed alpha-Activation of Heteroarenes
Naoto Chatani

1.3.2.2.1 Introduction and Fundamental Examples

The selective functionalization of heterocycles is of particular importance, because of the ubiquity of these structures in natural products and pharmaceutical agents. Direct utilization of a C–H bond [1] of heterocycles is a promising method for the preparation of heterocycles because no pre-functionalization is required. Although Friedel–Crafts acylation is the most commonly used method for introduction of keto functionality on an aromatic ring, it is not often applicable to *N*-heteroarenes because of deactivation of the Lewis acids by the coordination of *N*-heteroarenes and the electron-deficient aromatic character of *N*-heteroarenes.

Direct selective acylation of pyridine was achieved by use of olefins and CO, with $Ru_3(CO)_{12}$ as catalyst, as shown in Scheme 1 [2]. Reaction of pyridine **1** with CO and 1-hexene in the presence of $Ru_3(CO)_{12}$ gave hexanoylpyridine **2** with a 93:7 ratio of linear and branched isomers. Use of 2-hexene or 3-hexene in place of 1-hexene also gave **2** with exactly the same linear/branched product ratio as in the reaction with 1-hexene.

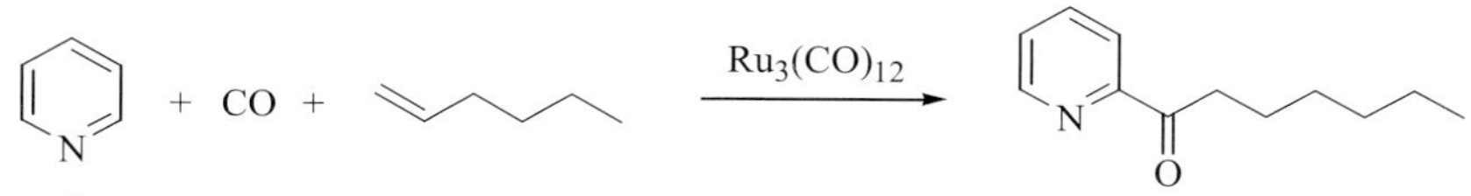

Scheme 1. C–H carbonylation at a C–H bond α to the nitrogen in pyridine; 1.3 mol% $Ru_3(CO)_{12}$, no solvent, 10 atm, 150 °C, 16 h.

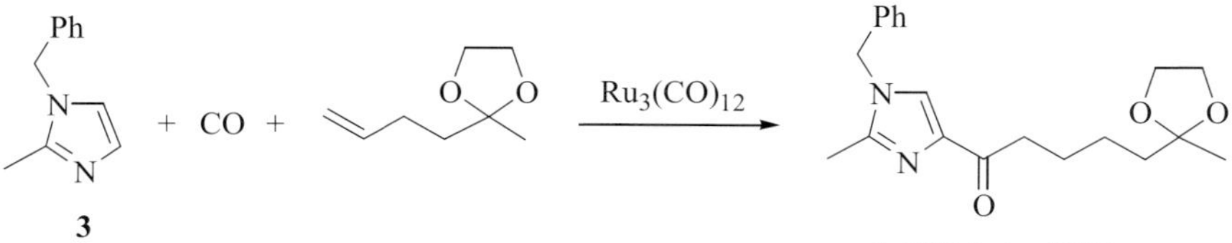

Scheme 2. C–H carbonylation at a C–H bond α to the nitrogen in imidazole; 4 mol-% $Ru_3(CO)_{12}$, toluene, 20 atm, 160 °C, 20 h.

This type of C–H carbonylation is also feasible at C–H bonds in five-membered *N*-heteroaromatics, such as imidazoles **3** (Scheme 2), thiazoles **5**, oxazoles **6**, and pyrazoles **7** [3]. Functional group compatibility was extensively studied in the reaction of **3**, and it was found that a variety of functional groups, for example ketone, ester, cyano, acetal, *N,O*-acetal, ketal, and silyl groups, were tolerated under the reaction conditions, indicating that C–H carbonylation reactions have now reached a satisfactory level in organic synthesis. The reactivity of the five-membered *N*-heteroaromatics is significantly affected by the pK_a of conjugate acid of the *N*-heteroaromatic compounds, as shown in Scheme 3. Thus, the higher is the pK_a, the greater the reactivity. This indicates that coordination of the substrates by the sp^2 nitrogen to the ruthenium center is a key step in C–H carbonylation of N-heteroaromatics. The substrates must compete with CO to coordinate with ruthenium. In fact, reaction of pyrazole **7**, which has the lowest pK_a among the substrates shown in Scheme 3, proceeded effectively only when the reaction was conducted under a lower pressure of CO (3 atm, 46%: 20 atm, trace). This observation emphasizes the importance of coordination of a nitrogen to ruthenium for the reaction to proceed.

Scheme 3. Order of reactivity of five-membered N-heteroaromatics in C–H carbonylation.

1.3.2.2.2 Mechanism

A trinuclear ruthenium cluster **9**, formed by coordination of the pyridine nitrogen to the ruthenium catalyst then specific activation of a C–H bond next to the nitrogen, is proposed as the key catalytic species. In fact, **9** had catalytic activity in the reaction shown in Scheme 1. Coordination of the olefin to the ruthenium center in **9** leading to **10**, insertion of the olefin into an H–Ru bond leading to **11**, and CO insertion gives the acyl ruthenium complex **12**, which undergoes reductive elimination to give the final product, with Ru being regenerated (Scheme 4). Coordination of pyridine through the sp^2 nitrogen is important. In fact, the monophosphine complex, Ru$_3$(CO)$_{11}$(PPh$_3$), was much less active than Ru$_3$(CO)$_{12}$, and the triphosphine complex, Ru$_3$(CO)$_9$(PPh$_3$)$_3$, had no catalytic activity.

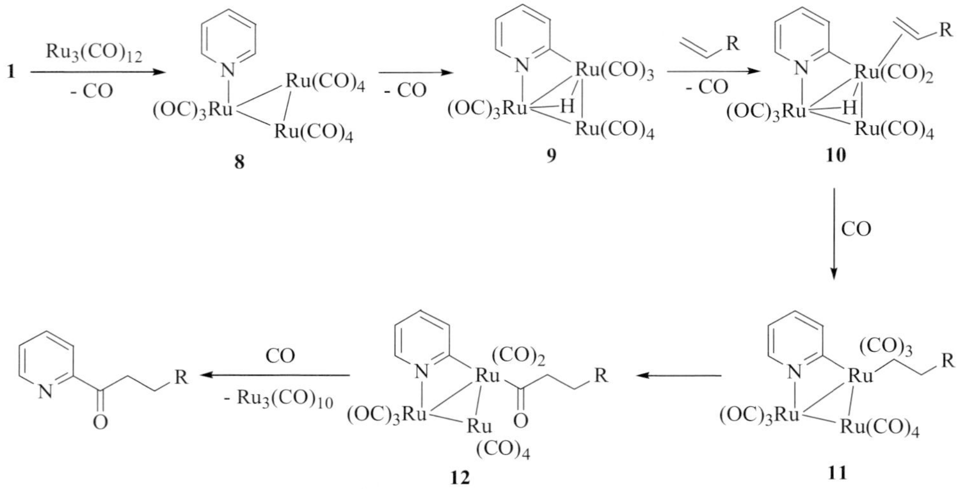

Scheme 4. Proposed reaction mechanism.

1.3.2.2.3 Scope and Limitations

In addition to five- and six-membered N-heteroaromatics, other N-heterocyclic compounds have been found to serve as substrates for C–H carbonylation reactions. Benzoimidazole **13** reacted in the same way with CO and olefins in the presence of $Ru_3(CO)_{12}$, but carbonylation occurred at the C–H bond β to the sp^2 nitrogen to give **14** (Table 1). The same relationship between reactivity and the pK_a values of conjugate acids of heterocycles was also observed in the β carbonylation [4].

Table 1. Examples of the structural diversity of C–H carbonylation reactions (conditions are different in each reaction).

Starting materials	Olefins	Products	Yield (%)	Ref.
13	Bu^t-CH=CH₂	**14**	77	4
15	$H_2C=CH_2$	**16**	80	5
17	$H_2C=CH_2$	**18**	91	6
19	$H_2C=CH_2$	**20**	94	7

Table 1. Continued

Starting materials	Olefins	Products	Yield (%)	Ref.
21	H$_2$C=CH$_2$	22	65	8
23	H$_2$C=CH$_2$	24	85	9

The 2-arylpyridine **15** did not undergo C–H carbonylation at the C–H bond in the pyridine ring, but instead at the ortho C–H bond (the C–H bond γ to the sp^2 nitrogen) in the aryl ring to give **16** [5]. This γ carbonylation was applicable to aryloxazolines and N-arylpyrazoles. Regioselective carbonylation occurred at the less hindered C–H bond when meta-substituted phenyloxazoline **17** was used as the substrate [6]. A pyrazole ring also functioned as the directing group, as in **19** [7].

Although examples are few, C–H carbonylation at C–H bonds δ to the sp^2 nitrogen was achieved when N-pyridylindoline **21** was used as the substrate [8]. The reaction is sensitive to solvent polarity. The choice of N,N-dimethylacetoamide as solvent is crucial for the reaction to proceed efficiently. In the γ- and δ-carbonylation, only ethylene works effectively as coupling partner.

C–H carbonylation catalyzed by Ru$_3$(CO)$_{12}$ is applicable to olefinic C–H bonds, as in **23** [9].

A high-throughput procedure using a mass spectrometric labeling strategy for examining applicable substrates in C–H carbonylation catalyzed by Ru$_3$(CO)$_{12}$ has recently been reported [10].

The directing group promoted C–H activation reaction is applicable to sp^3 C–H bonds adjacent to the nitrogen in alkylamines, as shown in Scheme 5. Alkylation occurred when reaction of **25** with CO and ethylene was conducted in the presence of Ru$_3$(CO)$_{12}$ as catalyst [11]. On the other hand, the use of a rhodium complex as catalyst resulted in C–H carbonylation [12].

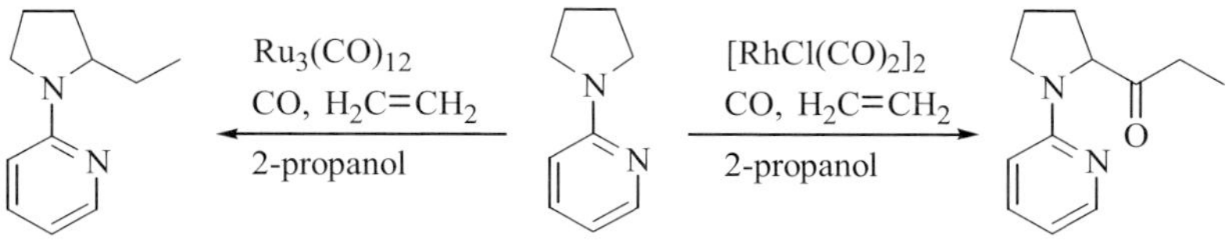

Scheme 5. C–H carbonylation and alkylation at the sp³ C–H bond α to the nitrogen.

Experimental

1-(1-Benzyl-2-methyl-1H-imidazol-4-yl)-5-(2-methyl-[1,3]dioxolan-2-yl)-pentan-1-one (4)

$Ru_3(CO)_{12}$ (25 mg, 0.04 mmol), 1-benzyl-2-methyl-1H-imidazole (3) (172 mg, 1 mmol), 2-but-3-enyl-2-methyl[1,3]dioxolane (568 mg, 4 mmol), and toluene (3 mL) were placed in a 50-mL stainless-steel autoclave. The system was flushed with 10 atm CO three times, then pressurized with CO to 20 atm, then immersed in an oil bath at 160 °C. After 20 h, it was removed from the oil bath and left to cool for ca. 1 h; the gases were then released. After evaporation, the resulting residue was subjected to column chromatography on silica gel with hexane–EtOAc as eluent to give 4 (233 mg, 68 % yield) as a colorless oil. An analytical sample was obtained by bulb-to-bulb distillation. Colorless oil; $R_F = 0.20$ (EtOAc); ^{1}H NMR (CDCl$_3$) δ 1.30 (s, 3H, CH$_3$C), 1.39–1.54, 1.60–1.80 (m, 6H, CH$_2$), 2.37 (s, 3H, 2-CH$_3$), 2.94 (t, $J = 7.6$ Hz, 2H, CH$_2$C(O), 3.83–3.96 (m, 4H, CH$_2$O), 5.06 (s, 2H, CH$_2$N, 7.00–7.13, 7.28–7.42 (m, 5H, Ph), 7.51 (s, 1H, 5-H); ^{13}C NMR (CDCl$_3$) δ 13.19 (2-CH$_3$), 23.70, 23.83, 24.30, 38.60, 38.96 (CH$_2$, CH$_3$), 50.33 (CH$_2$N), 64.53 (CH$_2$O), 110.01 (CH$_3$), 124.37 (5-C), 126.90, 128.37, 129.11, 134.99 (Ph), 140.31 (4-C), 145.62 (2-C), 196.17 (CO); IR (neat) 3130 w, 2930 s, 2878 s, 2646 w, 2198 w, 1812 w, 1667 s, 1539 s, 1499 m, 1457 s, 1414 s, 1375 s, 1312 m, 1206 s, 1163 s, 1061 s, 944 m, 853 m, 790 m, 730 m, 698 s, 668 m, 640 m; MS, m/z (rel. intensity) 342 (M$^+$, 3), 228 (28), 227 (20) 215 (14), 214 (81), 199 (36), 172 (14), 91 (100), 87 (78). Anal. Calcd for $C_{20}H_{26}N_2O_3$: C, 70.15; H, 7.65; N, 8.18. Found: C, 70.22; H, 7.55; N, 8.22.

1.3.2.3 Ruthenium(II)- and Iridium(III)-catalyzed Addition of Aromatic C–H Bonds to Olefins
T. Brent Gunnoe and Roy A. Periana

1.3.2.3.1 Introduction and Fundamental Examples

Ru(II) Catalysts

Additions of aromatic C–H bonds across olefin double bonds (olefin hydroarylation) are catalyzed by the Ru(II) complex TpRu(CO)(NCMe)(Ph) (**1**) (Tp = hydridotris(pyrazolyl)borate; Scheme 1) [1, 2]. The reactions proceed at moderate temperature (90 °C) with relatively low olefin concentrations. For the catalytic hydrophenylation of ethylene, approximately 50 turnovers are obtained after 4 h using 0.1 mol% of complex **1** in benzene and 25 psig ethylene. After 24 h, approximately 80 turnovers are observed after which catalytic activity ceases. At low ethylene pressures (<60 psig), ethylbenzene is the primary product with traces of 1,3- and 1,4-diethylbenzene observed. Increasing olefin pressure reduces the rate of catalysis. For example, at 250 psig ethylene pressure only 5 equiv. ethylbenzene are produced (per equiv. catalyst) after 4 h reaction at 90 °C. For reactions incorporating 1-hexene greater catalytic activity is observed with increasing olefin concentration up to approximately 60 equiv. olefin (based on Ru). Increasing the 1-hexene/catalyst ratio above 60:1 results in a decrease in the rate of catalysis. Hydrophenylation reactions of α-olefins (e.g. propene or 1-hexene) result in reduced catalytic activity compared with the hydrophenylation of ethylene. In addition, stoichiometric production of *trans*-β-alkyl styrenes results in termination of the catalysis after approximately 10 to 12 turnovers (Scheme 1). In the hydrophenylation of ethylene formation of styrene is not detected at low ethylene pressures; at higher ethylene pressure (250 psig), however, traces of styrene are produced with a small amount of butylbenzene, because of a second ethylene insertion step. The hydrophenylation of ethylbenzene produces 1,3-diethylbenzene and 1,4-diethylbenzene in an approximately 2:1 ratio.

These catalytic reactions provide a unique pathway for addition of aromatic C–H bonds across C=C bonds. In contrast with Friedel–Crafts catalysts for olefin hydroarylation, the Ru-catalyzed hydrophenylation reactions of α-olefins selectively produce linear alkyl arenes rather than branched products. Although the selectivity is mild, the formation of anti-Markovnikov products is a unique feature of the Ru(II) and Ir(III) catalysts discussed herein. Typically, the preferred route for incorporation of long-chain linear alkyl groups into aromatic substrates is Friedel–Crafts acylation then Clemmensen reduction, and the catalysts described herein provide a more direct route to linear alkyl arenes.

TpRu(CO)(NCMe)(Me) reacts with thiophene and furan to initiate stoichiometric C–H activation at the 2-position (Scheme 2). After both reactions the 2-aryl products have been isolated and fully characterized [3]. Extension of these stoichiometric reactions to catalytic transformations has been demonstrated for furan or thiophene and ethylene. Heating a combination of 1 mol% TpRu(CO)(NCMe)(2-furyl) in furan and mesitylene under 10–40 psig ethylene results in catalytic pro-

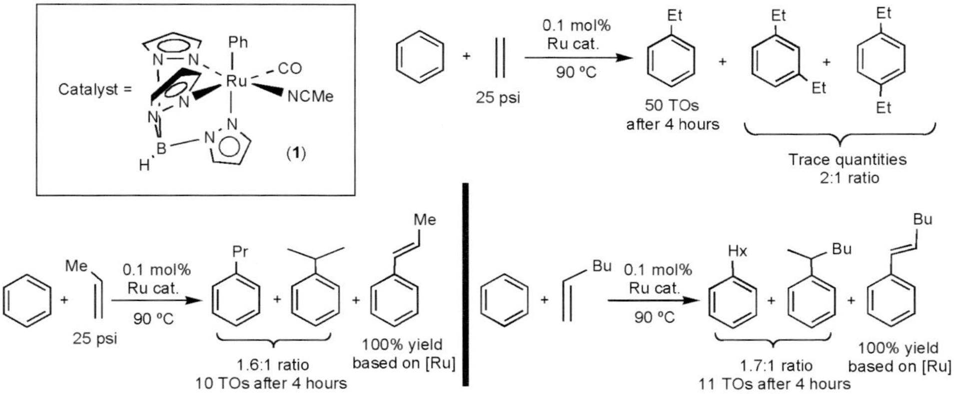

1.3 Alkylation and Vinylation of Arenes | 181

Scheme 1. TpRu(CO)(NCMe)(Ph) catalyzes the hydroarylation of olefins (Pr = propyl, Hx = hexyl, Bu = butyl).

duction of 2-ethylfuran (Scheme 2). In addition, 0.1 mol% TpRu(CO)(NCMe)(2-thienyl) in thiophene under 40 psig ethylene pressure at 90 °C produces 2-ethylthiophene with five catalytic turnovers after 12 h. Pyrrole is also activated on reaction with TpRu(CO)(NCMe)(Me), although C–C bond formation with the acetonitrile ligand occurs to produce a stable metallacycle (Scheme 2).

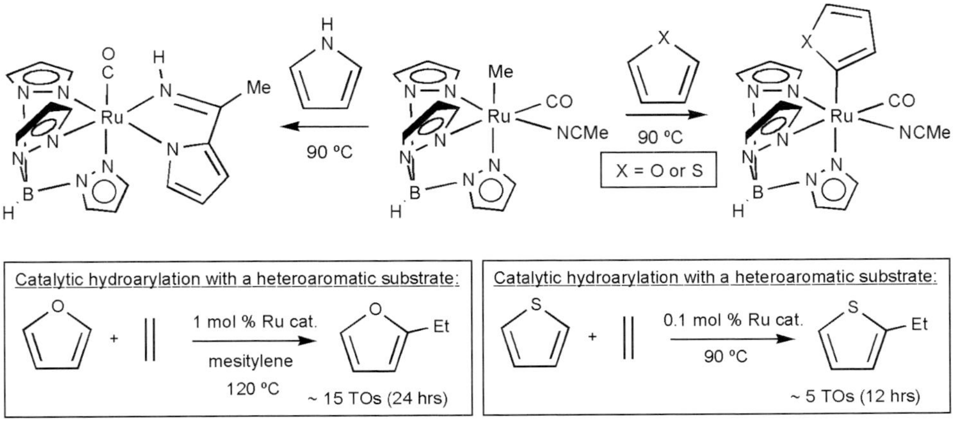

Scheme 2. Reactions of TpRu(CO)(NCMe)(Me) with five-membered heteroaromatic substrates.

The Ru(II) catalysts currently used for olefin hydroarylation reactions are limited in terms of incorporation of substituents into the olefin substrate. For example, attempted hydrophenylation of isobutylene with TpRu(CO)(NCMe)(Ph) as catalyst does not yield new organic products. In addition, extension of catalysis to hetero-functionalized olefins using the TpRu(CO)(NCMe)(aryl) systems has not been successful. For electron-deficient olefins (e.g. styrene, methyl methacrylate, acrylonitrile) the TpRu(II) complexes initiate radical polymerization of the olefin in transformations that probably involve a Ru(III/II) redox change [4]. The

{TpRu^{II}(CO)(R)} fragment reacts with olefins that bear electron-donating groups (e.g. ethyl vinyl ether or ethyl vinyl sulfide) to produce TpRu complexes that are not active in hydroarylation chemistry.

Ir(III) Catalysts

Ir(III) complexes can also be used to catalyze the hydroarylation of olefins [5, 6]. The catalyst precursors (acac-O,O)$_2$Ir(Ph)(L) (L= pyridine or H$_2$O) and [Ir(μ-acac-O,O,C^3)(acac-O,O)(acac-C^3)]$_2$ have been reported and are depicted in Scheme 3, with representative catalytic reactions. Among these catalyst precursors, (acac-O,O)$_2$Ir(Ph)(H$_2$O) has been demonstrated to be the most active system [5]. The linear-to-branched ratios observed for Ir(III) catalyzed reactions of α-olefins are remarkably similar to those obtained using TpRu(CO)(NCMe)(Ph). For example, hydrophenylation of propene using [Ir(μ-acac-O,O,C^3)(acac-O,O)(acac-C^3)]$_2$ as catalyst produces a 1.6:1 ratio of linear propylbenzene to cumene (identical to the ratio observed using the Ru(II) catalyst). Similar to the TpRu(CO)(NCMe)(R) catalysts, the regioselectivity of hydroarylation of ethylene that incorporates monoalkyl arenes produces meta- and para-disubstituted alkyl arenes in approximately 2:1 ratio without observation of ortho-disubstituted products.

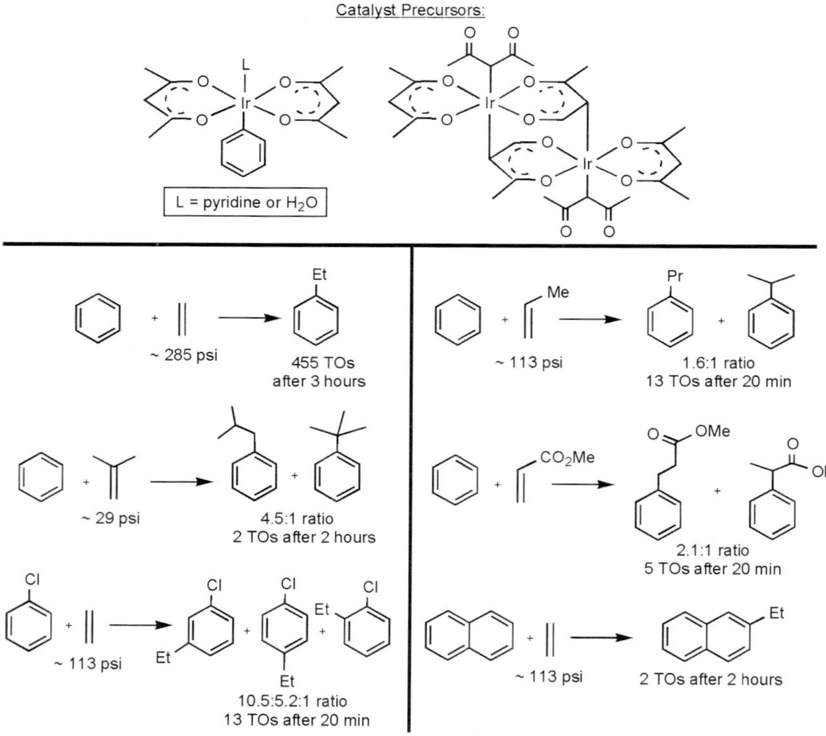

Scheme 3. Ir(III) catalyzed hydroarylation of olefins.

Advantages of the Ir(III)-catalyzed reactions over the Ru(II) systems include an increased number of turnovers as well as a more substantial range of olefins that are compatible with the catalyst. The hydrophenylation of ethylene using [Ir(μ-acac-O,O,C^3)(acac-O,O)(acac-C^3)]$_2$ as catalyst yields 455 turnovers after 3 hours at 180 °C whereas the TpRu(CO)(NCMe)(Ph) catalyst yields approximately 75 catalytic turnovers after 24 hours at 90 °C. The Ir(III) systems exhibit increased thermal stability relative to the TpRu(CO)(R)(NCMe) systems. In contrast to the Ru(II) catalytic systems, the hydrophenylation of α-olefins (e.g., propene or 1-hexene) catalyzed by Ir(III) does not yield observable quantities of β-alkyl styrenes. In addition, olefins that possess electron-withdrawing groups are compatible with the Ir(III) catalysts. For example, the hydrophenylation of methyl acrylate produces 3-phenylpropionic acid methyl ester and 2-phenylpropionic acid methyl ester (five turnovers) in a 2.1:1 ratio, and the hydrophenylation of styrene with 22 turnovers to produce 1,2-diphenylethane and 1,1-diphenylethane in a 45:1 ratio is observed. The Ir(III) catalysts exhibit reduced activity compared with the TpRu(II) systems requiring temperatures of approximately 180 °C {compared with approximately 90 °C for Ru(II)} to achieve sufficient rates of reaction.

1.3.2.3.2 Mechanism

The mechanisms of catalytic hydroarylation of olefins by TpRu(CO)(NCMe)(Ph) and the Ir(III) systems shown in Schemes 1 and 3 have been studied both experimentally and computationally [2, 5, 7–9], and the proposed pathways are closely related. Catalyst precursors in both cases are coordinatively and electronically saturated 18-electron complexes with at least one labile ligand. Initial exchange between the labile ligand and olefin likely precedes olefin insertion (Scheme 4). Olefin insertion into a metal–aryl bond creates an open coordination site that can bind and activate the C–H bond of the aromatic substrate, and release of alkyl arene regenerates the active catalyst. The intermediates labeled **I** would be expected to undergo facile β-hydride elimination to produce metal hydride complexes with coordinated olefins. Oxgaard and Goddard have calculated that the lack of production of olefinic products for hydroarylation reactions catalyzed by Ir(III) is because of the reversibility of the β-hydride elimination step [9]. Consistent with this proposal, experimental studies with TpRu(CO)(NCMe)(CH$_2$CH$_2$Ph) indicate that β-hydride elimination is kinetically facile and provide evidence for reversibility. For example, TpRu(CO)(NCMe)(CH$_2$CH$_2$Ph) rapidly produces free styrene, CHDCl$_2$, and TpRu(CO)(NCMe)(Cl) on heating to 90 °C in CDCl$_3$ (Scheme 5). In the hydrophenylation of α-olefins by TpRu(CO)(NCMe)(Ph), formation of trans-β-alkyl styrenes is observed, and it has been proposed that the steric influence of the alkyl group raises the ground-state energy of TpRu(CO)(H)(η^2-trans-(R)HC=CH(Ph)) relative to TpRu(CO)(H)(η^2-styrene) and increases the rate of dissociation [2].

At higher olefin concentrations the rate of catalysis, using either the Ru(II) or Ir(III) systems, decreases with increasing olefin concentration. For example, between 10 and 40 psig of ethylene pressure, the rate of hydrophenylation of ethyl-

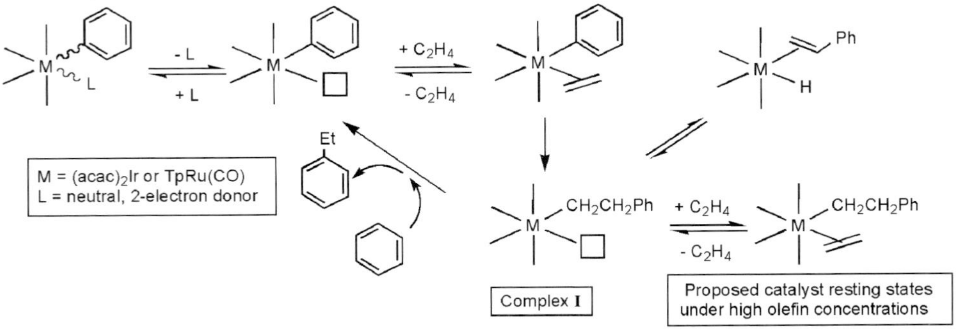

Scheme 4. Proposed mechanisms for catalytic hydroarylation of olefins (benzene and ethylene depicted as substrates; note: the catalyst precursors for Ir(III) exhibit trans disposition of L and Ph).

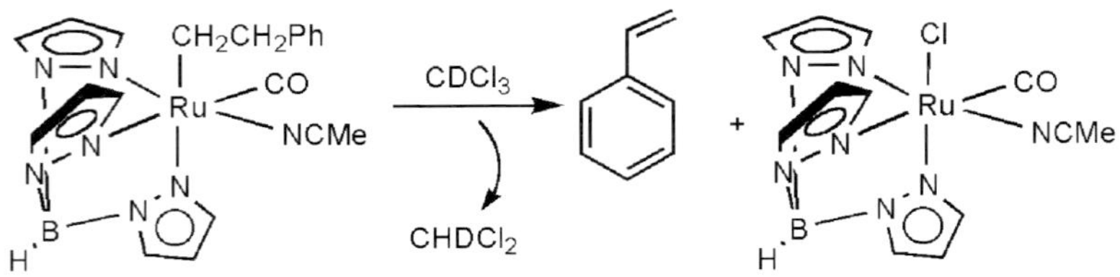

Scheme 5. Evidence for facile β-hydride elimination is obtained upon heating TpRu(CO)(NCMe)(CH$_2$CH$_2$Ph) in CDCl$_3$.

ene catalyzed by TpRu(CO)(NCMe)(Ph) decreases. For analogous catalysis by [Ir(μ-acac-O,O,C^3)(acac-O,O)(acac-C^3)]$_2$, the rate of catalysis increases with increasing ethylene pressure for ethylene/benzene ratios <0.2 whereas at ratios >0.2 the rate of ethylbenzene production decreases. These trends have been rationalized by invoking the trapping of catalytic intermediates after olefin insertion upon coordination of ethylene (Scheme 4). Thus, in order for the Ru(II) or Ir(III) systems to reenter the catalytic cycle, dissociation of dihapto-coordinated ethylene (or other olefin) must occur, and the reversible binding of ethylene results in inverse dependence of the rate of catalysis on olefin concentration. For the Ir(III) catalyst, lower concentrations of ethylene presumably shift the catalyst resting state such that olefin coordination is involved in the rate-determining step (or steps preceding the rate determining step) of the catalytic cycle, and a first-order dependence of catalytic rate on concentration of ethylene results.

1.3.2.3.3 Scope and Limitations

The Ru(II) and Ir(III) olefin hydroarylation catalysts have only recently been developed, and the scope and potential of these reactions have not been fully delineated. For example, the impact of altering ancillary ligand identity, metal oxidation state and the identity of the metal center has not been extensively probed. At this stage, the TpRu(CO)(NCMe)(R) catalysts seem to be limited to hydroarylation reactions of ethylene and simple α-olefins. Incorporation of electron-donating groups or electron-withdrawing groups into the olefin results in side reactions that disrupt the olefin hydroarylation chemistry. The introduction of more than one alkyl group inhibits catalyst activity, presumably because of steric issues that restrict the olefin coordination which precedes the insertion/C–C bond-forming step. Benzene, monoalkyl arenes, thiophene, and furan can be used as aromatic substrates. Although the detailed mechanism has not been reported, use of pyrroles with N–H bonds is complicated by C–C coupling with acetonitrile, while also providing a potential new avenue for catalytic C–C bond formation (Scheme 2). Tolerance of functionality on the arenes (e.g. anisoles, anilines, etc.) has not yet been determined. Currently, the primary advantage of TpRu(II) systems is sufficient activity to enable catalytic turnover under relatively ambient conditions (i.e. low olefin concentrations and 90 °C). The flexibility of the Tp ligand and the option of replacing CO with other neutral, two-electron donating ligands offer the possibility of tuning future catalysts for improved properties.

In comparison with the Ru(II) systems, Ir(III) catalysts have more tolerance of olefin functionality. For example, hydroarylation of styrenes, acrylates, and dialkyl olefins is possible, although the number of turnovers is significantly reduced compared with the hydroarylation of ethylene using Ir(III) catalysts. The compatibility of the Ir(III) systems with heteroaromatic substrates has not been reported; hydroarylation of ethylene with chlorobenzene is possible, however, and gives m-, p- and o-chloroethylbenzene in 10.5:5.2:1 ratio. Chlorobenzene reacts more slowly than toluene or ethylbenzene indicating that electron-withdrawing groups deactivate the arene toward the metal-mediated C–H activation step. Consistent with this result, attempted hydroarylation of ethylene using pentafluorobenzene and $[Ir(\mu\text{-acac-}O,O,C^3)(\text{acac-}O,O)(\text{acac-}C^3)]_2$ as catalyst results in no observed reaction.

Important questions for this class of catalyst include whether more active systems can be accessed and if increased selectivity/functional group tolerance can be achieved. Given the importance of C–C bond-forming reactions that involve metal-mediated C–X (X = halide or trifluoromethanesulfonate) activation of an aromatic substrate [10–14], the development of catalysts that mediate selective C–C bond formation of aromatic compounds based on C–H activation would complement current methods and be a valuable addition to the toolbox of synthetic organic chemists. The possibility of increasing catalyst activity has been addressed using computational studies [15]. The seminal result from this study is the conclusion that electron density at the metal center has an inverse effect on the rates of C–H activation and olefin insertion. Thus, it has been predicted that as the activation barrier for C–H activation is reduced, the activation barrier for olefin insertion will increase, thereby placing an upper limit on catalyst activity; it has, however, been suggested that alteration of the σ-framework could lead to more efficient catalysts.

Experimental

Preparation of TpRu(CO)(NCMe)(Ph)

TpRu(CO)$_2$(Me) [1] (0.5239 g, 1.36 mmol) and Me$_3$NO (0.2040 g, 2.72 mmol) were heated under reflux in approximately 25 mL acetonitrile for 40 min. The solution was cooled to room temperature, neutral alumina (5 g) was added, and the volatiles were removed under reduced pressure. The green–yellow crude reaction mixture was flashed on a plug of neutral alumina using dichloromethane (~200 mL). Volatiles were removed from the eluent and the resulting solid was washed with hexanes to give a white solid. The solid, TpRu(CO)(NCMe)(Me), was collected and dried under vacuum (0.3601 g, 67 % yield). TpRu(CO)(NCMe)(Me) (0.538 g, 1.35 mmol) was dissolved in a mixture of benzene (16 mL) and acetonitrile (7 mL, 0.135 mol) and the solution was heated in a screwcap pressure tube to 90 °C for approximately 40 h. The volatiles were removed in vacuo, and the remaining solid was dissolved in a small amount of dichloromethane followed by precipitation with pentane. Collection of the resulting solid yielded a white powder in 68 % yield that was pure by NMR spectroscopy. To obtain product pure by elemental analysis, the resulting solid was purified by flash chromatography through a plug of neutral alumina using dichloromethane as eluent. The product was collected and the solvent volume was reduced to approximately 10 mL under reduced pressure. Precipitation with pentane yielded the pure product as a white solid.

Hydrophenylation of Ethylene Using TpRu(CO)(NCMe)(Ph)

A representative catalytic reaction is described. TpRu(CO)(NCMe)(Ph) (0.021 g, 0.046 mmol) was dissolved in 4.1 mL (0.0456 mol) distilled benzene and decane (0.269 mL, 1.38 mmol) was added to the homogeneous solution as internal standard. The solution was placed in a thick-walled glass reaction vessel and charged with 25 psig ethylene. The tube was then placed in an oil bath heated to 90 °C. Periodically the tube was removed from the oil bath and plunged into an ice bath. A sample (0.1 mL) of the reaction solution was removed under a purge of dinitrogen and the tube was quickly returned to the oil bath and ethylene pressure was restored. The removed samples (~1 µL) were analyzed by GC–FID.

Preparation of Ir(O,O-acac)$_2$(C-acac)(H$_2$O)

IrCl$_3$(H$_2$O)$_x$ (54.11 % Ir, 1 g, 2.82 mmol), 2,5-pentadione (10 mL, 9.75 mmol), and NaHCO$_3$ (1 g, 11.9 mmol) were added to a round-bottomed flask equipped with a reflux condenser vented to an oil bubbler. The mixture was heated to gentle reflux with stirring for 48 h. During this time a yellow solid precipitated. The reaction mixture was cooled to room temperature and the solid was collected as crude product. The yellow solid was dissolved in 200 mL H$_2$O at room temperature with vigorous stirring and the solution was filtered and concentrated under vacuum to give 650 mg (45 %) Ir(O,O-acac)$_2$(C-acac)(H$_2$O) as a yellow microcrystalline solid.

Hydrophenylation of Propene Using Ir(III)
A 3-mL stainless steel autoclave equipped with a glass insert and a magnetic stir bar was charged with 1 mL distilled benzene and 3–5 mg (5 mmol, ~0.1 mol%) catalyst. The reactor was degassed with nitrogen and pressurized with 0.96 MPa propylene and 2.96 MPa nitrogen. The autoclave was heated for 30 min in a well stirred heating bath maintained at 180 °C. The liquid phase was sampled and product yield was determined by GC–MS using methyl cyclohexane, introduced into the reaction solution after the reaction, as internal standard.

1.3.2.4 Catalytic Functionalization of N-Heterocycles via their Rhodium–Carbene Complexes
Sean H. Wiedemann, Jonathan A. Ellman, and Robert G. Bergman

1.3.2.4.1 Introduction and Fundamental Examples
The o-functionalization of arenes directed by an embedded nitrogen atom ranks among the first examples of a directed catalytic C–H transformation [1]. Moore et al. demonstrated that pyridine can be o-acylated by CO and an olefin using $Ru_3(CO)_{12}$ as catalyst [2]. Later, imidazoles, among other N-heterocycles, were also found to be active substrates for this reaction [3]. Other than early Zr-catalyzed studies [1], however, at the beginning of this work we were unable to uncover any examples of C-*alkylation* of N-heterocycles by olefins. The $RhCl(PPh_3)_3$-catalyzed cyclization of olefin tethered benzimidazoles was our first effort in this area (Scheme 1) [4]. In prior work, Wilkinson's catalyst had proven effective for C–H transformations involving a *pendant* directing group [5]. Despite this precedent, mechanistic evidence suggests that a unique mechanism is operative in our reaction.

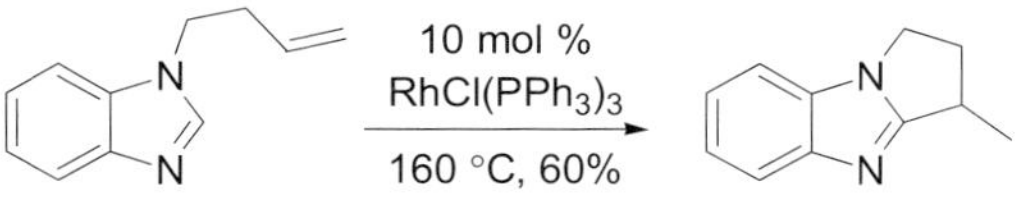

Scheme 1. Initial cyclization result.

When benzimidazole **1** was treated with a stoichiometric amount of an optimized Rh-catalyst mixture ($[RhCl(coe)_2]_2$ and PCy_3), a single catalyst–substrate complex (**2**) was isolated (Scheme 2) [6]. Structural studies of **2** revealed an unusual ligand binding mode in which the organic fragment is bound to rhodium as an N-heterocyclic carbene (NHC). Kinetic studies showed our catalytic cycle to be the first known example involving a reactive M-NHC intermediate. This new-found mechanistic understanding illuminates general structural requirements for Rh-NHC-mediated arene addition (Scheme 3). These specifications parallel those for the formation of stable NHCs having two donating heteroatoms flanking an sp^2-hybridized carbon [7].

This chapter covers the intramolecular cyclization of olefin-tethered benzimidazoles and the intermolecular C–H addition of azoles to olefins [8]. These reactions

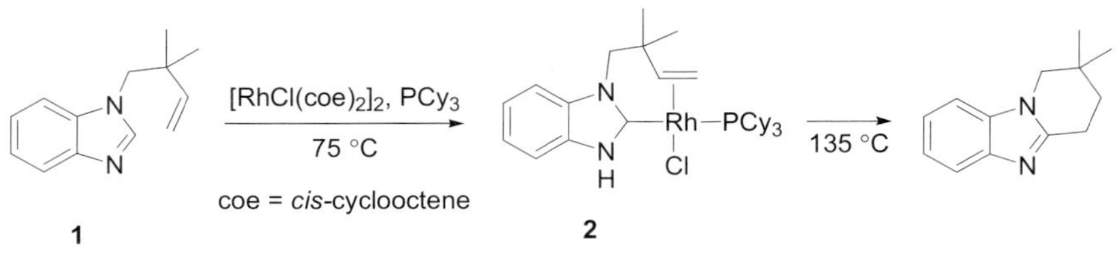

Scheme 2. Discovery of catalyst resting state.

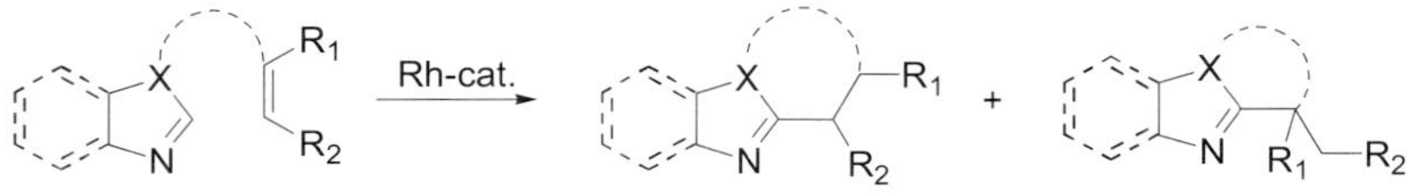

Scheme 3. General reaction.

can be conducted using a catalyst mixture of commercially available materials at neutral pH. Mild Lewis acids or, more conveniently, Brønsted acids have been shown to promote the desired reaction. Microwave heating can also significantly reduce reaction times [9].

Some variants of our central reaction have been discovered. Consistent with the notion that NHC ligands are stable in the absence of aromatic π-stabilization, a non-aromatic N-heterocycle, 4,4-dimethyl-2-oxazoline (**3**), can be added to olefins in the presence of a rhodium catalyst [10]. The optimized reaction conditions are especially mild, making this an effective method for preparation of protected carboxylic acids.

Close inspection of **2** reveals that it can be thought of as a metal–aryl complex in a low oxidation state. Thus it meets the conditions for another type of C–H transformation – cross-coupling. Various N-heterocycles bearing the N–CH–X motif will react with aryl iodides, under the action of rhodium catalysis, to give arylated products in moderate yield [11].

1.3.2.4.2 Mechanism

The proposed catalytic cycle for the annulation of **1** has been chosen as a representative example of the C–H transformations discussed in this chapter (Scheme 4). Formation of the isolable catalyst resting state **2** occurs at a temperature well below that required for catalytic activity, suggesting that the initial C–H activation steps are rapid [6]. The formation of **2** from **1** and Rh/PCy$_3$ probably begins with coordination of a nitrogen lone-pair to Rh(I) [12]. Studies are in progress to define a specific microscopic pathway in the subsequent tautomerization/coordination to give **2**, but C–H to N–H tautomerization is certainly crucial to catalyst activity because all suitable substrates feature an unsaturated nitrogen atom adjacent to the site of coupling.

DFT calculations performed on a simplified version of **2** indicate that the transition state for carbometallation of the pendant olefin is the highest energy barrier of the overall reaction [13]. Subsequent rapid steps, including proton-transfer and

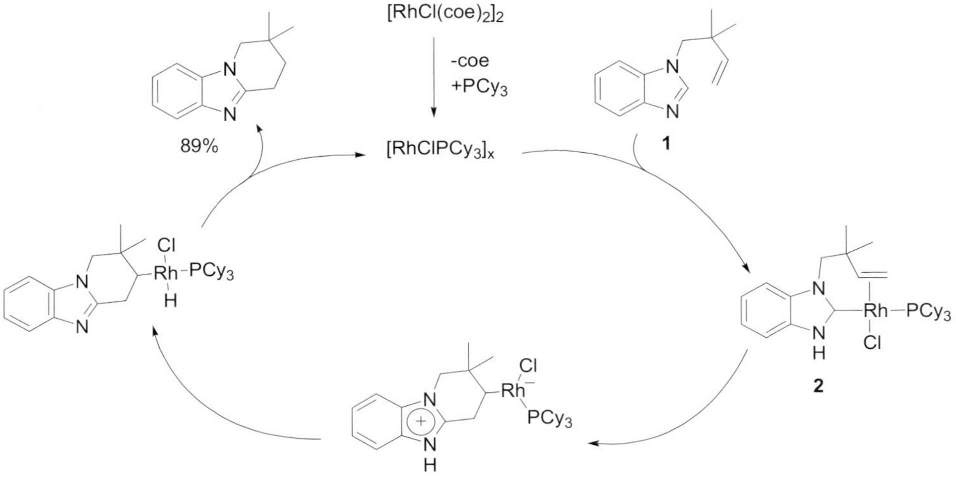

Scheme 4. Representative mechanism of olefin arylation.

reductive elimination, are needed to liberate unsaturated Rh, which is rapidly reconverted to resting-state **2**.

This Rh(I)-based mechanism does not properly explain the pronounced rate acceleration observed on addition of HCl. Rather, we suggest that the well-known propensity of Rh(I) to oxidatively add HCl evokes a different, Rh(III)-based mechanism. The elementary steps along such a pathway (β-hydride insertion then C–C reductive elimination) differ substantially from those depicted in Scheme 4. Despite this uncertainty, our current mechanistic understanding of the non-H$^+$ mediated reaction constitutes a satisfactory model because it has led to the successful refinement and expansion of our initial findings.

1.3.2.4.3 Scope and Limitations

The intramolecular coupling of benzimidazoles using a Rh(I)/PCy$_3$ catalyst is quite general with respect to substitution on the pendant olefin (Table 1, Column A) [4]. Alkenes bearing up to three substituents can be cyclized successfully. When cyclization to different ring sizes is possible, the reaction tends to be selective for formation of five-membered rings (Table 1, Entry 2), although steric factors can override this preference (Table 1, Entry 5). The best yields and selectivity are obtained when competing Rh-catalyzed positional isomerization of the pendant olefin is slow or impossible.

Table 1. Substrate generality of intramolecular coupling.

Entry	Substrate	Product	Yield (%)		
			Column A Heating: conventional Catalyst: Rh/PCy$_3$[a]	Column B microwave Rh/PCy$_3$·HCl[b]	Column C microwave RhCl(PPh$_3$)$_3$[c]
1			70[d]	50[d]	59[d]
2			79	76	55
3			–	62	42
4			82	97	64
5			71[d]	78	59
6			75[d]	–	–

[a] Conditions: 5 mol % ½[RhCl(coe)$_2$]$_2$, 7.5 mol % PCy$_3$, THF or toluene, 160–180 °C, 1–3 d. [4]
[b] Conditions: 5 mol % ½[RhCl(coe)$_2$]$_2$, 5 mol % PCy$_3$, o-dichlorobenzene/acetone, 225–250 °C, 6–12 min. [9]
[c] Conditions: 10 mol % RhCl(PPh$_3$)$_3$, 250 °C, 20 min. [9]
[d] Reactions run with 2x the normal catalyst loading.

The greatest limitations of our initial method in respect to, for example, industrial applications, are the air-sensitive nature of the catalyst and the high temperatures and long reaction times required. Two revised methods utilizing microwave heating have been developed to overcome these difficulties [9]. Microwave heating as a strategy for increasing reaction rate and yield has become well-accepted in both academic and industrial laboratories for a variety of chemical transformations [14].

The first revised method (Table 1, Column B) offers reduced air sensitivity and takes advantage of beneficial Brønsted acid additive effects by making use of PCy$_3$

protected as its HCl salt. The combined benefits of the acid additive and microwave heating lead to yields after 12 min of heating that generally exceed those obtained from 20 h of conventional heating without additives.

The second revised method (Table 1, Column C) offers the most operationally simple reaction conditions. Commercially available, air-stable Wilkinson's catalyst is used instead of a Rh/phosphine mixture, and the reactions are complete in less than 20 min. Isolated yields continue to be in the useful range without the need for special air-sensitive techniques.

The additive-modified catalyst mixture developed for intramolecular olefin coupling is sufficiently active to catalyze the analogous intermolecular reaction. A variety of five-membered N-heterocycles can be catalytically alkylated at the C-2 position (Table 2) [8]. Additional functional groups on the heterocycle[15] (Table 2, Entry 1) and on the olefin (Table 3) are well tolerated. Products corresponding to linear addition are usually obtained exclusively, even when the olefin is rapidly isomerized under the reaction conditions.

Table 2. Heterocycle generality of intermolecular coupling.[a]

Entry	Heterocycle	Product	Substitution	Yield (%)
1			R = H R = OMe R = COOMe R = Cl	96 94[b] 89[b] 93[b]
2			X = NMe X = S X = O	67 97 77
3			–	99
4			–	86[c]

[a] Conditions: 5 mol % ½[RhCl(coe)$_2$]$_2$, 7.5 mol % PCy$_3$, 5 mol % 2,6-lutidine·HCl, 5 equiv 3,3-dimethylbutene, THF, 150 °C. [8]
[b] Conditions: 5 mol % ½[RhCl(coe)$_2$]$_2$, 5 mol % PCy$_3$·HCl, 5 equiv 3,3-dimethylbutene, THF, 150 °C. [10]
[c] Conditions: 5 mol % ½[RhCl(coe)$_2$]$_2$, 5 mol % PCy$_3$·HCl, 2.5 mol % PCy$_3$, 5 equiv 3,3-dimethylbutene, THF, 45 °C. [10]

Table 3. Alkene generality of intermolecular coupling of azoles and **3**.

Entry	Alkene	Azole product[a]	Yield (%)	Azoline product[b]	Yield (%)
1	(pentenyl)	benzimidazole-C5H11	80	oxazoline-C5H11	74
2	methylenecyclohexane	–	–	oxazoline-CH2-cyclohexyl	79
3	cyclohexene	–	–	oxazoline-cyclohexyl	57
4	CH2=CHCH2C(O)O-i-Bu	benzimidazole-(CH2)3C(O)O-i-Bu	87[c]	oxazoline-(CH2)3C(O)O-i-Bu	71[d]
5	vinyl pinacol acetal	benzimidazole-(CH2)2-pinacol acetal	60[c]	oxazoline-(CH2)2-pinacol acetal	58[d]
6	CH2=CHC(O)O-t-Bu	4,5-dimethylthiazole-(CH2)2C(O)O-t-Bu	93[c]	oxazoline-(CH2)2C(O)O-t-Bu	59[d]

[a] Conditions: 1 equiv benzimidazole or 4,5-dimethylthiazole, 5 mol % ½[RhCl(coe)$_2$]$_2$, 7.5 mol % PCy$_3$, 5 mol % 2,6-lutidine·HCl, 5 equiv alkene, THF, 150 °C. [8]
[b] Conditions: 1 equiv 4,4-dimethyl-2-oxazoline (**3**), 5 mol % ½[RhCl(coe)$_2$]$_2$, 5 mol % PCy$_3$·HCl, 2.5 mol % PCy$_3$, 5 equiv alkene, THF, 45 °C. [10]
[c] Reactions run with 2x the normal catalyst loading.
[d] Reactions run at 105 °C.

Olefins react with 4,4-dimethyl-2-oxazoline (**3**) much more efficiently (Table 3, azoline column) than they do with other (aromatic) substrates [10]. Rather than requiring reaction temperatures of 135–180 °C for the reaction to occur, addition of **3** to olefins is optimally conducted between 45 and 105 °C. The substrate scope of this reaction includes, for the first time, disubstituted alkenes (Table 3, azoline column, Entries 2 and 3). The coupling products of **3** with olefins can be deprotected to reveal carboxylic acids or elaborated further using well-known methods [16].

A variety of N-heterocycles can be catalytically phenylated with iodobenzene (Table 4) [11]. This reaction provides substrate scope complementary to that of existing Pd-catalyzed N-heterocycle arylation methods [17]. Both aromatic and non-aromatic heterocycles serve as suitable substrates. Dihydroquinazoline (Table 4,

Entry 4), in addition to being active toward arylation, is a good substrate for addition to alkenes [18].

Table 4. Heterocycle arylation with iodobezene.[a]

Entry	Heterocycle	Product	Temp. (°C)	Yield (%)
1	benzimidazole (NH)	2-Ph-benzimidazole (NH)	150	56
2	N-methylbenzimidazole	2-Ph-N-methylbenzimidazole	135	51
3	benzoxazole	2-Ph-benzoxazole	135	79
4	3,4-dihydroisoquinoline (NH)	1-Ph-3,4-dihydroisoquinoline	150	78
5	4,4-dimethyl-2-oxazoline	2-Ph-4,4-dimethyl-2-oxazoline	105	51

[a] Conditions: 10 mol % ½[RhCl(coe)$_2$]$_2$, 40 mol % PCy$_3$, 4 equiv Et$_3$N, 2 equiv iodobenzene, THF, 150 °C. [11]

Experimental

2,3-Dihydro-2-methyl-1H-pyrrolo[1,2-a]benzimidazole (Table 1, Column C, Entry 4)

To a 2–5 mL process vial was added 103 mg (0.60 mmol) 1-(2-methylallyl)-1H-benzimidazole and 56 mg (0.060 mmol) RhCl(PPh$_3$)$_3$. The reaction vessel was flushed with N$_2$ for 1 min and then capped with a septum. A 3:1 o-dichlorobenzene–acetone mixture (3 mL, untreated) was added via syringe and the vial was placed in a Smith Synthesizer cavity and subjected to microwave radiation (300 W, normal absorption) with stirring. A pre-set temperature of 250 °C was reached rapidly (<1 min) and sustained for 20 min. On cooling with 40 psig compressed air the vessel was uncapped and the contents evaporated to dryness in a Savant evaporator. The residue was then taken up in ethyl acetate (20 mL) and washed with 1 M HCl (3 × 20 mL). The acidic layers were combined and washed with ethyl acetate, and then made alkaline to pH 9 by slow addition of solid K$_2$CO$_3$. The resulting basic solution was then extracted with CH$_2$Cl$_2$ (3 × 20 mL). The organic layers were combined, dried over MgSO$_4$, filtered, and evaporated to dryness; TLC (silica, hexanes–ethyl acetate, 1:1, with 1 % Et$_3$N): R_F = 0.20. The fraction with R_F = 0.20 was isolated by flash chromatography: 66 mg (64 %) as a pale orange solid with m.p. 88 °C. ^{1}H NMR (CDCl$_3$, 500 MHz) δ = 7.70–7.68 (m, 1H, Ar-H), 7.27–7.17 (m, 3H,

Ar-H), 4.23 (dd, 1H, $J = 10$, 7.5 Hz, 1-H), 3.63 (dd, 1H, $J = 10$, 6 Hz, 1-H), 3.25–3.13 (m, 2H, 2,3-H), 2.67 (dd, 1H, $J = 16$, 6, 3-H), 1.32 (d, 3H, $J = 6.5$ Hz, 3-CH$_3$).

2-Hexyl-4,4-dimethyl-2-oxazoline (Table 3, Azoline column, Entry 1)

In a nitrogen atmosphere glove-box, a glass-walled vessel equipped with a stopcock (20 mL) was filled with a solution of 1-hexene (420 mg, 5.0 mmol), [RhCl(coe)$_2$]$_2$ (18.0 mg, 0.050 mmol relative to monomer), PCy$_3$.HCl (16.0 mg, 0.050 mmol) [19], and PCy$_3$ (7.0 mg, 0.025 mmol) in THF (6.7 mL). Immediately before closing the Kontes seal, 99.1 mg (1.0 mmol) 4,4-dimethyl-2-oxazoline (**3**) was added. The reaction mixture was heated at 45 °C for 18 h and then cooled to 25 °C. The reactor was opened in air and excess acid was neutralized by addition of Et$_3$N (0.5 mL); TLC (silica, pentane–diethyl ether, 1:1, I$_2$ stain): $R_F = 0.3$. The reaction suspension was mixed with silica, concentrated to dryness using a rotary evaporator, and the fraction with $R_F = 0.3$ was isolated by flash chromatography: 136 mg (74 %) as colorless oil. IR: 1668 cm^{-1} (C=N). ^{1}H NMR (CDCl$_3$, 500 MHz,): $\delta = 3.88$ (s, 2H, 3-H), 2.22 (t, 2H, $J = 7.8$ Hz, 1′-H), 1.60 (m, 2H, 2′-H), 1.4–1.2 (m, 6H, 3′-5′-H), 1.21 (s, 6H, 4-CH$_3$), 0.86 (t, 3H, $J = 6.8$ Hz, 6′-H).

1.3.2.5 Fujiwara Reaction: Palladium-catalyzed Hydroarylations of Alkynes and Alkenes

Yuzo Fujiwara and Tsugio Kitamura

1.3.2.5.1 Introduction and Fundamental Examples

The catalytic activation of aromatic C–H bonds leading to useful addition with new C–C bond formation is of considerable interest in organic synthesis and industrial processes, because it would provide simple and economical methods for producing aryl-substituted compounds directly from simple arenes, as shown in Scheme 1.

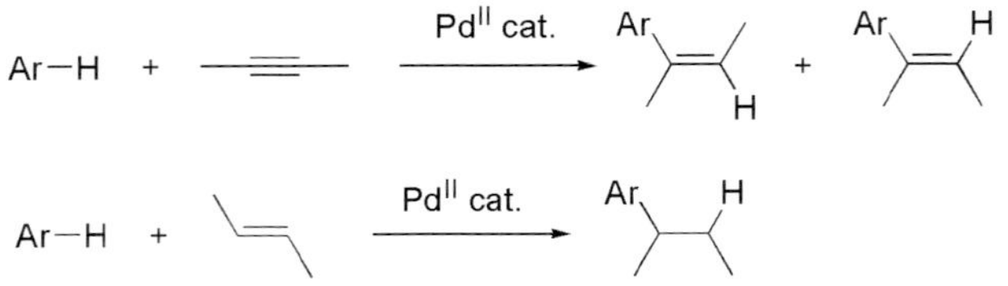

Ar–H + ≡ $\xrightarrow{\text{Pd}^{II} \text{ cat.}}$ Ar\=/H + Ar\=H/

Ar–H + \\ $\xrightarrow{\text{Pd}^{II} \text{ cat.}}$ Ar\\H/

Scheme 1. General equations for hydroarylation of alkynes or alkenes.

Study of the reactivity of aromatic C–H bonds in the presence of transition metal compounds began in the 1960s despite the quite early discovery of Friedel–Crafts alkylation and acylation reactions with Lewis acid catalysts. In 1967, we reported Pd(II)-mediated coupling of arenes with olefins in acetic acid under reflux [1]. The reaction involves the electrophilic substitution of aromatic C–H bonds by a Pd(II) species, as shown in Scheme 2, and this is one of the earliest examples of aromatic C–H bond activation by transition metal compounds. Although the direct use of aromatic compounds in synthesis is generally restricted to activated groups (e.g. C–Br, C–Cl bonds) other than the C–H bonds, manufacture of aryl halides generally is not an environmentally friendly process and thus the future of bulk synthesis of aromatic compounds may lie in the direct transformation of C–H bonds. Activation of aromatic C–H bonds by ortho-chelating-assisted oxidative addition to low-valent transition metal compounds (Scheme 3), leading to addition to C–C multiple bonds, has been recognized as another promising route (see Chapter III.1.3.2.1) [2].

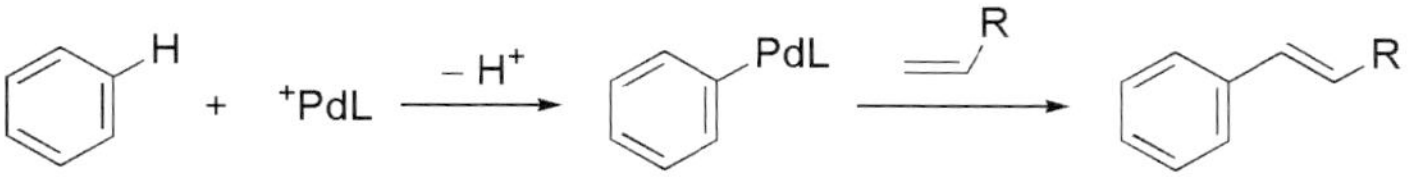

Scheme 2. C–H Bond transformation by electrophilic substitution.

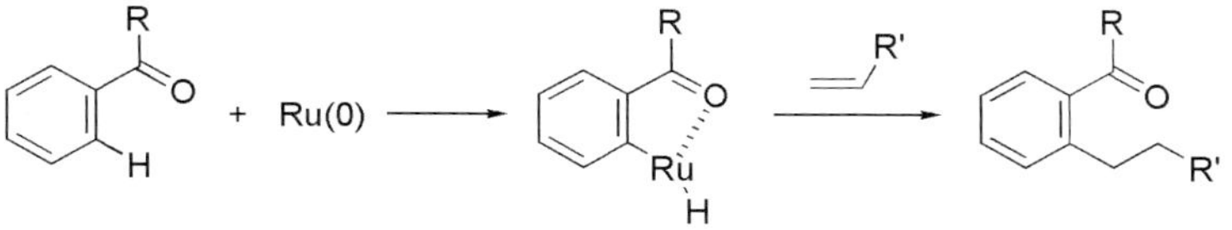

Scheme 3. C–H bond transformation by ortho-chelating assistance (see Chapter III.1.3.2.1).

In our attempt to extend the coupling reaction of arenes with alkenes to the coupling with alkynes, as shown in Scheme 4, it was found that the reaction of arenes with ethyl propiolate in TFA (trifluoroacetic acid) gave addition products instead of a coupling product [3]. This addition reaction has been extended to various alkynes and various arenes and also to intramolecular reactions for synthesis of heterocycles such as coumarins, quinolines, and thiocoumarins.

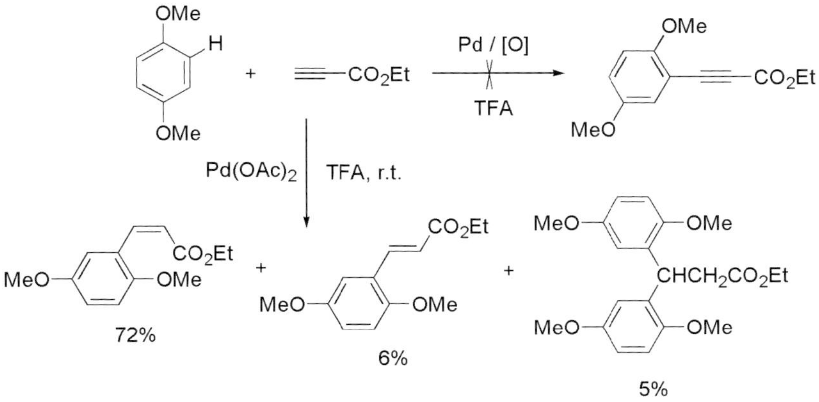

Scheme 4. Reactions of arenes with alkynes in the presence of a Pd catalyst.

1.3.2.5.2 Mechanism

Isotope experiments revealed that D-atoms were incorporated to the vinyl position of adducts either in inter- or intramolecular reactions in d-TFA (Scheme 5) [3b, c]. Reaction of heteroarenes with alkynoates in AcOD gave similar results [4]. Also, addition of heteroaromatic C–D bonds to the C–C multiple bonds and the large isotope effect ($k_H/k_D = 3$) between pyrrole and d_5-pyrrole in the reaction with ethyl phenylpropiolate were observed.

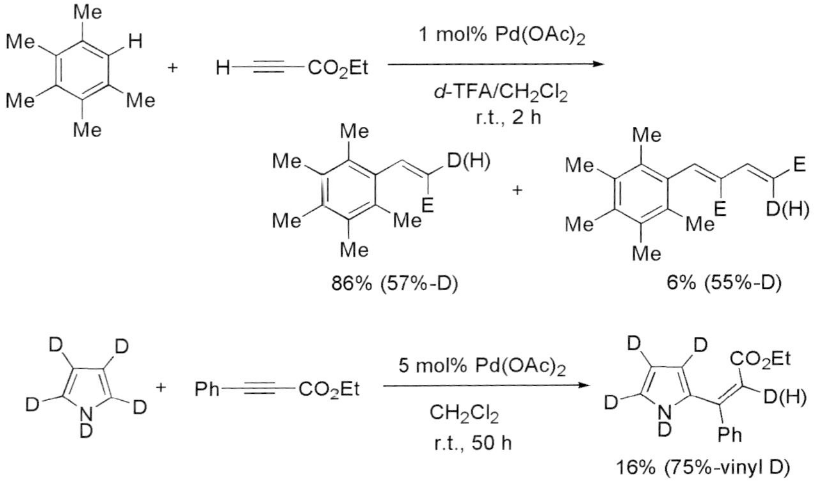

Scheme 5. Isotope experiments for hydroarylations of alkynes.

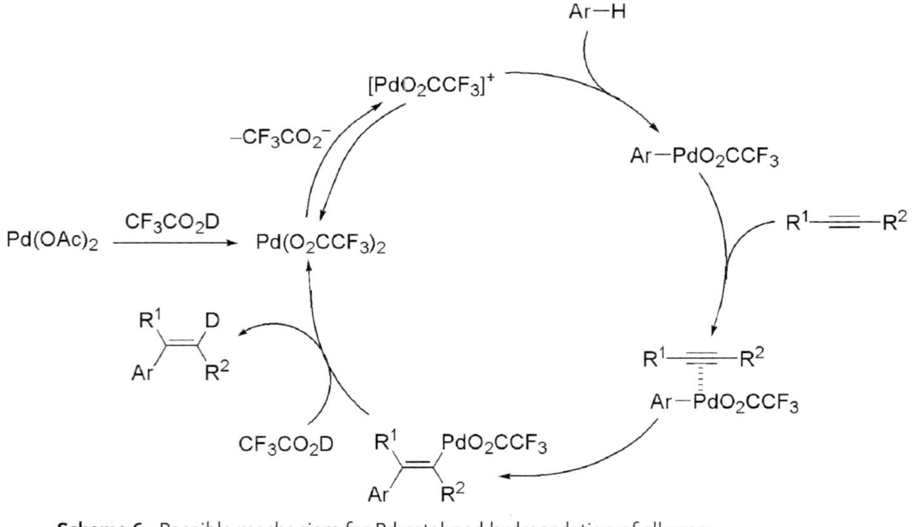

Scheme 6. Possible mechanism for Pd-catalyzed hydroarylation of alkynes.

Thus, a possible mechanism involving σ-aryl–Pd complexes similar to those involved in the coupling reaction of arenes with olefins has been suggested. Scheme 6 shows the possible mechanism of the hydroarylations. The facile formation of similar aryl–Pd complexes from Pd(II) and arenes in TFA has been indicated by the coupling of arenes with arenes, and also demonstrated by formation of aromatic acids from simple arenes with carbon monoxide [5]. The formation of vinyl–Pd complexes has been suggested by the formation of adducts of two alkynes and one arene. Use of TFA as solvent facilitates the generation of a highly cationic $[Pd(II)O_2CCF_3]^+$ species to form σ-aryl–Pd complexes by electrophilic substitution of aromatic C–H bonds.

It is considered that the hydroarylation of alkenes proceeds with a mechanism similar to that depicted in Scheme 6.

1.3.2.5.3 Scope and Limitations

Table 1 shows that a variety of commercially available aromatic compounds undergo Pd-catalyzed hydroarylations. Generally electron-rich aromatics give good results [3], consistent with a mechanism involving electrophilic aromatic C–H bond activation. This type of hydroarylation has high chemoselectivity with Br and OH substituents on aromatic compounds. Besides ethyl propiolate, terminal and internal alkynes including substituted propiolates, arylalkynes, and 3-butyn-2-ones are used for this hydroarylation. In addition to $Pd(OAc)_2$, a Pt(II) catalyst, $PtCl_2/2AgOAc/TFA$, has catalytic activity with a higher selectivity.

As shown in Table 2, reaction of heteroaromatic compounds with alkynoates occurs under very mild conditions [4, 6]. Heteroaromatic compounds such as pyrroles, furans, and indoles readily hydroarylate alkynoates at room temperature in the presence of a catalytic amount of $Pd(OAc)_2$ in acetic acid or CH_2Cl_2, usually affording cis-heteroarylalkenes. This reaction provides a synthetic route to heteroarylalkenes, especially cis-alkenes, from simple heteroaromatic compounds.

The intermolecular reaction of phenols with propiolates in TFA in the presence of a catalytic amount of $Pd(OAc)_2$ affords coumarin derivatives [3c, 7]. The results are summarized in Table 3. The coumarins are obtained in high yields from electron-rich phenols such as 3,4,5-trimethoxyphenol, 3,5-dimethoxyphenol, 3-methoxyphenol, 2-naphthol, and 3,4-methylenedioxyphenol. Various alkynoates, including diethyl acetylenedicarboxylate, give the corresponding coumarins in good-to-high yields.

The intramolecular version of this reaction provides a general method for synthesis of biologically active heterocycles, for example coumarins, quinolinones, and thiocoumarins, as shown in Schemes 7 and 8 [3a, c]. The reaction tolerates a variety of functional groups, for example Br, CHO, etc.

Very recently, it was found that montmorillonite-enwrapped Pd(II) catalyst had high activity in hydroarylations of alkynes or alkenes [8]. The results are summarized in Table 4. The catalyst could be reused without any loss of activity.

Table 1. Representative examples of the hydroarylation of alkynes.

Arenes	Alkynes	Products	Yield (%)
pentamethylbenzene	HC≡C-CO$_2$Et	(E)-pentamethylstyryl CO$_2$Et	85
3,5-dimethylbenzene	HC≡C-CO$_2$Et	2,4-dimethylstyryl CO$_2$Et	49
naphthalene	HC≡C-CO$_2$Et	1-naphthyl CH=CH-CO$_2$Et	45
2,4,6-trimethylphenol	HC≡C-CO$_2$Et	3,5-dimethyl-4-hydroxy styryl CO$_2$Et	57
2-bromo-1,3,5-trimethylbenzene	HC≡C-CO$_2$Et	bromo-trimethyl styryl CO$_2$Et	46
pentamethylbenzene	Me-C≡C-CO$_2$Et	pentamethyl-(Me)C=C(Me)-CO$_2$Et	78
pentamethylbenzene	Ph-C≡C-Ph	pentamethyl-(Ph)C=C(Ph)	58
pentamethylbenzene	HC≡C-COMe	pentamethylstyryl COMe	92

Table 2. Representative examples of hydroarylation of alkynoates by heteroaromatic compounds.

Heteroaromatics	Alkynoates	Products	Yield (%)
pyrrole (NH)	Ph—≡—CO$_2$Et	2-(1-phenyl-2-ethoxycarbonylvinyl)pyrrole	78
pyrrole (NH)	≡—CO$_2$Et	2-(2-ethoxycarbonylvinyl)pyrrole	52
N-Me pyrrole	Ph—≡—CO$_2$Et	N-Me-2-(1-phenyl-2-ethoxycarbonylvinyl)pyrrole	83
2,5-dimethylpyrrole	Ph—≡—CO$_2$Et	3-substituted 2,5-dimethylpyrrole	71 (E/Z = 2/1)
2-EtO$_2$C-3-Me-5-Me pyrrole	Ph—≡—CO$_2$Et	4-substituted product	66
2-methylfuran	Ph—≡—CO$_2$Et	5-substituted 2-methylfuran	72
3,4-dimethylfuran	Ph—≡—CO$_2$Et	2-substituted 3,4-dimethylfuran	66
indole	Ph—≡—CO$_2$Et	3-substituted indole	78
3-methylindole	Ph—≡—CO$_2$Et	2-substituted 3-methylindole	69

Table 3. Pd(II)-catalyzed reaction of phenols with alkynoates.

Phenols	Alkynoates	Products	Yield (%)
3,4,5-trimethoxyphenol	Ph—≡—CO$_2$Et	5,6,7-trimethoxy-4-phenylcoumarin	87
3,4-methylenedioxyphenol	Ph—≡—CO$_2$Et	6,7-methylenedioxy-4-phenylcoumarin	93
3,5-dimethoxyphenol	Ph—≡—CO$_2$Et	5,7-dimethoxy-4-phenylcoumarin	56
3-methoxyphenol	Ph—≡—CO$_2$Et	7-methoxy-4-phenylcoumarin	85
2-naphthol	Ph—≡—CO$_2$Et	4-phenyl-benzo[f]coumarin	88
3,5-dimethoxyphenol	H—≡—CO$_2$Et	5,7-dimethoxycoumarin	59
3,5-dimethoxyphenol	Me—≡—CO$_2$Et	5,7-dimethoxy-4-methylcoumarin	97
3,5-dimethoxyphenol	EtO$_2$C—≡—CO$_2$Et	5,7-dimethoxy-4-ethoxycarbonylcoumarin	47
4-^{t}Bu-phenol	H—≡—CC$_2$Et	6-^{t}Bu-coumarin	51

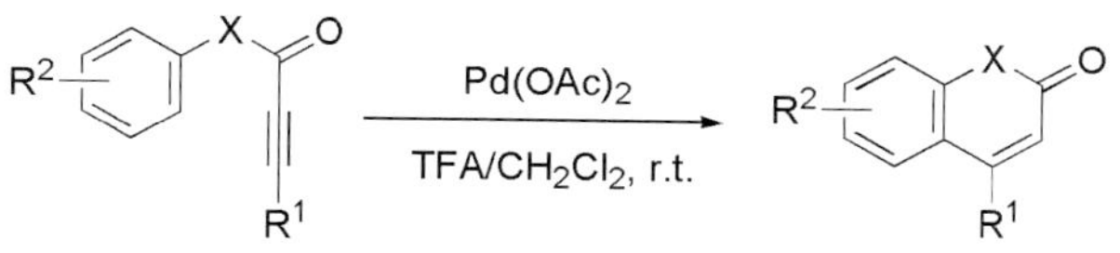

Scheme 7. Intramolecular hydroarylation of alkynes.

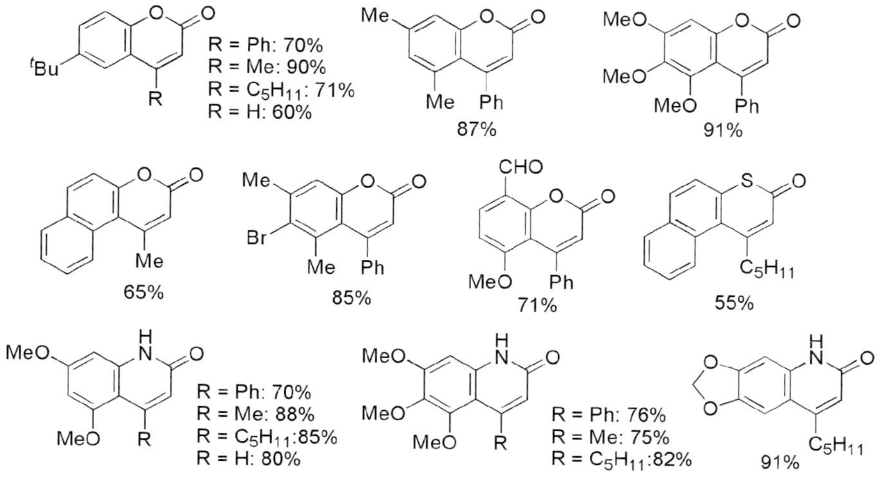

Scheme 8. Typical heterocycles from intramolecular hydroarylation.

Table 4. Pd(II)-Montmorillonite-catalyzed hydroarylation of alkynes or alkenes.

Aromatics	Alkynes or alkenes	Products	Yield (%)
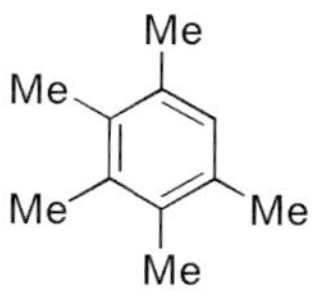 Me,Me,Me,Me,Me-benzene	Ph—≡—CO$_2$Et	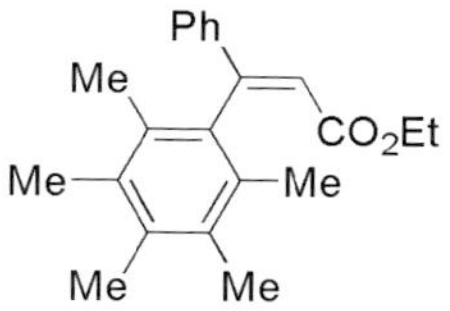 tetramethylbenzene with C(Ph)=CH-CO$_2$Et	83
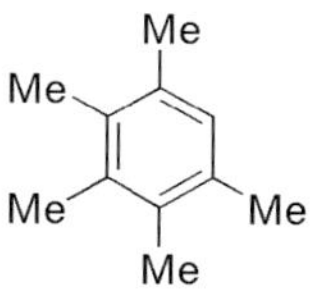 Me,Me,Me,Me,Me-benzene	Ph—≡—Ph	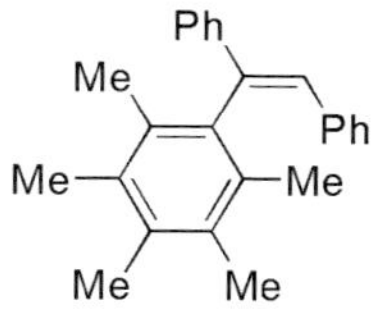 tetramethylbenzene with C(Ph)=CH-Ph	78
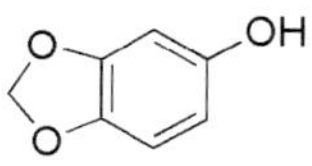 methylenedioxyphenol	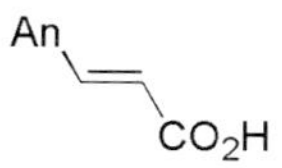 An—CH=CH—CO$_2$H	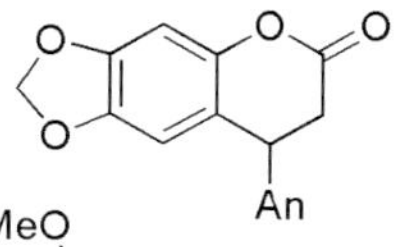 methylenedioxy-chromanone with An	100
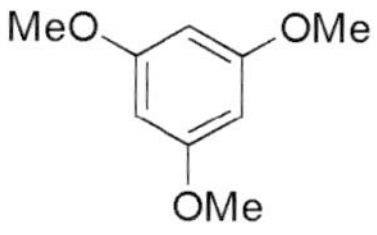 3,5-dimethoxy-methoxybenzene	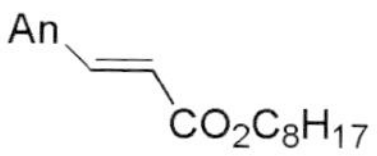 An—CH=CH—CO$_2$C$_8$H$_{17}$	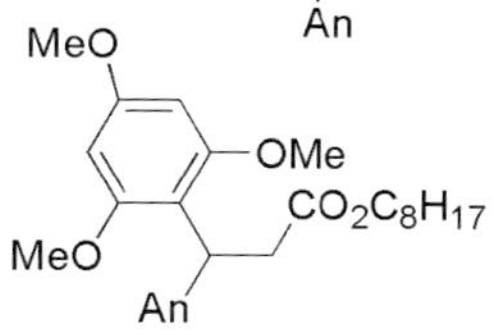 trimethoxybenzene with CH(An)-CH$_2$-CO$_2$C$_8$H$_{17}$	78

Experimental

Hydroarylation of Ethyl Propiolate by Pentamethylbenzene. Ethyl (2Z)-3-(pentamethylphenyl)propenoate

Ethyl propiolate (0.475 g, 4.9 mmol) was added, with stirring, to a cold mixture of pentamethylbenzene (1.51 g, 10.2 mmol), Pd(OAc)$_2$ (10 mg, 0.045 mmol), TFA (4 mL), and CH$_2$Cl$_2$ (1 mL) on an ice–water bath and the mixture was stirred for 5 min. The mixture was then warmed to room temperature and stirred for 3 h. The reaction mixture was poured into saturated NaCl solution and extracted with ether. The extract was washed with saturated NaCl, neutralized with Na$_2$CO$_3$ solution, dried over anhydrous Na$_2$SO$_4$, and the solvent was evaporated in vacuo. Two compounds were isolated by column chromatography on silica gel with 8:1 hexane–AcOEt as eluent to give ethyl (2Z)-3-(pentamethylphenyl)propenoate, 1.05 g (88%), as white crystals, mp 71.8–72.3 °C: ^{1}H NMR (300 MHz, CDCl$_3$) δ 1.09 (t, J = 7.2 Hz, 3H, CH$_3$), 2.14 (s, 6H, 2CH$_3$), 2.19 (s, 6H, 2CH$_3$), 2.22 (s, 3H, CH$_3$), 4.01 (q, J = 7.2 Hz, 2H, OCH$_2$), 6.12 (d, J = 12.0 Hz, 1H, vinyl), 7.12 (d, J = 12.0 Hz, 1H, vinyl). ^{13}C NMR (75 MHz, CDCl$_3$) δ 13.84, 16.22, 16.60, 17.46, 59.61, 122.01, 129.61, 131.72, 133.11, 133.79, 146.32, 165.23. IR (CHCl$_3$, cm^{-1}) 1724 (C=O). Ethyl (2E,4Z)-4-(ethoxycarbonyl)-5-(pentamethylphenyl)-2,4-pentadienoate, 0.05 g (6%), as colorless crystals with mp 60.8–62.0 °C: ^{1}H NMR (300 MHz, CDCl$_3$) δ 0.87 (t, J = 6.9 Hz, 3H, CH$_3$), 1.32 (t, J = 6.9 Hz, 3H, CH$_3$), 2.12 (s, 6H, 2CH$_3$), 2.18 (d, 6H, 2CH$_3$), 2.22 (s, 3H, CH$_3$), 3.97 (q, J = 6.9 Hz, 2H, OCH$_2$), 4.25 (q, J = 6.9 Hz, 2H, OCH$_2$), 6.17 (d, J = 15.9 Hz, 1H, vinyl), 7.25 (d, 1H, vinyl), 7.50 (d, J = 15.9 Hz, 1H, vinyl). ^{13}C NMR (75 MHz, CDCl$_3$) δ 13.40, 14.26, 16.19, 16.67, 17.78, 60.47, 60.53, 120.41, 130.37, 132.04, 132.44, 133.98, 134.44, 141.48, 145.25, 166.01. IR (CHCl$_3$, cm^{-1}) 1734 (C=O), 1716 (C=O).

Pd(II)-catalyzed Reaction of 3,4,5-Trimethoxyphenol with Ethyl Phenylpropiolate. 5,6,7-Trimethoxy-4-phenylcoumarin

Ethyl phenylpropiolate (0.238 g, 2 mmol) was added to a cold mixture of Pd(OAc)$_2$ (5.6 mg, 0.025 mmol), 3,4,5-trimethoxyphenol (0.184 g, 1 mmol), and TFA (1.5 mL) on an ice–water bath. After stirring at the same temperature for 5 min, the mixture was stirred at room temperature for 6 h. The reaction mixture was then neutralized with aqueous NaHCO$_3$ solution, extracted with CH$_2$Cl$_2$, and the extract was dried over anhydrous Na$_2$SO$_4$ and concentrated. Flash column chromatography on silica gel with hexane–AcOEt as eluent gave 0.258 g (87%) 5,6,7-trimethoxy-4-phenylcoumarin as colorless crystals, mp 146.8–147.7 °C: ^{1}H NMR (300 MHz, CDCl$_3$) δ 3.26 (s, 3H, OCH$_3$), 3.79 (s, 3H, OCH$_3$), 3.94 (s, 3H, OCH$_3$), 6.07 (s, 1H, vinyl), 6.73 (s, 1H, ArH), 7.34 (m, 2H, Ph), 7.40 (m, 3H, Ph). ^{13}C NMR (75 MHz, CDCl$_3$) δ 56.22, 60.88, 61.01, 96.23, 107.23, 114.02, 127.17, 127.46, 127.98, 138.98, 139.41, 151.04, 151.64, 155.35, 156.86, 160.58. IR (KBr, cm^{-1}) 1726 (C=O).

1.3.2.6 Palladium-catalyzed Oxidative Vinylation
Piet W. N. M. van Leeuwen and Johannes G. de Vries

1.3.2.6.1 Introduction
The reaction sequence in the vinylation of aromatic halides and vinyl halides, i.e. the Heck reaction, is oxidative addition of the alkyl halide to a zerovalent palladium complex, then insertion of an alkene and completed by β-hydride elimination and HX elimination. Initially though, C–H activation of a C–H alkene bond had also been taken into consideration. Although the Heck reaction reduces the formation of salt by-products by half compared with cross-coupling reactions, salts are still formed in stoichiometric amounts. Further reduction of salt production by a proper choice of aryl precursors has been reported (Chapter III.2.1) [1]. In these examples aromatic carboxylic anhydrides were used instead of halides and the co-produced acid can be recycled and one molecule of carbon monoxide is sacrificed. Catalytic activation of aromatic C–H bonds and subsequent insertion of alkenes leads to new C–C bond formation without production of halide salt byproducts, as shown in Scheme 1. When the hydroarylation reaction is performed with alkynes one obtains arylalkenes, the products of the Heck reaction, which now are synthesized without the co-production of salts. No reoxidation of the metal is required, because palladium(II) is regenerated.

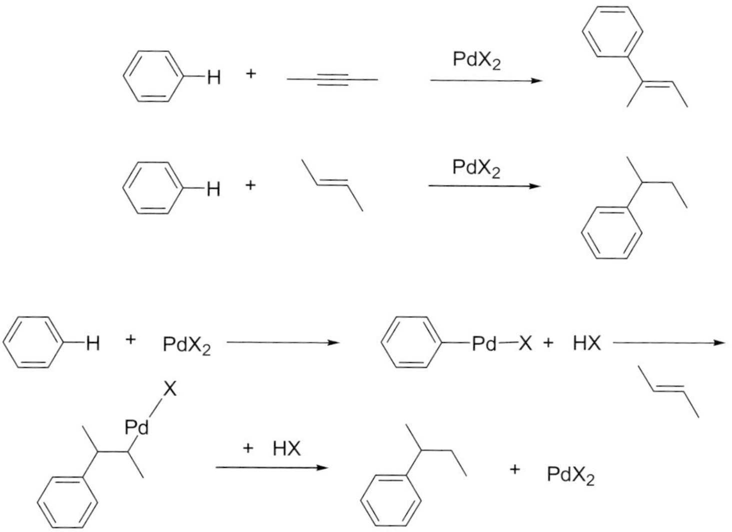

Scheme 1. Hydroarylation of alkenes and alkynes (see Chapter III.1.3.2.5).

The reactions shown in Scheme 1 require activation of the aromatic C–H bond by a metal and subsequent insertion of an alkene or alkyne in the aryl–carbon palladium bond (Chapter III.1.3.2.5). C–H activation has been the topic of many studies since the 1960s and several metal complex systems are known to induce this reaction [2]. Here we will focus on activation by electrophilic metal systems based on palladium.

1.3.2.6.2 Palladium Catalyzed Oxidative Vinylation of Arenes

Initially, a stoichiometric reaction was observed. This involved arylation of alkenes giving alkenes, rather than alkanes as presented in Scheme 1. The sequence is shown in Scheme 2 in more detail. After insertion of the alkene β-H elimination occurs giving the arylalkene product and palladium(0). This reaction was first reported by Fujiwara in 1967 [3]. The catalytic version requires the re-oxidation of palladium(0) to palladium(II) by an oxidant [4a]. This oxidative, catalytic coupling of arenes and alkenes was published in 1969 by Fujiwara. Copper and silver acetate combined with dioxygen as the reoxidizing agents gave turnover numbers up to 5 for the formation of *trans*-stilbene from styrene and benzene [4a]. Turnover numbers up to 14 were achieved by Tsuji in 1984 by using *t*-butyl perbenzoate as the oxidant [4b]. Palladium benzoate is used as the catalyst, and acetic acid and benzene as the solvent; typically the reaction is performed at 100 °C. Similarly, furans were coupled with acrylates $H_2C=CHCO_2R$ (R = Me, Et) to give 3-(*E*)-furyl propenoates. In the absence of olefins, Pd-catalyzed benzoxylation of C_6H_6, PhMe, and PhCl occurred to give the corresponding aryl benzoates. The regioselectivity showed that metalation was involved rather than a radical reaction. With the use of benzoquinone as stabilizing agent and *t*-BuOOH as the oxidant TON as high as 280 (90 °C, 15 h, AcOH solvent) have been reported [5]. Other oxidants such as H_2O_2, MnO_2, or AgO_2CPh were less effective.

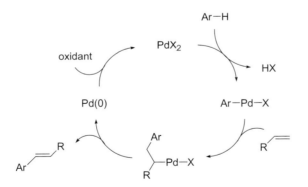

Scheme 2. Oxidative arylation of alkenes by electrophilic palladium(II).

1.3.2.6.3 ortho-Palladation Followed by Vinylation or Alkylation

Palladation of an arene is a very facile reaction when, before the C–H activation step, coordination of palladium to a nearby ligand functionality in the molecule occurs. The first report of a stoichiometric intramolecular palladation is probably the reaction of diazobenzene and palladium chloride by Cope in 1967 [6a]. Intramolecular palladation is a widespread reaction that has often been used as a starting point for synthesizing new molecules using insertion reactions in the arylpalladium bond of unsaturated substrates [6b, c]. When the metal-to-ligand interaction is strong, as in imine and pyridine complexes, the reaction remains stoichiometric and catalysis is hard to achieve. Aromatic amides lead to weaker interactions between the amide ligand and palladium and under these conditions catalytic conversions have been reported. Orthovinylation of aromatic amides as a stoichiometric reaction was reported by Horino in 1981 [7]. The cyclopalladated intermediate has been isolated and characterized (Scheme 3).

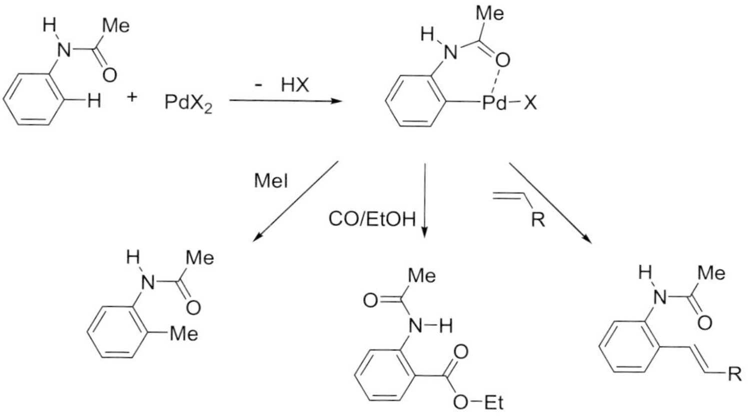

Scheme 3. Stoichiometric palladation and further reactions [7].

The highest yields for this reaction were obtained with an unsubstituted amide and with one containing a *p*-chloro substituent, with vinyl methyl ketone as the alkene. Use of styrene led to lower yields. The reaction of the intermediate acetanilidopalladacycle with methyl iodide, which produces *N-o*-tolylacetamide is interesting mechanistically [7b]. Stoichiometrically, full conversion was achieved, but attempts with catalysts gave only 10 turnovers and required the use of silver acetate to convert palladium iodide into the acetate, because palladium iodide does not have enough activity for metalation.

Catalytic results for the vinylation reaction were obtained by us by using rapid screening experiments with a parallel synthesis apparatus [8]. This setup enabled us to perform systematically a large number of experiments in a short time. From these experiments, in which catalyst, Ar–H, olefin, solvent, and oxidant were varied, a single lead emerged known already as the stoichiometric reaction reported above, the oxidative vinylation of acetanilides. Relevant results are summarized in Table 1. Aniline derivatives **1** and **2** are not reactive under the conditions tested (entries 1 and 2). In contrast, the reaction of acetanilide (**3**) with *n*-butyl acrylate proceeds smoothly and selectively at 80 °C using 2 mol% Pd(OAc)$_2$ and benzoquinone (BQ) as oxidant to yield *n*-butyl *E*-(2-acetamido)cinnamate **4** (Eq. 2) exclusively, albeit in moderate yield. No formation of the 3- or 4-substituted product was observed in any of the reactions performed, showing the importance of the ortho-directing and stabilizing effect of the amide group, as observed by Horino.

Table 1. Coupling of aniline derivatives with n-butyl acrylate using Pd(OAc)$_2$[a].

Entry	Substrate		Temp (°C)	Solvent	Additives	Yield (%)
1	3,5-dimethyl-N-methylaniline	1	80	HOAc	None	0
2	3,5-dimethoxyaniline	2	80	HOAc	None	0
3	acetanilide	3	80	HOAc	None	35
4			20	HOAc	None	54
5			20	HOAc/Toluene[b]	TsOH[c]	72
6			20	HOAc/NMP[b]	TsOH[c]	23
7			20	CF$_3$COOH	None	61
8			20	HOAc/Toluene[b]	NaCl[c]	0
9[d]			20	HOAc/Toluene[b]	TsOH[c]	29

[a] Substrate (3.0 mmol), n-butyl acrylate (3.3 mmol), Pd(OAc)$_2$ (0.06 mmol), BQ (3.0 mmol) in 6.75 mL of solvent. Yields are isolated yields
[b] 2:1 ratio (v/v). NMP = N-methylpyrrolidinone
[c] 1.5 mmol
[d] Reaction performed using H$_2$O$_2$ as the oxidant

Remarkably, reducing the reaction temperature to 20 °C results in higher yields (Table 1, entries 3 and 4). Acetic acid is the solvent of choice and using mixtures (up to 1:1 v/v) of HOAc with CH$_2$Cl$_2$ or THF does not affect the yields to a large extent. No C–H activation of aromatic solvents, e.g. toluene, is observed under these conditions. Coordinating solvents, for example NMP, hamper the catalytic reaction (entry 6), because the solvent competes successfully for the coordination site with the substrate. The presence of a substoichiometric amount (0.5–1.0 equiv.) of para-toluenesulfonic acid (TsOH) has a large beneficial effect, resulting in 72% isolated yield when acetanilide is used as the substrate (Table 1, entry 5). Fujiwara and coworkers showed that the acidity of the solvent can have a large influence on the reaction rate in their Pd(II)-catalyzed hydroarylation of alkenes, CF$_3$COOH being much more effective than HOAc. These observations were explained in terms of greater electrophilicity of the Pd(II) center when OAc$^-$ was replaced by CF$_3$COO$^-$, resulting in faster metalation of the aromatic C–H bond [9]. In our work, employing CF$_3$COOH as solvent (without added TsOH) results in similar catalyst performance to HOAc/TsOH (Table 1, entry 7).

Addition of inorganic acids (HCl, H$_2$SO$_4$) has a large detrimental effect on catalyst performance. Apparently, halides block the catalytic cycle by coordinating to the Pd(II) center, thereby reducing its electrophilicity (entry 8). Furthermore, the presence of halide anions can favor protonolysis of Pd–alkyl bonds over β-H elimination, as was studied by Zhang et al. [10]. The role of halide ions as a ligand were studied in stoichiometric reactions of arylpalladium reagents and Pd(II)-catalyzed reaction of phenylmercuric acetate with a variety of allylic compounds. The halide ion was found to inhibit β-H elimination and to promote β-heteroatom elimination in acidic media. In this reaction a C–Pd intermediate with a β-heteroatom (including halogen, acetoxy, alkoxy, and hydroxyl groups) gives only the β-heteroatom elimination product in the presence of halide ions, and β-H elimination is effectively blocked [10].

Use of other oxidants, e.g. hydrogen peroxide (Table 1, entry 9) or Cu(II)(OAc)$_2$, give conversions that are significantly lower than when benzoquinone is used. The role of the BQ can be twofold. It serves as the oxidant, resulting in the formation of Pd(II) and hydroquinone. This reaction is known to be accelerated by acid [11]. In addition, BQ can act as a ligand, stabilizing the low-valent Pd species present during the catalytic cycle [12]. Addition of more than one equivalent of BQ does not improve the yield significantly.

Table 2. Coupling of substituted anilide derivatives with *n*-butyl acrylate using Pd(OAc)$_2$ in acetic acid.[a]

Entry	Substrate		Product		Yield (%)
1	(4-methyl anilide)	5	(product with COOBu)	13	85
2	(3-methyl anilide)	6	(product with COOBu)	14	91
3	(2-methyl anilide)	7	(product with COOBu)	15	38[b]
4	(4-MeO anilide)	8	(product with COOBu)	16	62
5	(4-CF$_3$ anilide)	9	(product with COOBu)	17	(29)

Table 2. Continued

Entry	Substrate		Product		Yield (%)	
6	[N-methyl-N-phenylacetamide]	10	[ortho-alkenylated product with COOBu]	18	0	
7	[formanilide]	11	[ortho-alkenylated formanilide with COOBu]	19	(26)	
8	[benzanilide]	12	[ortho-alkenylated benzanilide with COOBu]	20	55	

[a] Reactions performed as in Table 1, entry 5. Yields in parentheses were determined by GC
[b] Determined by ^{1}H NMR

Substituents on the aromatic moiety of the acetanilide substrate substantially affect the efficiency of the coupling reaction (Table 2). As expected, ortho-substitution hampers the reaction. It has been reported that the palladation of ortho-substituted acetanilides does not occur [7] or occurs at elevated temperatures only [7b]. With N-m-tolylacetamide (6) the reaction efficiency is enhanced to give a 91% yield of 14. Interestingly, reaction of N-methylanilide (10) gave no conversion at all. Formanilide and benzanilide (entries 7, 8) can be applied, although the yields are low to moderate.

The reaction also exhibits a large electronic dependence. Competition experiments with a series of 4-substituted acetanilides showed that electron-rich arenes react significantly faster ($k_{obs}(5) > k_{obs}(8) \approx k_{obs}(3) >> k_{obs}(9)$), although linear Hammett plots could not be obtained. A kinetic isotope effect is observed for the reaction ($k_H/k_D = 3$). Studies performed by Ryabov et al. have revealed that stoichiometric reaction of [(C$_6$H$_4$NHC(O)CH$_3$)Pd(II)(OAc)]$_2$ complexes with styrenes can be acid-catalyzed, and that protonation is likely to occur at the bridging acetate ligands [13]. This protonation is followed by alkene coordination and then rate-limiting insertion of alkene into the palladium-carbon bond. We tested the dimeric ortho-palladated anilide complexes in the reaction with n-butyl acrylate and found that the rate of the reaction is at least an order of magnitude higher when using the preformed complexes compared with the catalysts generated in situ. These results suggest a reaction pathway via slow electrophilic attack of cationic [PdOAc]$^+$ species on the π-system of the arenes [5].

1.3.2.6.4 Additives

Heteropolyoxametalates are often used in combination with palladium salts as catalysts in oxidation processes using dioxygen as the oxidant. Indeed, the oxidative coupling reaction of benzenes with alkenes was also successfully achieved by use of the Pd(OAc)$_2$/molybdovanadophosphoric acid (HPMoV)/O$_2$ system [14a]. For example, reaction of benzene with ethyl acrylate using this catalytic system in acetic acid afforded ethyl cinnamate as a major product in satisfactory yield. Typically, the reaction is conducted in acetic acid at 90 °C under 1 bar of O$_2$. After 6 h the TON is 15. This number was recently improved to 121 [14b].

In a recent review it was argued that such additives of copper, benzoquinone, and HPMOV are not really needed; all that is needed is the presence of oxidation-resistant ligands that prevent palladium metal formation [15]. Indeed, activation of the C–H bond is not as slow as, for example, the Wacker reaction of ethene in which reoxidation of palladium must be performed by copper oxidation, although in this catalytic system the additives may still play a role in stabilizing the intermediate low-valent palladium species and thus prevent catalyst decomposition. This thesis was corroborated by the work of de Vos and Jacobs, who showed that addition of benzoic acid to the oxidative arylation reaction in the presence of oxygen led to superior results in the coupling of a variety of substituted arenes with acrylates, cinnamates, and α,β-unsaturated ketones. Very good yields and TON up to 762 were obtained at 90 °C. A mixture of the o, m, and p isomers is obtained if substituted arenes are used [16].

1.3.2.6.5 Intramolecular Reactions

In an intramolecular version, oxidative coupling of indoles carrying butenyl groups at the 3-position resulted in cyclization to give annulated derivatives (Scheme 4) [17]. Stoltz reported use of oxygen or benzoquinone as the most effective oxidants, ethyl nicotinate as a stabilizing agent (perhaps as a ligand), and 10 % of palladium acetate as the catalyst. Turnover numbers are modest but, interestingly, the use of chiral centers in the side chain leads to stereospecific conversions with the formation of quaternary, chiral carbon centers. Likewise, allyl aryl ethers of electron-rich arenes lead to benzofurans (Scheme 4) [18]. Because the arene is very electron-rich one might consider a different mechanism – nucleophilic attack of the arene at the alkene activated by coordination to palladium. The stereospecificity of such a reaction was used to show that the reaction indeed involves C–H activation as the first productive step.

1.3.2.6.6 Recent Developments

A few recent examples of related C–C bond-forming reactions, all involving a palladium-catalyzed C–H activation step at arenes, will be mentioned. Salts are produced in these reactions, or acetic acid, as in the first example. Allylation of indoles at the 3-position was achieved by using palladium acetate, and bipyridine and allylic acetates as the reactants (Scheme 5) [19].

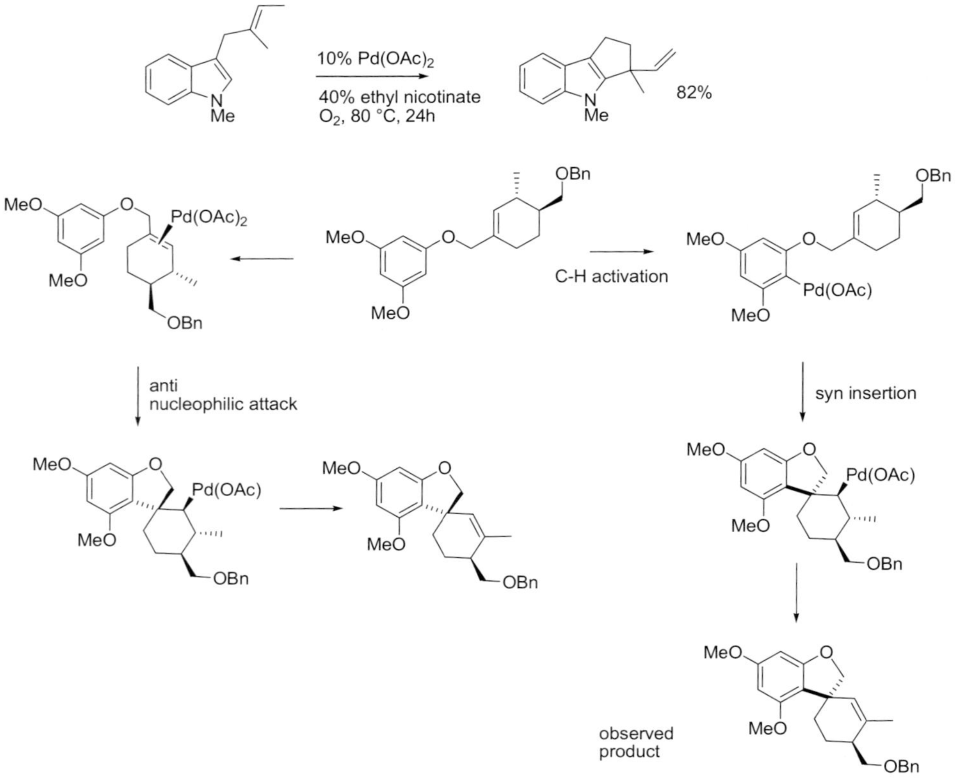

Scheme 4. Annulation of indole and the stereochemistry of oxidative dihydrobenzofuran formation.

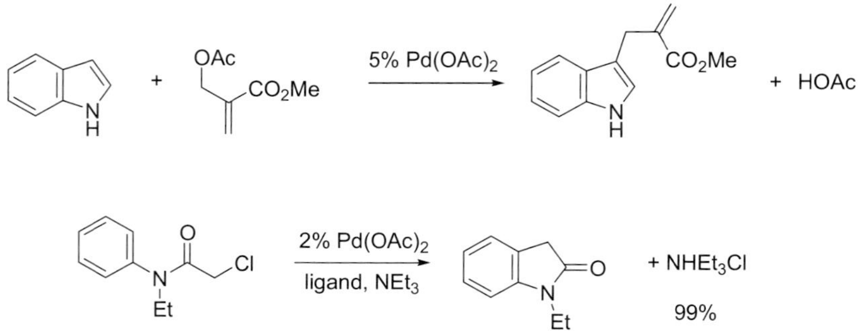

Scheme 5. C–H transformations not requiring oxidants, with acid or salt formation.

The cyclization of α-chloroacetanilides reported by Buchwald may also involve C–H activation (also Scheme 5). As mentioned above, acetanilides are among the most reactive substrates in this reaction. The presence of a base (Et$_3$N), and 2-PhC$_6$H$_4$PBu$_2$ as the ligand, are conditions more typical of cross-coupling of halides and therefore the sequence may start this way, followed by intramolecular

C–H activation. Turnovers up to 50 were achieved [20]. The cyclization is highly regioselective, obviating the need for prefunctionalized arenes.

A normal Heck reaction, thus producing salt, may be accompanied by an additional C–H activation step, as found by Gallagher et al. [21]. Biaryl bromides undergo the Heck reaction to give both the expected products ("retention") and "crossover" products derived from migration of palladium and net transfer of reactivity from one aryl ring to the other (Scheme 6). Under the conditions used crossover is increasingly favored when electron-deficient arenes are involved. Crossover products derived from transfer on to the pyridine ring have also been observed.

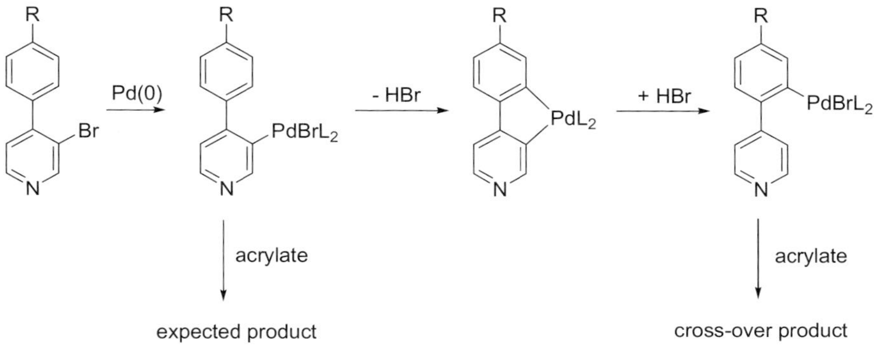

Scheme 6. "Cross-over" via intramolecular C–H activation.

A reaction involving C–H activation requiring neither reoxidation nor salt separation has been reported by Larock (Scheme 7) [22]. Addition to nitriles of arylpalladium intermediates obtained via C–H activation gives palladium ketimidates which react with acid to regenerate palladium salt, the catalyst, and ketimines (or their protonated conjugates). The co-produced ammonium trifluoroacetate formed after hydrolysis to liberate diaryl ketones does require regeneration. The maximum turnover number was 7. The example shows that in the absence of β-hydrogen atoms, maintaining palladium in a divalent state is also a promising approach to new catalytic conversions.

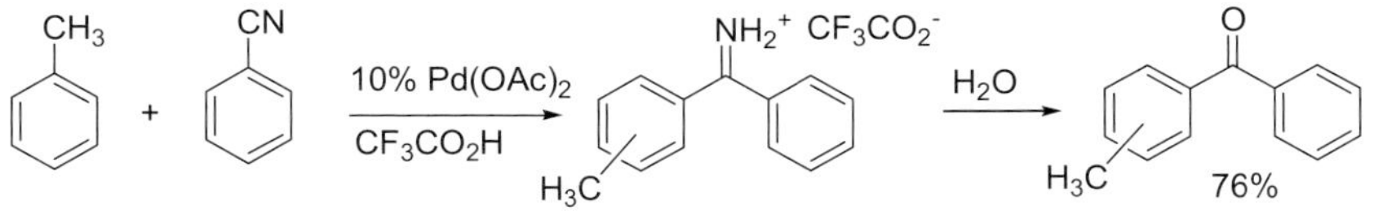

Scheme 7. Diaryl ketones via C–H activation.

1.3.2.6.7 Conclusions

From the examples above we can conclude that palladium-catalyzed oxidative vinylation of arenes has developed well beyond its inception by Fujiwara in the sixties. Quite good turnovers with O_2 as oxidant have now been achieved by use of additives such as benzoic acid. Although the reaction itself is not regioselective, leading to ortho, meta, and para mixtures with substituted benzenes, use of donating groups on the arene can induce highly selective orthometalation reactions leading to single products. Intramolecular versions are also high selective. Although some mechanistic work has been performed [23] the picture is not entirely clear. The need for additives such as benzoquinone, benzoic acid or ethyl nicotinate suggests that even under these oxidative conditions Pd(0) may form, leading to the formation of clusters that need stabilization to prevent formation of palladium black. This reaction is an obvious example of green chemistry, water being the only side-product. Thus, one can hope that in the near future this chemistry will be used for ton-scale production. In contrast, the alkylation reaction is much less well-developed. Here, the limitation would seem to be the choice of leaving group, because many leaving groups may inhibit the catalysis. We expect many new developments in these areas.

Experimental

General Procedure for Coupling Acetanilide Derivatives to *n*-Butyl Acrylate
In a typical experiment, 3.0 mmol anilide, 13.5 mg (0.06 mmol) $Pd(OAc)_2$, 324 mg (3.0 mmol) benzoquinone, and 286 mg (1.5 mmol) *p*-toluenesulfonic acid monohydrate are weighed into a one-necked round-bottomed flask equipped with a stirring bar. Next, 4.5 mL acetic acid is added, followed by a solution of 0.42 mL (3.0 mmol) *n*-butyl acrylate in 2.25 mL toluene. The flask is capped with a rubber septum and the mixture is stirred overnight. Samples of the mixture are taken, diluted with diethyl ether, washed with a saturated $NaHCO_3$ solution, dried over $MgSO_4$, and analyzed by GC or GC–MS. After 16 h the reaction mixture is diluted with 15 mL ether and carefully neutralized with 2.5 M NaOH solution. After extraction of the aqueous phase with 15 mL ether the combined organic phases are washed with water (15 mL), dried ($MgSO_4$), and evaporated in vacuo. The resulting solids are purified by column chromatography to yield the corresponding product as a white powder.

1.3.3
Minisci Radical Alkylation and Acylation
Ombretta Porta and Francesco Minisci

1.3.3.1 Introduction
The Friedel–Crafts alkylation and acylation are of very little, if any, synthetic interest when applied to heterocyclic aromatic bases; the substitution of protonated heterocycles by nucleophilic carbon-centered radicals is instead successful. This reaction, because of the dominant polar effect which is mainly related to the charge-transfer character of the transition state (Scheme 1), reproduces most of the aspects of the Friedel–Crafts aromatic substitution, but reactivity and selectivity are the opposite.

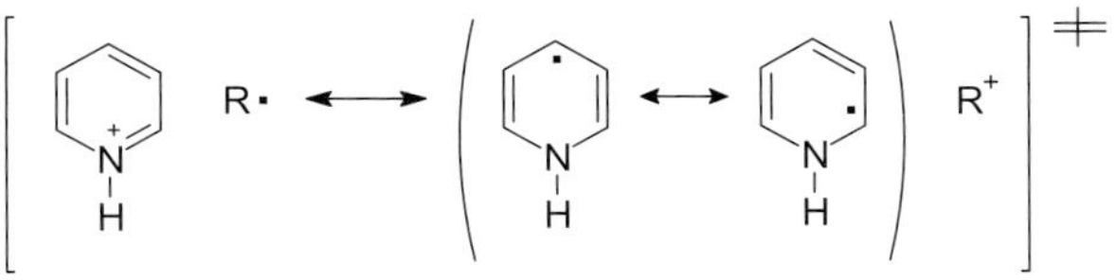

$$\left[\underset{\underset{H}{|}}{\underset{N^+}{\bigcirc}} \quad R\cdot \longleftrightarrow \left(\underset{\underset{H}{|}}{\underset{N}{\bigcirc}}^{\cdot} \longleftrightarrow \underset{\underset{H}{|}}{\underset{N}{\bigcirc}}_{\cdot} \right) \quad R^+ \right]^{\ddagger}$$

Scheme 1. Charge-transfer character of the transition state.

The parallelism with the Friedel-Crafts aromatic substitution arises because the more stable the carbocation, the more nucleophilic the corresponding radical will usually be. Thus, in principle, all the electrophilic species employed in the Friedel–Crafts reaction, when used as the corresponding radicals, should behave as nucleophiles in the selective substitution of heteroaromatic bases.

This reaction [1] is of synthetic interest for several reasons:

1. almost all carbonyl (R–Ċ=O, RO–Ċ=O, RNH–Ċ=O) and alkyl radicals without electron-withdrawing groups directly bonded to the radical center can be successfully introduced;
2. a wide variety of simple and cheap sources of these radicals are available starting from several of the most important classes of organic compounds;
1. all heteroaromatic bases (pyridines, quinolines, isoquinolines, acridines, diazines, benzodiazines, triazines, imidazoles, benzoimidazoles and benzothiazoles) and compounds of great biological interest (nucleosides, purines, and pteridines) with at least an α or γ position free are suitable substrates, and when all these positions are functionalized, ipso-substitution has occasionally been observed;
4. the reaction is highly chemo- and regioselective – substitution usually occurs at α and γ positions only.

Several reviews [2] have been published, covering the overall synthetic and mechanistic evolution of the reaction, starting from 1973.

1.3.3.2 Mechanism

The general reaction mechanism is shown in Scheme 2. The high rate constants (from 10^5 to 10^8 $M^{-1} s^{-1}$) for addition of nucleophilic radicals to protonated heterocycles [3] reduce other possible competitive side reactions of the radicals involved, strongly contributing to the synthetic interest.

With unprotonated heterocyclic bases, nucleophilic radicals either do not react (t-alkyl, benzyl, acyl, α-oxoalkyl, α-N-amidoalkyl) or react more slowly (3 to 6 orders of magnitude less) leading poor synthetic value (low yields and low chemo- and regioselectivity).

According to the FMO theory, the LUMO of the pyridinium cation has the highest coefficients at the carbon atoms in the α and γ positions and the dominant interaction in the radical addition occurs between the SOMO of the nucleophilic radical and the LUMO of the protonated heteroarene [3]. The rearomatization step of the radical cation **2** is very effective. In line with the mechanism of Scheme 2,

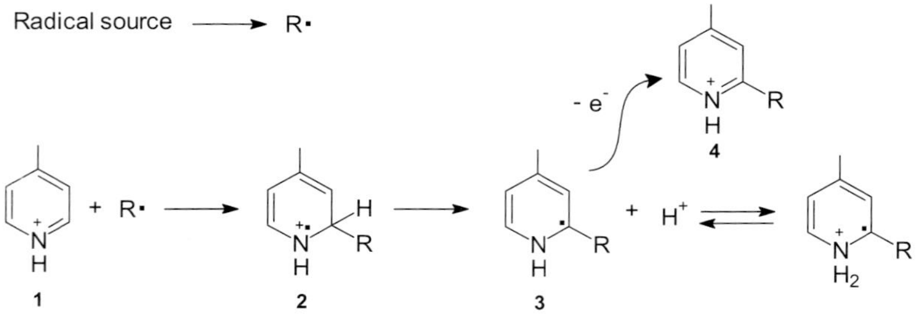

Scheme 2. Reaction mechanism of homolytic heteroaromatic substitution.

adduct **2** irreversibly loses a proton from the α C–H position affording the pyridinyl radical **3**, which being a weaker base (pK_a 2–3) than the corresponding dihydropyridine ($pK_a \approx 7$) has an high equilibrium concentration in not strongly acidic medium. Because α-aminoalkyl radicals are good reducing agents with ionization potentials (~5 eV) close to those of Li and Na, very mild oxidants, for example Ti(IV) [4], can rapidly and selectively oxidize pyridinyl radicals **3** to the final rearomatized product **4**. A great variety of thermal, photochemical and redox radical sources has been successfully utilized [2].

1.3.3.3 Scope, Limitations and Fundamental Examples

1.3.3.3.1 Radical Sources Involving Simultaneous Transformations at sp^2 and sp^3-Hybridized Carbon Atoms

A quite general method of synthesis is depicted in Scheme 3, in which heterocyclic aromatic C–H and aliphatic C–H transformations are simultaneously involved.

Scheme 3. Simultaneous sp^2 and sp^3 transformation.

1 is the protonated heteroaromatic base and **5** can be an alkane, alkene, alkylaromatic, alcohol, ether, aldehyde, amine, amide, etc. The oxidant is usually a species which generates oxygen- or nitrogen-centered radicals suitable for hydrogen abstraction from C–H bonds, e.g. O_2, H_2O_2, RO-OH, $(ArCOO)_2$, $ArCO_2$-OR, $(ROO)_2CO$, $S_2O_8^{2-}$, NH_2OH, NH_2OSO_3H, R_2NCl, R_3NO, $PhI(OAc)_2/N_3^-$. The redox metal salt catalysis, particularly suitable for the general Scheme 3, involves the couples Fe(III)/Fe(II), Cu(II)/Cu(I), Ag(II)/Ag(I), Ti(IV)/Ti(III), Co(III)/Co(II), and Ce(IV)/Ce(III).

Fundamental examples are shown in Eqs (1)–(8) of Table 1.

Table 1. Representative examples of simultaneous sp^2 and sp^3 transformation.

Table 1. Continued

$$\text{4-methylquinoline-N-H}^+ + CH_3CH_2OH + NH_2OH \xrightarrow{Ti(III)} \text{4-methyl-2-(1-hydroxyethyl)quinoline-N-H}^+ + NH_3 + H_2O \quad (7)$$

70% [6]

$$\text{quinoline-N-H}^+ + CH_3(CH_2)_3\text{-N(CH}_3)_2 \text{ (with C=O)} \xrightarrow{Fe(II)} \text{2-substituted quinoline-N-H}^+\text{-(CH}_2)_4\text{-N(CH}_3)_2 \quad (8)$$

35% [10]

+ 4-[(CH$_2$)$_4$-N(CH$_3$)$_2$]quinoline-N-H$^+$ +   2,4-bis[(CH$_2$)$_4$-N(CH$_3$)$_2$]quinoline-N-H$^+$

55%   10%

The yields reported in Eqs. (1)–(8) depend, with exception of Eqs. (6) and (8), on the starting heteroaromatic bases, but selectivity is higher because conversions are not always quantitative. Conversions can be increased by increasing the amounts of reagents. In Eq. (1) the aerobic oxidation is catalyzed by N-hydroxyphthalimide which generates the N-oxyl radical responsible for abstraction of hydrogen from benzaldehyde [5]. In Eqs. (2)–(4), the OH· and t-BuO· radicals, formed in classic Fenton-type processes, generate the nucleophilic radicals by hydrogen abstraction from formamide or s-trioxane. In Eq. (5) the Ph· and PhCOO· radicals abstract a hydrogen atom from cyclohexane producing the cyclohexyl radical. In Eq. (7) the $NH_3^{+\cdot}$ radical, formed by Ti(III) reduction of $^+NH_3OH$, abstracts a hydrogen atom from ethanol generating the ketyl radical.

The regioselectivity observed in Eq. (6) with diprotonated quinoxaline (reaction performed in 96% H_2SO_4) is of particular interest. Whereas in monoprotonated quinoxaline (Eq. 1) the C-2 carbon atom has the lowest electron density and substitution occurs at C-2 position only, in diprotonated quinoxaline (Eq. 6) the electron density of the equivalent C-2 and C-3 is as low as that of the equivalent C-6 and C-7 carbon atoms (NMR and INDO calculations) [9] and substitution occurs at both C-2 and C-6 positions. To minimize polysubstitution, the conversions in Eq. (6) were limited to about 50% (selectivity, based on the reacted bases, is >90% in any case). Another feature of Eq. (6) is the exceptional selectivity of hydrogen abstraction from the C-5 position of n-hexyl derivatives by the aminium radical i-Bu$_2$NH$^+$, generated from i-Bu$_2$NHCl$^+$ and Fe(II) [2].

A synthetic variation of the methodology reported in Eq. (6) is shown in Eq. (8), in which the selective intramolecular hydrogen abstraction by aminium radical occurs. Results similar to those of Eq. (8) were obtained with n-hexyl-N-chloroamine and Fe(II). Noteworthy is the selective intramolecular hydrogen abstraction from the C–H at the 4-position (Eq. (8), entropic effect) versus the selective intermolecular hydrogen abstraction from the C–H at the 5-position (Eq. (6), polar and enthalpic effect) [2]. With diprotonated quinoxaline, substitution occurs at both C-2 and C-6-positions.

Another method consists in generating an electrophilic carbon-centered radical (e.g. the $CH_3COCH_2\cdot$ radical from acetone, peroxydisulfate and Ag(I)) which, instead of reacting with the protonated heteroarene, readily adds to simple alkenes forming a radical adduct that, owing to its nucleophilic character, selectively reacts with the heterocyclic ring (Scheme 4) [2].

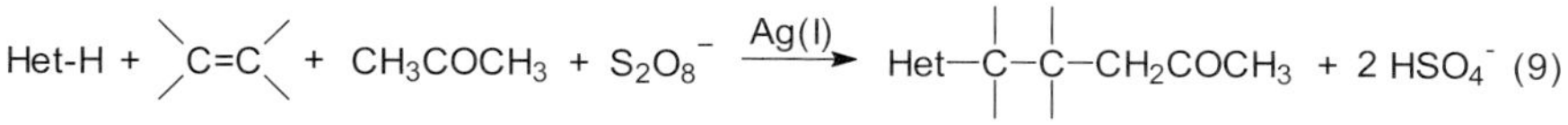

$$\text{Het-H} + \underset{\diagdown}{\overset{\diagup}{C}}=\underset{\diagup}{\overset{\diagdown}{C}} + CH_3COCH_3 + S_2O_8^- \xrightarrow{Ag(I)} \text{Het}-\underset{|}{\overset{|}{C}}-\underset{|}{\overset{|}{C}}-CH_2COCH_3 + 2\ HSO_4^- \quad (9)$$

Scheme 4. Heteroaromatic substitution by electrophilic radicals and alkenes.

1.3.3.3.2 Carboxylic Acids and Alcohols as Radical Sources

Carboxylic acids are the most general, versatile and useful source of carbon-centered radicals successfully used for selective alkylation and acylation of protonated heteroarenes. Alkyl, acyl, carbamoyl, and alkoxycarbonyl radicals have been obtained by oxidative decarboxylation of the corresponding acids with peroxydisulfate as an oxidant and Ag(I) as catalyst.

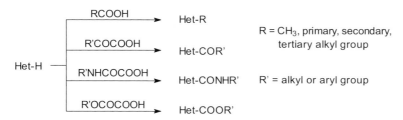

Scheme 5. Carboxylic acids as a sources of radicals.

Several other processes have been developed, however, to accomplish the oxidative decarboxylation of carboxylic acids: oxidation by Pb(OAc)$_4$, by iodosobenzenediacetate, and by Ag(II) salt generated in situ in a catalytic cycle from a variety of peroxides (benzoyl peroxide, percarbonate, perborate) [2] other than the already mentioned peroxydisulfate. Representative examples are shown in Eqs (9)–(12) of Table 2.

The heteroaromatic substitution reflects the Friedel–Crafts reaction with the opposite reactivity and selectivity. The synthetic advantages and disadvantages are also opposite to those of concern for the selectivity of monosubstitution – whereas introduction of a carbonyl group deactivates the aromatic ring toward further substitution in the electrophilic process, in contrast it activates the heteroaromatic ring in the homolytic reaction, favoring polysubstitution when the α and γ positions are free (Eq. 3); the opposite occurs in alkylation.

Selective monosubstitution with carbonyl-centered radicals is, however, possible by taking advantage of the reduced basicity because of introduction of a carbonyl group and use of a suitable acidic medium to make the starting base mainly protonated and the monosubstituted derivative much less protonated. The reac-

Table 2. Representative examples of acids and alcohols as radical sources.

quinolinium-H + $(CH_3)_3CCOOH$ + $S_2O_8^-$ $\xrightarrow{Ag(I)}$ 2-tBu-quinolinium-H + $2\,HSO_4^-$ + CO_2 (9)

9 (96%) [11]

pyrazinium-H + PhCOCOOH + $S_2O_8^-$ $\xrightarrow{Ag(I)}$ 2-COPh-pyrazinium-H + $2\,HSO_4^-$ + CO_2 (10)

100% [12]

benzothiazolium-NH + cy-C_6H_{11}NHCOCOOH + $S_2O_8^-$ $\xrightarrow{Ag(I)}$

2-CONHcy C_6H_{11}-benzothiazolium-NH + $2\,HSO_4^-$ + CO_2 (11)

71% [13]

pyrazinium-H + EtOCOCOOH + $S_2O_8^-$ $\xrightarrow{Ag(I)}$ 2-COOEt-pyrazinium-H + $2\,HSO_4^-$ + CO_2 (12)

83% [14]

4-methylquinolinium-H + CH_2OH-CH_2OH + $S_2O_8^-$ $\xrightarrow{Ag(I)}$ 2-CH_2OH-4-methylquinolinium-H + CH_2O + $2\,HSO_4^-$ (13)

10 (92%) [15]

tions in Eqs (10)–(12) were conducted in a two-phase system (i.e. H_2O–CH_2Cl_2) which favors monosubstitution by continuous extraction of the reaction products from the aqueous solution.

Conversions of the starting bases are almost quantitative for Eqs (9), (10), and (12) whereas for Eq. (11) it is 78 % (selectivity is >90 %).

Alcohols are not only source of ketyl radicals generated by hydrogen abstraction from the α-C–H position (Eq. (7), Table 1). Oxidation of alcohols with $Pb(OAc)_4$, $PhI(OAc)_2$, and $S_2O_8^{2-}$ with Ag(I) as catalyst produces alkoxy radicals (RO·) which may further undergo β-scission (Eq. 13), intramolecular hydrogen abstraction, or intra- and intermolecular addition to alkenes, generating a nucleophilic carbon-centered radical useful for heteroaromatic substitution (Scheme 6) [2].

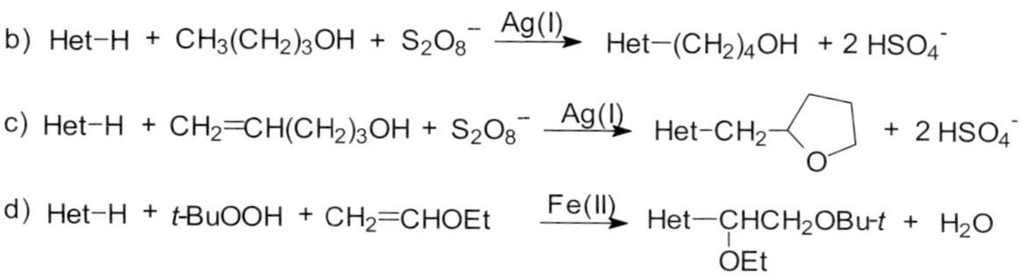

b) Het–H + $CH_3(CH_2)_3OH$ + $S_2O_8^-$ $\xrightarrow{Ag(I)}$ Het–$(CH_2)_4OH$ + 2 HSO_4^-

c) Het–H + $CH_2{=}CH(CH_2)_3OH$ + $S_2O_8^-$ $\xrightarrow{Ag(I)}$ Het–CH_2–⟨O⟩ + 2 HSO_4^-

d) Het–H + t-BuOOH + $CH_2{=}CHOEt$ $\xrightarrow{Fe(II)}$ Het–$\underset{OEt}{CHCH_2OBu\text{-}t}$ + H_2O

Scheme 6. Alcohols as sources of carbon-centered radicals in heteroaromatic substitution.

1.3.3.3.3 Alkyl Halides as Radical Sources

Alkyl iodides have been widely used for selective alkylation of heteroaromatic bases. The method is based on rapid iodine abstraction by aryl radicals (obtained from benzoyl peroxide or diazonium salts) or by a methyl radical (obtained from MeCOOH, t-BuOH, t-BuOOH, $(t\text{-}BuO)_2$, $(MeCOO)_2$, $MeSOMe/H_2O_2$, or $MeCOMe/H_2O_2$) [2]. An example is depicted in Eq. (14) of Table 3.

Perfluoroalkyl radicals are readily obtained by the same procedures but, owing to their electrophilic character, they are not suitable for selective heteroaromatic substitution. In the presence of simple alkenes, perfluoroalkyl radicals add very rapidly to the olefinic bond and the radical adduct, which has nucleophilic character, selectively reacts with the heterocyclic ring (Eq. 15).

Alkyl bromides have also been used as sources of alkyl radicals for selective heteroaromatic alkylation by using Bu_3SnH in combination with azobisisobutyronitrile or a variety of silanes and peroxides (Scheme 7) [2].

Het–H + R–Br + R'$_3$SiH + t-BuOOH $\longrightarrow$ Het–R + R'$_3$SiBr + t-BuOH + H_2O

Scheme 7. Alkyl bromides as sources of radicals in the heteroaromatic substitution

1.3.3.3.4 Various Sources of Radicals

A variety of different sources of radicals have been used in several heteroaromatic substitution reactions [2]; these include acyl peroxides, oxaziridines, thiohydroxamic Barton esters, the Gif reaction, alkyl xanthates, and ketones/H_2O_2 (Scheme 8).

Table 3. Representative examples of alkyl iodides as sources of radicals.

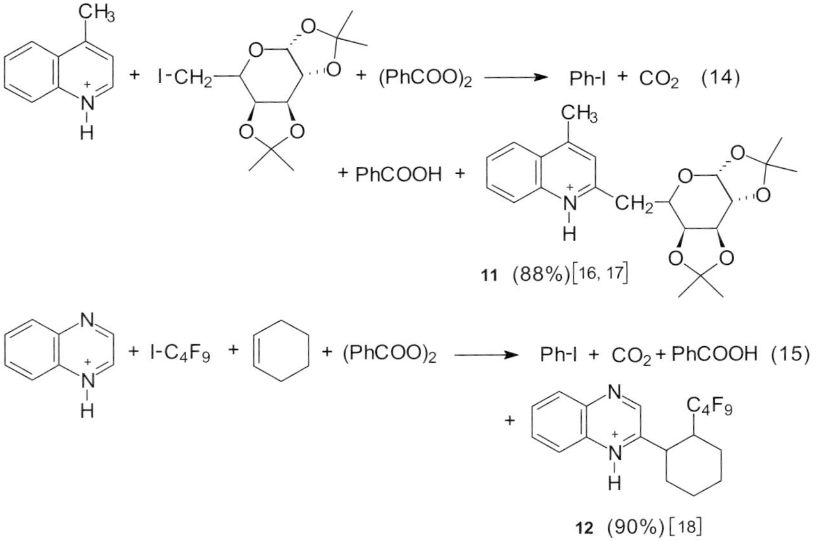

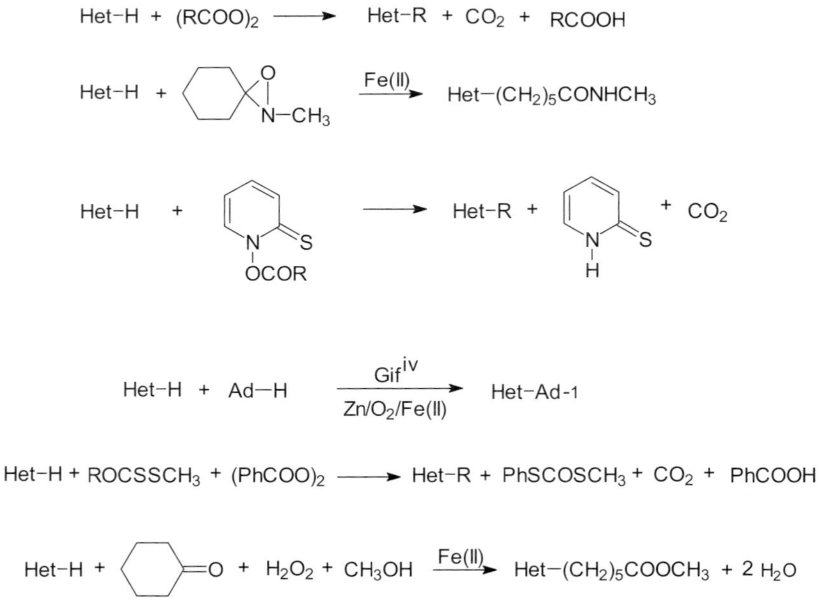

[Reaction 14: 4-methylquinolinium + I-CH$_2$-(sugar) + (PhCOO)$_2$ → Ph-I + CO$_2$ + PhCOOH + product **11** (88%) [16, 17]]

[Reaction 15: quinoxalinium + I-C$_4$F$_9$ + cyclohexene + (PhCOO)$_2$ → Ph-I + CO$_2$ + PhCOOH + product **12** (90%) [18]]

Het–H + (RCOO)$_2$ ⟶ Het–R + CO$_2$ + RCOOH

Het–H + (spirocyclohexane-oxaziridine with N–CH$_3$) $\xrightarrow{\text{Fe(II)}}$ Het–(CH$_2$)$_5$CONHCH$_3$

Het–H + (N-acyloxy-pyridinethione, OCOR) ⟶ Het–R + (pyridinethione-NH) + CO$_2$

Het–H + Ad–H $\xrightarrow[\text{Zn/O}_2\text{/Fe(II)}]{\text{Gif}^{iv}}$ Het–Ad-1

Het–H + ROCSSCH$_3$ + (PhCOO)$_2$ ⟶ Het–R + PhSCOSCH$_3$ + CO$_2$ + PhCOOH

Het–H + (cyclohexanone) + H$_2$O$_2$ + CH$_3$OH $\xrightarrow{\text{Fe(II)}}$ Het–(CH$_2$)$_5$COOCH$_3$ + 2 H$_2$O

Scheme 8. Different sources of radicals for the heteroaromatic substitution.

Experimental

Quinoline-2,4-dicarboxyamide [6] (6) (Eq. 3)

A solution of 10 mmol quinoline (1.3 g), 10 mmol H_2SO_4, 25 mmol H_2O_2 and 0.3 mmol of $FeSO_4 \cdot 7H_2O$ in 100 mL $HCONH_2$ was warmed, with stirring, for 4 h at 60 °C. The resulting solution was diluted with water and made basic with a 10 % NaOH solution. The solid formed was separated and the aqueous solution was extracted with CH_2Cl_2. On work-up of the organic extract 2.21 g of a solid were obtained (m.p. 271–274 °C). GC analysis (with lepidine-2-carboxyamide as internal standard) revealed 97 % yield of quinoline-2,4-dicarboxyamide. On crystallization from EtOH the reaction product (m.p. 278–279 °C) was identical with an authentic sample of **6** (m.p., GC, ^{1}H NMR and MS).

Reaction of Quinoxaline with 1-Hexylamine and N-Chlorodiisobutylamine [9] (7 and 8, X = NH_2) (Eq. 6)

A solution of N-chlorodiisobutylamine (50 mmol) in 50 mL 96 % H_2SO_4 was added dropwise (3 h) with vigorous stirring at r.t. to a mixture of quinoxaline (50 mmol), finely powdered $FeSO_4 \cdot 7H_2O$ (5 mmol), and 1-hexylamine (150 mmol) in 60 mL 96 % H_2SO_4. of Additional $FeSO_4 \cdot 7H_2O$ (10 mmol) was added in portions during the reaction. After 2 h the solution was poured on to ice (500 g), made alkaline with a 12 M NaOH solution, extracted with $CHCl_3$ and analyzed by GC. Unreacted quinoxaline (50 %), isomer **7** (12 %), and isomer **8** (33 %) were subsequently isolated by flash column chromatography on silica gel (EtOAc–MeOH, 7:3). Isomer **7**: b.p. 150 °C (0.5 mmHg); ^{1}H NMR ($CDCl_3$) δ 1.42 (d, 3H, CH_3), 1.3–2.2 (m, 6H, $-(CH_2)_3-$), 3.17 (sext, 1H, CH), 3.42 (t, 2H, CH_2N), 7.7–8.2 (m, 4H, H-5, H-6, H-7, H-8), 8.76 (s, 1H, H-3); MS, m/e 248 (M^+), 213, 185, 171, 158, 144, 131. Isomer **8**: b.p. 160 °C (0.5 mmHg); ^{1}H NMR ($CDCl_3$) δ 1.40 (d, 3H, CH_3), 1.2–2.2 (m, 6H, $-(CH_2)_3-$), 2.97 (sext, 1H, CH), 3.43 (t, 2H, CH_2N), 7.65 (d, 1H, H-7), 7.90 (d, 1H, H-5); 8.10 (d, 1H, H-8), 8.82 (s, 2H, H-2, H-3); MS, m/e 248 (M^+), 213, 185, 171, 158, 144, 136.

2-tert-Butylquinoline (9) [11] (Eq. 9)

A mixture of 2.5 mmol quinoline, 7 mmol pivalic acid, 0.2 mmol $AgNO_3$, 5 mmol $(NH_4)_2S_2O_8$, and 5 mmol H_2SO_4 in 25 mL of water and 25 mL of chlorobenzene was heated under reflux for 2 h. The aqueous solution was made basic with NaOH, the organic solvent was separated, and the aqueous solution was extracted with CH_2Cl_2. GC analysis of the organic extract (with quinaldine as internal standard) revealed the presence of 2.4 mmol **9** and 0.1 mmol unreacted quinoline. Compound **9** was isolated as pure liquid by flash column chromatography on silica gel (hexane–EtOAc, 6:1). ^{1}H NMR (500 MHz, $CDCl_3$) δ 1.51 (s, 9H, $3CH_3$), 7.28 (d, 1H, J = 8.2 Hz, H-3), 7.50 (1H, ddd, J = 1.4, 6.8, 8.2 Hz, H-6), 7.70 (1H, ddd, J = 1.4, 6.9, 8.3 Hz, H-7), 7.79 (1H, dd, J = 1.6, 8.3 Hz, H-5), 8.06 (1H, d, J = 8.3 Hz, H-8), 8.09 (1H, d, J = 8.3 Hz, H-4); MS, m/e 185 (M^+), 170, 155, 125.

2-Hydroxymethyllepidine (10) [15] (Eq. 13)

A solution of 20 mmol lepidine, 20 mmol CF_3COOH, 80 mmol $(NH_4)_2S_2O_8$, and 2 mmol $AgNO_3$ in 75 mL water and 75 mL $(CH_2OH)_2$ was heated under reflux for 3 h. The aqueous solution was made basic with NaOH and extracted with CH_2Cl_2. GC analysis (with quinoline as internal standard) of the organic extract revealed the presence of 1.3 mmol unreacted lepidine and 18.5 mmol **10**. Flash column chromatography on silica gel (hexane–EtOAc, 2:1) gave 17.1 mmol pure **10**: m.p. 85 °C; ^1H NMR (CDCl$_3$) δ 2.60 (s, 3H, 4-CH$_3$), 4.91 (s, 2H, CH$_2$O), 7.16 (s, 1H, H-3), 7.4–7.7 (m, 2H, H-5 and H-7), 7.9–8.1 (m, 2H, H-6 and H-8); MS, m/e 173 (M$^+$), 144, 128, 115.

6-Deoxy-6-(4′-methyl-2′-quinolyl)-1,2:5,6-di-O-isopropyline-α-d-galactopyranose (11) [16] (Eq. 14)

A solution of 10 mmol lepidine, 15 mmol iodosugar, 27 mmol $(PhCOO)_2$, and 15 mmol CF_3COOH in 100 mL of CH_3CN was heated under reflux for 8 h. The solution was poured into cold 10% aqueous Na_2CO_3 solution and extracted with CH_2Cl_2. After evaporation of the solvent, 8.1 mmol of **11** were isolated as a liquid by flash column chromatography on silica gel (EtOAc–hexane, 6:4): ^1H NMR (300 MHz, CDCl$_3$) δ 1.25, 1.35, 1.55 (3s, 3H, 3H, 6H, C-CH$_3$), 2.70 (s, 3H, Het-CH$_3$), 3.15–3.30 (m, 2H, H-6a, 6b), 4.29 (dd, 1H, H-4, J_{4-5} = 1.5 Hz, J_{3-4} = 7.5 Hz), 4.31 (dd, 1H, H-2, J_{1-2} = 5.0 Hz, J_{2-3} = 2.5 Hz), 4.59 (m, 1H, H-5), 4.64 (dd, 1H, H-3), 5.47 (d, 1H, H-1), 7.25 (s, 1H, H-3′), 7.4–7.7 (2t, 2H, H-5′, H-6′), 7.9–8.1 (2d, 2H, H-7′, H-8′); MS, m/e 385 (M$^+$, 45), 370 (M$^+$ − 15, 60), 298 (100), 157 (95); HRMS 385.188 (calcd for $C_{22}H_{25}NO_5$, 385.189).

Reaction of Quinoxaline with Perfluoro-*n*-butyl Iodide and Cyclohexene (12) [18] (Eq. 15)

A solution of 2 mmol quinoxaline, 4 mmol cyclohexene, 3 mmol perfluoro-*n*-butyl iodide, 2 mmol CF_3COOH, and 3 mmol benzoylperoxide in 10 mL CH_3COOH was heated under reflux for 6 h. On removal of CH_3COOH under vacuum, a solution of 10% aqueous NaOH was added and the mixture was extracted with EtOAc. Compound **12** was isolated by flash column chromatography on silica gel (hexane–EtOAc, 95:5) (1.8 mmol, 90%): ^1H NMR (500 MHz, CDCl$_3$) δ 1.37–1.63 (m, 4H), 1.69–2.04 (m, 3H), 2.20 (m, 1H, H-3/H-6), 3.14–3.30 (m, 2H, H-2 and H-1), 7.70 (ddd, 1H, J = 1.7, 7.0, 8.3 Hz, H-7′/6′), 7.74 (ddd, 1H, J = 1.7, 7.0, 8.3 Hz, H′-6′/7′), 8.03 (dd, 1H, J = 1.7, 8.3 Hz, H-8′/5′), 8.08 (dd, 1H, J = 1.7, 8.3 Hz, H-5′/8′), 8.71 (s, 1H, H-3′); ^{19}F NMR (470 Hz, CFCl$_3$) −127.1 (dm, 1F, ^2J = 293 Hz), −124.7 (dm, 1F, ^2J = 293 Hz), −121.6 (dm, 1F, ^2J = 296 Hz), −120.2 (dm, 1F, ^2J = 296 Hz, ^3J = 12 Hz, F$_\beta$-F$_\gamma$), −117.5 (d, 1F, ^2J = 281 Hz, F$_a$), −104.7 (d, 1F, ^2J = 281 Hz, , F$_a$), −80.8 (t, 3F, J = 9.0 Hz, F$_\gamma$); MS, m/e 430 (M$^+$), 411 (M$^+$-F), 157, 144, 129, 69, 57, 43.

1.4
Aryl–Aryl Coupling Reactions

1.4.1
Intermolecular Arylation Reactions

1.4.1.1 Intermolecular Arylation Reactions of Phenols and Aromatic Carbonyl Compounds

Masahiro Miura and Tetsuya Satoh

1.4.1.1.1 Introduction

Transition metal-catalyzed cross-coupling is now recognized to be one of the most powerful carbon–carbon bond-formation reactions [1]. The palladium-catalyzed coupling of aryl halides or their synthetic equivalents, for example aryl triflates, with arylmetals is very often employed in the synthesis of biaryl molecules, whose skeletons are found in a wide range of important compounds including natural products and organic functional materials [1–3].

Appropriately functionalized aromatic substrates such as phenols and aromatic carbonyl compounds have recently been found to undergo intermolecular arylation directly and regioselectively on treatment with aryl halides in the presence of transition metal catalysts such as Pd, Rh, and Ru [3, 4] (for intramolecular reaction, see Chapter III.1.4.2). As illustrated in Scheme 1, which is a general catalytic sequence with palladium, coordination of a given functional group to a metal center is the key to effective coupling by C–H bond cleavage. Clearly the reaction has a significant advantage – the stoichiometric metalation of aromatic substrates is not required. Representative examples and some related reactions are summarized in this section.

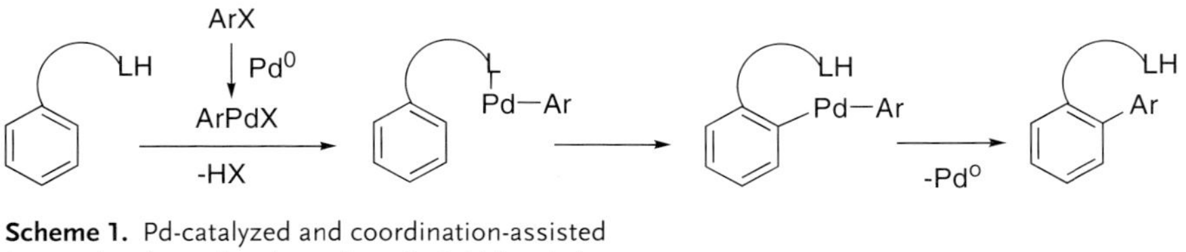

Scheme 1. Pd-catalyzed and coordination-assisted intermolecular aryl-aryl coupling via C–H bond cleavage

1.4.1.1.2 Reaction of Phenols

Arylation of 2-phenylphenols with aryl iodides (Scheme 2) is one of the first examples to proceed by the sequence given in Scheme 1 [5, 6]. The reaction proceeds via (1) oxidative addition of iodobenzene to Pd^0 to give PhPdI, (2) coordination of the phenolic oxygen to Pd, (3) palladation at the 2′-position to form a diarylpalladium intermediate, and (4) reductive elimination of the product. The use of a relatively strong inorganic base, for example Cs_2CO_3, is important for a smooth coupling. K_2CO_3 and Na_2CO_3 are less effective, in this order. Aromatic palladation via C–H bond cleavage may involve a Friedel–Crafts type electrophilic substitution or oxidative addition [7]. That the reactions of substrates with an electron-donating substituent in the 5′-position proceed smoothly seems to be consistent with the former mechanism. The strongly electron-withdrawing nitro group does not inhibit the reaction, however. Although the mechanism is not definitive, these results may imply that both pathways may participate in the reaction. Anyway, the base appears to promote the reaction effectively.

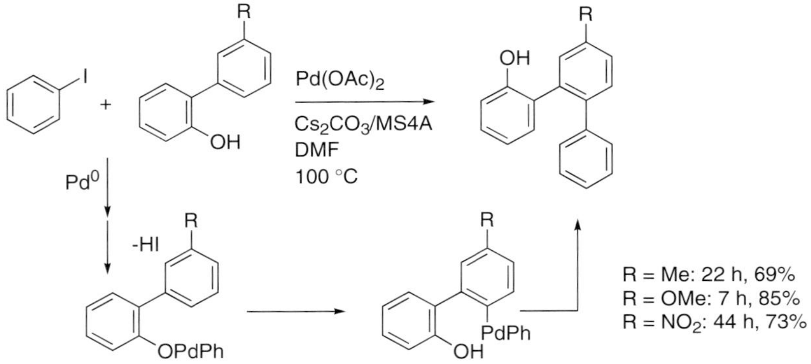

R = Me: 22 h, 69%
R = OMe: 7 h, 85%
R = NO$_2$: 44 h, 73%

Scheme 2. Pd-catalyzed regioselective arylation of 2-phenylphenols.

Simple 2-phenylphenol with less steric restriction can undergo diarylation to give a sterically congested 1,2,3-triphenylbenzene derivative (Eq. 1). 1-Naphthol is arylated at the peri position selectively (Eq. 2). Interestingly, 2-naphthol undergoes successive diarylation on treatment with bromobenzenes in the presence of PPh$_3$

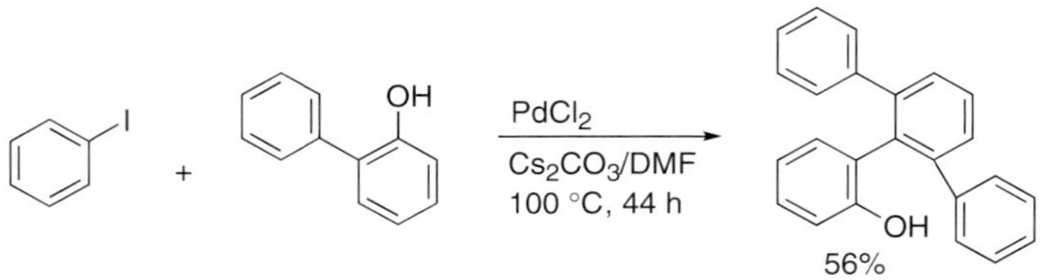

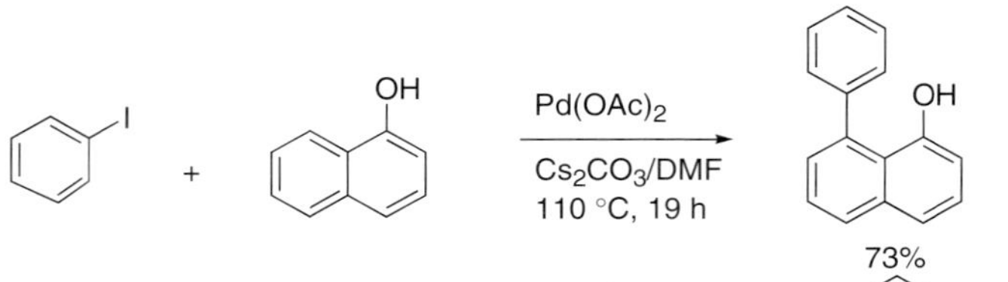

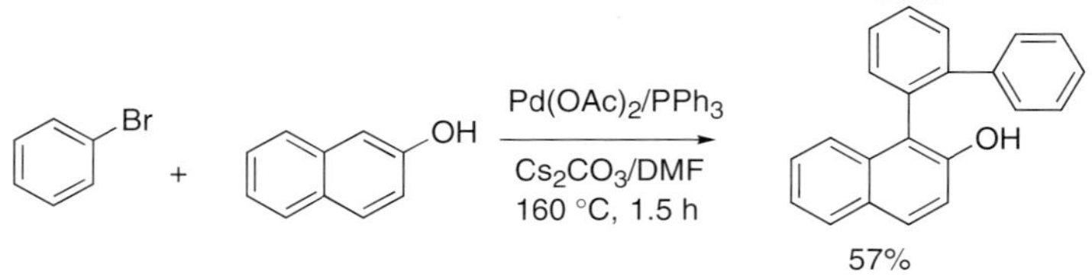

as ligand (Eq. 3). The reaction is believed to involve two different mechanisms. The first arylation at the 1-position may occur in a similar sequence to the α-arylation of ketones via enolates (paths a and c in Scheme 3) [4, 9]. The second may proceed in the same manner with the reaction in Scheme 2.

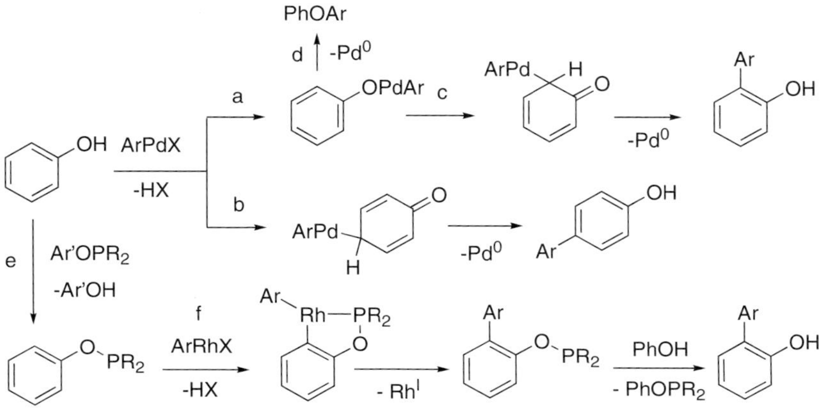

Scheme 3. Possible mechanistic sequences for the catalytic arylation of phenol.

Phenol itself can be arylated multiply around the oxygen up to five times by use of excess bromobenzene (Eq. 4) [8]. The use of a less polar solvent such as o-xylene is important; no reaction of phenol occurs in DMF. The lack of hexa-arylated product may be attributed to steric reasons. When the 2- and 6-positions of phenol are masked by *tert*-butyl groups, the 4-position is arylated (Eq. (5) and path b in Scheme 3) [10]. It is worth noting that a diaryl ether is formed by reductive elimination of the alkoxyarylpalladium intermediate when a bulky phosphine ligand is used (path d) [11].

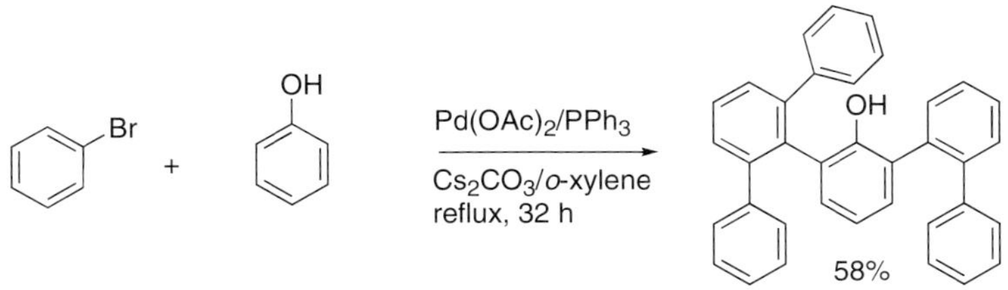

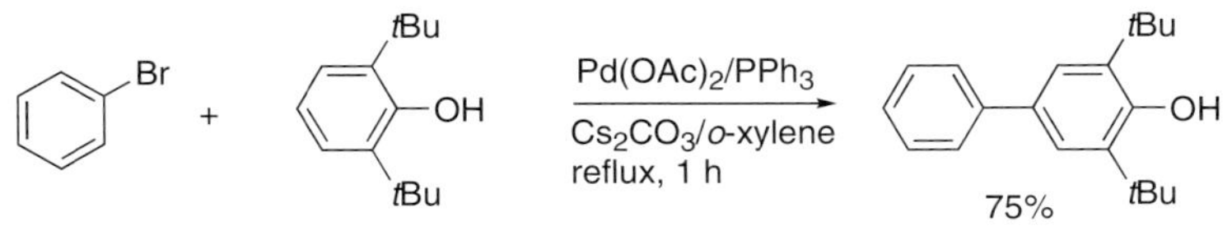

The ortho arylation of phenols with aryl bromides has also been found to occur when rhodium catalysts are used [12–14]. The key to the reaction is to employ phosphinites or phosphites as ligands. An example using a phosphinite ligand is given in Eq. (6) [13]. The reaction is believed to proceed by coordination of a phosphinite to ArRh(III) generated by oxidative addition (paths e and f); this is followed by cyclometalation and reductive elimination. The arylated phosphinite then undergoes transesterification with the starting phenol to give the product. In this reaction a phosphinite having the same aryloxy group as the phenol substrate must be prepared. This can, however, be avoided by using $P(NMe_2)_3$, which reacts with phenols in situ to give the corresponding phosphites (Eq. 7) [14]. It is noted that under the rhodium-catalyzed conditions no Heck type reaction occurs (Eq. 6) and minor amounts of the products arylated at the 2′-position, as in the palladium-catalyzed reaction (Scheme 2), are observed.

1.4.1.1.3 Reaction of Aromatic Carbonyl Compounds and Related Substrates

Alkyl aryl ketones are known to be arylated at the α-position of the alkyl groups, via the corresponding enolates, by treatment with aryl halides in the presence of palladium catalysts [4, 9]. The ortho arylation of alkyl aryl ketones is also possible. For example, in the reaction of benzyl phenyl ketones with bromobenzenes, the arylation first occurs at the benzylic position; the ortho positions are then arylated via C–H bond cleavage (Eq. 8) [15]. The ortho arylation is believed to occur after coordination of the enol oxygen to ArPd(II), which is followed by ortho palladation as in the reaction of 2-phenylphenols shown in Scheme 2.

Benzanilides also undergo ortho arylation (Eqs. 9 and 10) [16]. For these reactions, aryl triflates as arylation reagents are more effective than aryl bromides. The reaction is also believed to proceed via coordination of amide anion to ArPd(II), because no reaction occurs with secondary amides.

Benzylidene anilines formed from benzaldehydes and anilines have been found to undergo ortho arylation effectively when a ruthenium catalyst such as [RuCl$_2$(η^6-C$_6$H$_6$)]$_2$ is used in the presence of K$_2$CO$_3$ as base (Eqs. 11 and 12) [17]. A polar solvent such as NMP is used. In this reaction the substrates have no acidic hydrogen and thus ortho metalation seems to occur via coordination of the neutral nitrogen to the metal center, as in the reaction of phosphinite (Scheme 3). Similarly, 2-phenylpyridine [18] (Eq. 13) and 2-phenyl-1H-imidazole (Eq. 14) [19] are arylated.

It should be noted that introduction of a substituent to these substrates at an appropriate position enables the exclusive formation of the mono-arylated product as represented by Eqs. (10) and (12) (see also Scheme 2). The use of a limited amount of arylating reagent is sometimes successful for obtaining mono-arylated product selectively. In the reaction of Eq. (14), the Cp ligand on the ruthenium catalyst is believed to enable selective mono-arylation [19].

Direct aromatic arylation with arylmetal reagents has also been developed. Alkyl aryl ketones undergo arylation on treatment with arylboronates in the presence of a ruthenium catalyst (Scheme 4) [20]. As expected, depending on the bulkiness of the alkyl groups, either mono- or diarylation can be performed selectively. The reaction proceeds via initial C–H bond cleavage at the ortho position to give a metallacycle; this is followed by the reaction with another ketone molecule. Successive transmetalation and reductive elimination close the catalytic cycle. Thus, an equimolar amount of ketone is reduced during the reaction.

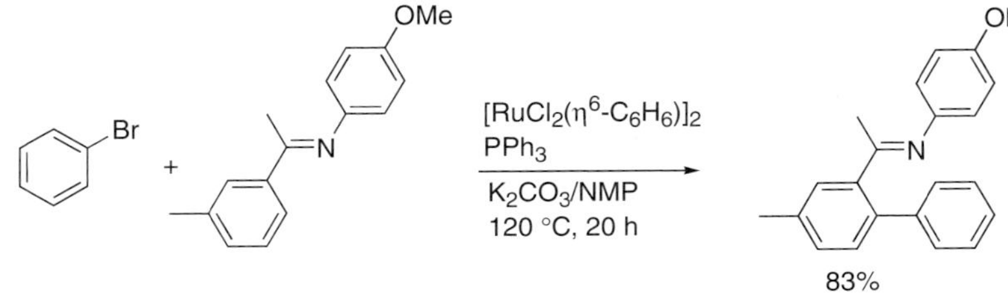

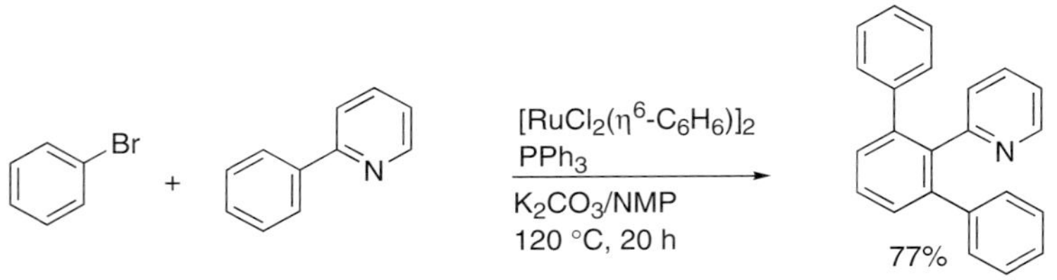

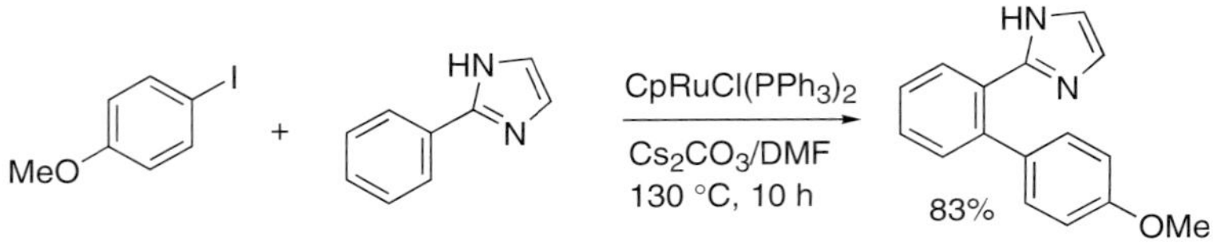

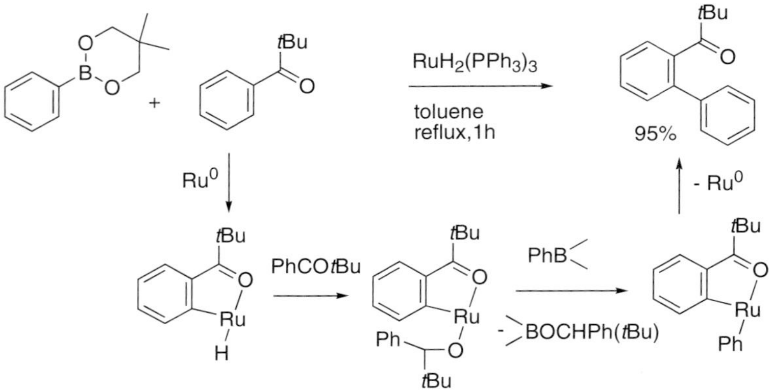

Scheme 4. Rhodium-catalyzed *ortho*-arylation of *t*-butyl phenyl ketone with 5,5-dimethyl-2-phenyl-[1,3,2]dioxaborinane.

2-Phenylpyridine is arylated by using aryltin reagents such as Ph$_4$Sn in the presence of a rhodium catalyst (Eq. 15) [21]. The use of a halogenated solvent is important for obtaining a satisfactory yield. Although the precise mechanism is not yet clear, it is likely that a cyclometalated intermediate participates.

$$\text{PhSnPh}_3 + \text{2-phenylpyridine} \xrightarrow[\text{Cl}_2\text{CHCHCl}_2]{\text{RhCl(PPh}_3)_3} \text{product (56\%)} \quad (15)$$

120 °C, 20 h

Experimental

2,6-(Diphenyl)benzoylaniline (Eq. 9)

A mixture of benzanilide (197 mg, 1 mmol), phenyltriflate (904 mg, 4 mmol), Pd(OAc)$_2$ (11.2 mg, 0.05 mmol), PPh$_3$ (78 mg, 0.3 mmol), Cs$_2$CO$_3$ (1.30 g, 4 mmol), and toluene (8 mL) was stirred under nitrogen at 110 °C for 24 h. After cooling, the reaction mixture was poured into dilute HCl, extracted with ethyl acetate, and the extract was dried over sodium sulfate. Evaporation of the solvent and washing of the residual solid well with hexane gave product 2,6-(diphenyl)benzoylaniline (335 mg, 96%). m.p. 279–280 °C; ^{1}H NMR (400 MHz, DMSO-d$_6$) $\delta = 6.95$ (t, $J = 7.3$ Hz, 1H), 7.15 (t, $J = 7.3$ Hz, 2H), 7.20 (d, $J = 7.3$ Hz, 2H), 7.29 (d, $J = 7.3$ Hz, 2H), 7.36 (t, $J = 7.3$ Hz, 4H), 7.43 (d, $J = 7.8$ Hz, 2H), 7.50 (d, $J = 7.3$ Hz, 4H), 7.60 (t, $J = 7.8$ Hz, 1H), 10.14 (s, 1H); ^{13}C NMR (100 MHz, DMSO-d$_6$) $\delta = 120.01$, 123.66, 127.46, 128.25, 128.53, 128.63, 129.06, 129.11, 136.49, 138.62, 139.53, 140.38, 166.76; MS m/z 349 (M$^+$).

1.4.1.2 Palladium-Catalyzed Arylation of Heteroarenes

Masahiro Miura and Tetsuya Satoh

1.4.1.2.1 Introduction

As described in Section III.1.4.1.1, the catalytic direct arylation reactions of aromatic compounds occurs effectively via C–H bond cleavage when the substrates are appropriately functionalized. On the other hand, various five-membered heteroaromatic compounds involving one or two heteroatoms, even without a functional group, are known to undergo arylation, usually at their 2- and/or 5-position(s), on treatment with aryl halides under the action of palladium catalysis. The coupling has recently been developed significantly [1, 2]. Representative examples with some mechanistic discussion are summarized in this section.

1.4.1.2.2 Reaction of Pyrroles, Furans, and Thiophenes

The arylation of these substrates, which are susceptible to electrophiles, may be regarded as proceeding through a Friedel–Crafts type mechanism involving attack of ArPd(II) species, as judged by the usual substitution pattern (path a in Scheme

1) [3]. However, either a Heck-type mechanism via carbopalladation (path b) or a coordination-assisted mechanism (path c) seems to be capable of taking part [4]. Thus, it is likely there is not one mechanism only and these pathways occur depending on the substituents on each nucleus and reaction conditions as described below.

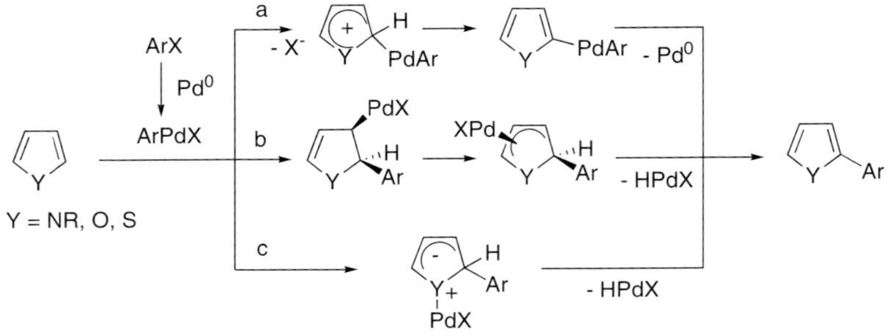

Scheme 1. Possible catalytic sequences for the Pd-catalyzed intermolecular arylation of five-membered heteroaromatic compounds involving one heteroatom.

An early significant example is the coupling of 1-substituted indoles with 2-chloro-3,6-dialkylpyrazines (Eq. 1) [5, 6]. Depending on the 1-substituents, the reaction occurs selectively at 2- or 3-position. Substitution at the 3-position is a rare occurrence and seems to be caused by the electron-withdrawing tosyl group, although the precise mechanism is still not clear.

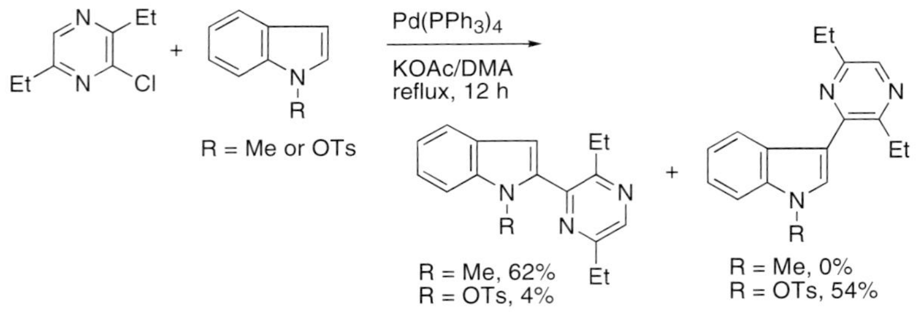

The reaction of pyrrole with iodobenzene proceeds selectively at the 2-position when MgO is used as base, even without N-protection (Eq. 2) [7, 8]. The strong N–Mg interaction has been proposed as the key to the effective coupling. As expected, N-unprotected indoles are arylated selectively at the 2-position in the presence of MgO (Eq. 3). The arylation of indolidines occurs under conditions similar to those in Eq. (1) (Eq. 4) [9].

Furan [10] and its 2-formyl derivative (Eq. 5) [11] are arylated at 2- and 5-positions, respectively. Arylation of 3-ethoxycarbonylfuran and its thiophene analogue occurs selectively at the 2-position in toluene, whereas the 5-position is attacked preferentially in NMP (Eq. 6) [12]. It has been proposed that paths b and a, respectively, predominantly participate in the former and latter reactions.

Selective 2-monoarylation of thiophene itself, and furan, can be achieved by use of excess of the substrate (Eq. 7) [10]. It has been demonstrated that use of AgF and DMSO as base and solvent, respectively, enables the reaction to proceed at 60 °C [13]. Various 2- or 3-substituted thiophenes and benzothiophenes have been subjected to catalytic arylation [2–4, 12–16]. 2,2′-Bithiophene can be diarylated at the 5,5′-positions (Eq. 8) [16]. Interestingly, N-phenyl-2-thiophenecarboxamides undergo 2,3,5-triarylation accompanied by a formal decarbamoylation on treatment with excess bromobenzenes (Eq. 9) [15]. The reaction involves initial coordination-assisted 3-arylation and successive decarbamoylation promoted by Pd(II) species and base in the medium. Introduction of an electron-withdrawing group to the 3-position of thiophene makes the 4-arylation possible, while reaction at the 2- and 5-positions precedes (Eq. 10) [15]. Thus, the reaction of 3-cyanothiophene affords the corresponding 2,4,5-triarylated products. Although the mechanism of the 4-arylation is not definitive, it has been proposed that the 2-arylation of 3-cyanobenzo[b]thiophene proceeds by path c in Scheme 1, whereas path a predominates in the reaction of 3-methoxybenzo[b]thiophene (Eq. 11) [4]. The polymerization of 3-alkyl-2-iodothiophenes has been reported (Eq. 12) [17].

1.4.1.2.3 Reaction of Imidazoles, Oxazoles, and Thiazoles

The order of reactivity for the reaction sites of azole compounds in electrophilic reactions is known to be 5 > 4 > 2 [18]. Thus arylation at the relatively electron-rich 5-position may be regarded as similar to that of pyrroles, furans, and thiophenes (Scheme 1). In contrast, the reaction at the 2-position may be regarded as proceeding differently [3]. Although the precise mechanism is still unclear, it may involve base-assisted deprotonative palladation with the ArPd(II) species (path d in Scheme 2). Insertion of the C=N double bond into the Ar–Pd bond is also a possible pathway (path e). Anyhow, the 5-arylation usually occurs faster than the 2-arylation, although the order of reactivity can be changed by use of an additive, for example the Cu(I) species described below.

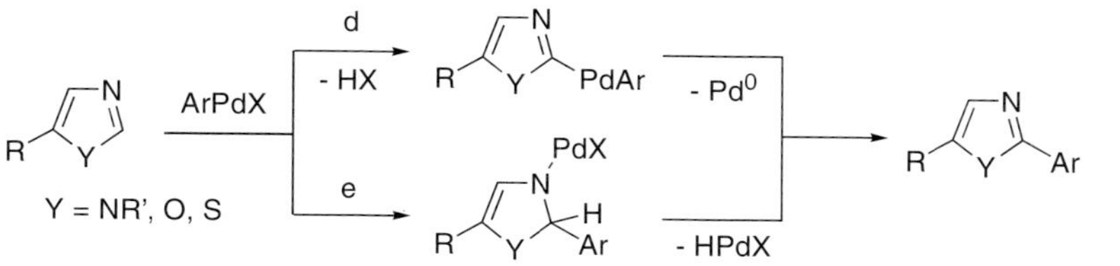

Scheme 2. Possible catalytic sequences for the Pd-catalyzed intermolecular arylation of azole compounds at the 2-position.

The reaction of 1-methyl-1H-imidazole with 2 equiv. iodobenzene gives its 5-phenyl derivative as the major product, with the 2,5-diphenyl derivative. Interestingly, addition of 2 equiv. CuI as promoter produces the 2-phenyl derivative together with the 2,5-diphenyl derivative; no 5-phenyl derivative is formed (Eq. 13) [3].

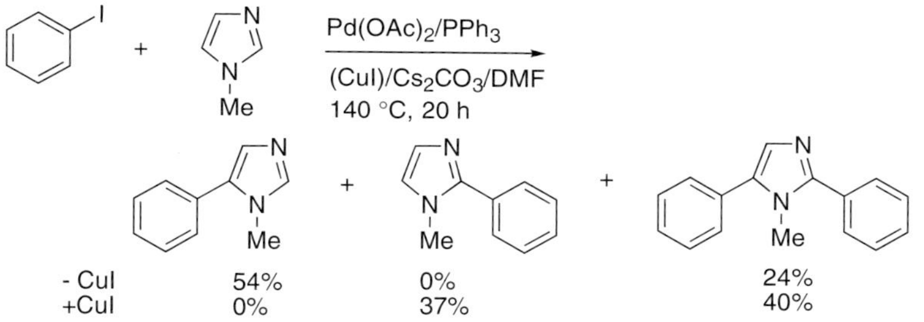

	5-phenyl	2-phenyl	2,5-diphenyl
- CuI	54%	0%	24%
+ CuI	0%	37%	40%

In MgO is used as base, 4(5)-phenyl- and 2-phenyl-1H-imidazoles can be obtained selectively by the reaction of 1H-imidazole itself in the absence and presence of CuI, respectively (Eq. 14) [7, 8]. The reaction of 1H-benzimidazole with iodobenzene also occurs at the 2-position. The same reaction can be performed by using a rhodium catalyst (Eq. 15) [19]. The arylation of imidazo[1,2-a]pyrimidine has also been reported (Eq. 16) [20].

Although examples for the arylation of oxazole itself are limited, reaction with 2-chloro-3,6-dialkylpyradines at the 5-position is known [6]. 2-Phenyloxazole [3] and benzoxoazole [3, 10, 19, 21] are good substrates for the direct arylation (Eqs. 17 and 18) [3].

The 2,5-diarylation of thiazole can be performed effectively with a bulky phosphine ligand. In this reaction, no mono-arylated product is observed, even in the early stage of the reaction, suggesting that the second arylation proceeds relatively fast (Eq. 19) [22]. The selective 2-arylation is accomplished by using CuI and Bu$_4$NF as cocatalyst and base, respectively (Eq. 20) [23]. By using a catalyst system of Co(OAc)$_2$-IMes (IMes = 1,3-bis-mesitylimidazolyl carbene) the 5-position is arylated selectively (Eq. 21) [20]. The Pd-catalyzed arylation of thiazole and 1-methylpyrrole with a polymer-linked aryl iodide has been reported [24].

Experimental

2,5-Di(p-anisyl)thiazole (Eq. 18)

Cs_2CO_3 (4.8 mmol, 1.56 g) was placed in a 100-mL two-necked flask and then dried at 150 °C in vacuo for 2 h. $Pd(OAc)_2$ (0.1 mmol, 22.4 mg), $P(t-Bu)_3$ (0.2 mmol, 40.5 mg), p-bromoanisole (4.8 mmol, 898 mg), thiazole (2 mmol, 170 mg), 1-methylnaphthalene (ca. 100 mg) as internal standard, and DMF (5 mL) were then added. The resulting mixture was stirred under N_2 at 150 °C for 8 h. After cooling, the reaction mixture was extracted with ethyl acetate and the extract was dried over sodium sulfate. The yield of 2,5-di(p-anisyl)thiazole determined by GC analysis was 76%. Column chromatography on silica gel using hexane–ethyl acetate, 95:5, as eluent gave the product (410 mg, 69%): m.p. 174–175 °C; ^{1}H NMR δ 3.84 (s, 3H), 3.86 (s, 3H), 6.93–6.97 (m, 4H), 7.51 (d, J = 8.7 Hz, 2H), 7.85–7.90 (m, 3H); ^{13}C NMR δ 55.37, 55.40, 114.29, 114.49, 124.17, 126.73, 127.73, 127.84, 137.88, 138.22, 159.63, 161.03, 166.24; MS m/z 297 (M$^+$). Anal. calcd for $C_{17}H_{15}NO_2S$: C, 68.66; H, 5.08; N, 4.71; S, 10.78. Found: C, 68.47; H, 5.10; N, 4.68; S, 10.41.

1.4.1.3 Palladium-Catalyzed Arylation of Cyclopentadienyl Compounds
Gerald Dyker

1.4.1.3.1 Introduction and Fundamental Examples

Arylated cyclopentadienes, especially pentaarylcyclopentadienes **6**, have interesting properties as ligands for transition metals [1] or as electroluminescent materials [2]. The classical methods for their preparation are multistep procedures [3], which are somewhat tedious and usually less suitable for sterically demanding aryl groups. In contrast, the palladium-catalyzed arylation of either metallocenes such as zirconocene dichloride (**1**) or simply cyclopentadiene (**3**) with aryl halides leads to the target compounds **4–6** in a single preparative step (Scheme 1) [4–6].

With the low-boiling cyclopentadiene (**3**) as starting material the reaction is preferably performed in a sealed tube. The product ratio of the multifold arylation, whether starting from **1** or from **3**, depends on several factors, for instance on the polarity of the solvent and on the electronic properties and bulkiness of the aryl halide **2**. Electron-withdrawing substituents on the aryl halide **2** usually somewhat favor the tetraaryl-substituted product **5**. The phosphine ligand also affects the outcome of the reaction – for arylation with meta- or para-substituted aryl halides tris(*tert*-butyl)phosphine [7] is recommended as an appropriate ligand, mainly to avoid ligand scrambling [8], which is occasionally observed with triphenylphosphine, causing minor impurities of phenyl-substituted cyclopentadienes in the crude product (so a recrystallization step is required for purification). With sterically demanding aryl halides triphenylphosphine nevertheless gives better yields than tris(*tert*-butyl)phosphine.

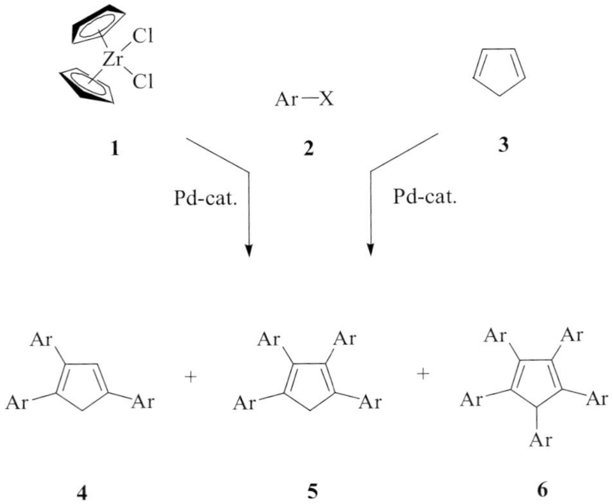

1 C–H Transformation at Arenes

Scheme 1. Multiple palladium-catalyzed arylation of cyclopentadiene units; Ar = aryl, X = Br, Cl.

1.4.1.3.2 Mechanism

For the C–C bond-forming step coordination of an electrophilic aryl palladium halide to a cyclopentadienyl anion is assumed, followed by reductive elimination. Presumably the Pd catalyst is not involved in the C–H bond-breaking step, which is interpreted as an apparently simple deprotonation with cesium carbonate as base. The overall process is similar to the arylation of other soft nucleophiles [9].

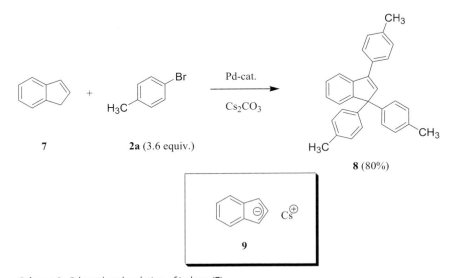

Scheme 2. Pd-catalyzed arylation of indene (**7**).

1.4 Aryl–Aryl Coupling Reactions | 237

This mechanistic consideration is in accord with the result from arylation of indene (**7**) [5]. The arylation occurs exclusively at the 1,3-positions, those with the highest electron density, of intermediary indenyl anions (**9**). The related arylation of azulene also proceeds exclusively at 1,3-positions, interpreted as electrophilic aromatic substitution, again by an electrophilic aryl-Pd species [10].

1.4.1.3.3 Scope and Limitations

Simple pentaarylcyclopentadienes, for example the pentatolyl-substituted product **6a** with substituents in the para or meta position, are obtained in excellent yield. Surprisingly, rather crowded pentaarylcyclopentadienes, for example the pentaxylyl

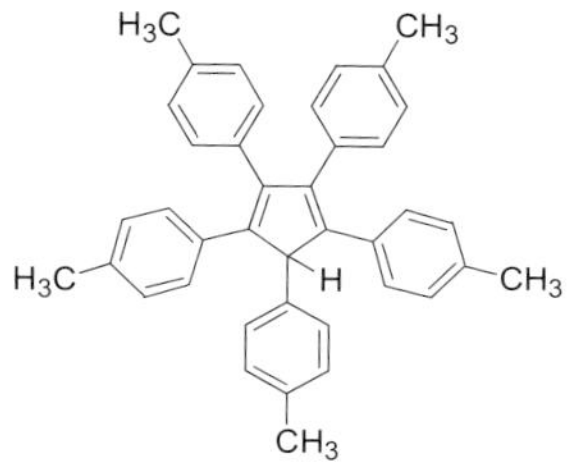

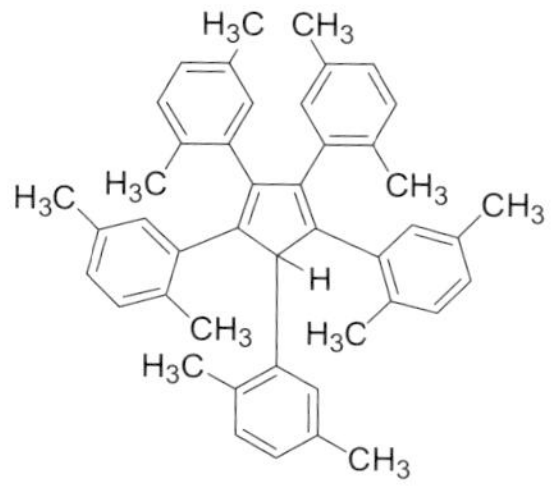

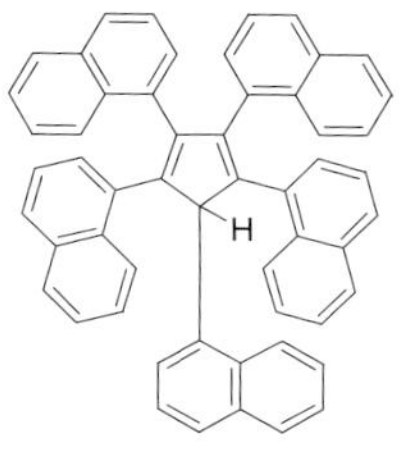

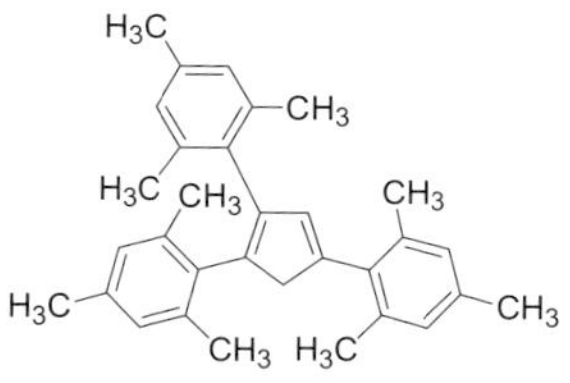

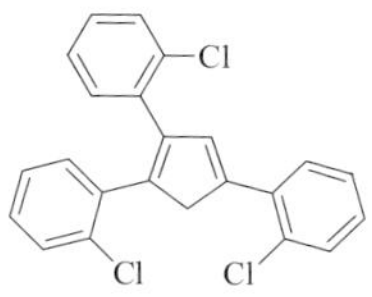

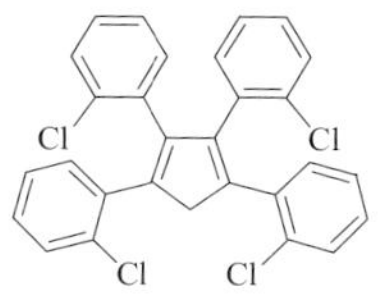

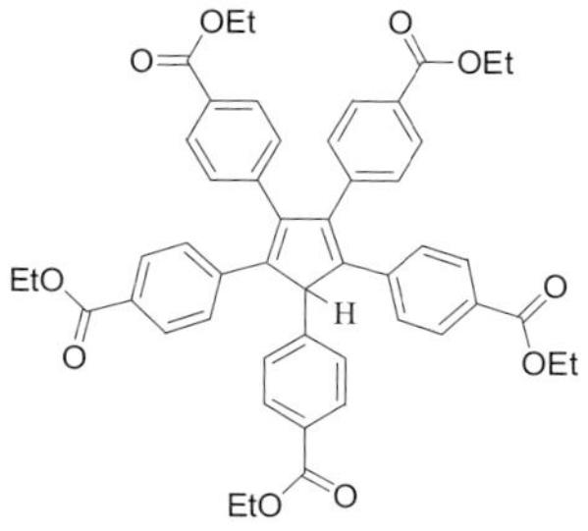

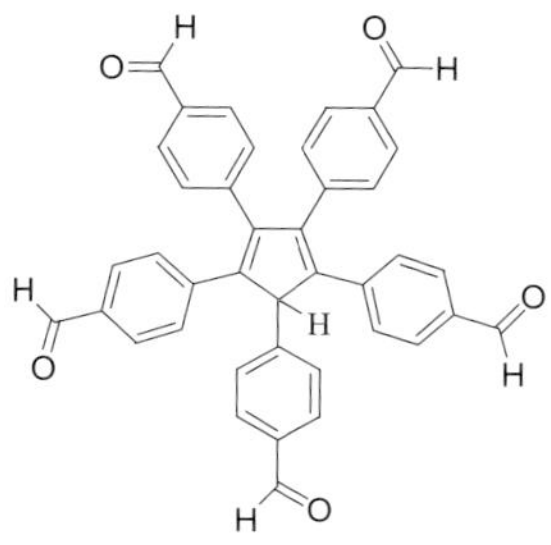

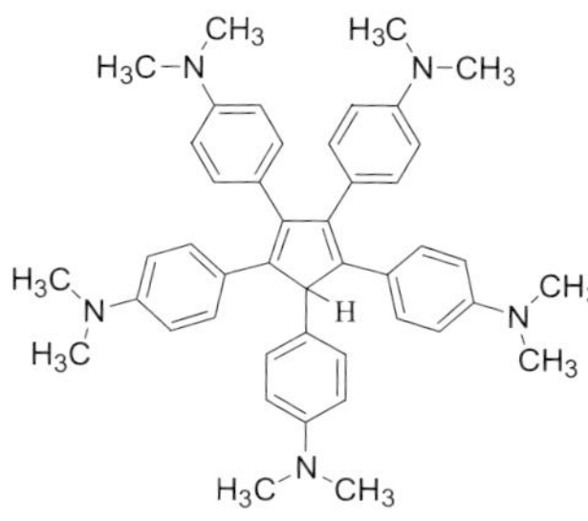

Scheme 3. Products from the multifold arylation of cyclopentadiene.

derivative **6b** and the pentanaphthyl compound **6c**, are also accessible in good yields, as mixtures of rotamers, of course. With a combination of a bulky aryl bromide, for example *para*-xylyl bromide, and a bulky phosphine ligand, for example tris(*ortho*-tolyl)phosphine, the corresponding triarylated product **4** can be obtained selectively. The even bulkier 2,4,6-trimethylbromobenzene (**2d**) also participates in the arylation process – the three-fold product **4d** and some tetraarylated compound are isolated under these conditions.

The arylation of cyclopentadiene (**3**) with 2-chlorobromobenzene using tris(*tert*-butyl)phosphine as ligand leads to a mixture of the tetra-arylated product **5e** and its pentaarylated congener, both isolated in 17–18 % yield. With the less reactive *ortho*-dichlorobenzene the process stops at the stage of the triarylated species **4e**, which can then be further arylated with 2-chlorobromobenzene.

Both, esters and tertiary amino groups are tolerated, giving rise to products **6f** and **6h** which are useful as centers for dendrimeric structures. The pentakisaldehyde **6g** was obtained from 4-bromobenzaldehyde protected as the ethylene glycol acetal. Attempts to use heterocyclic bromides for arylation were hitherto unsuccessful.

Experimental

1,2,3,4,5-Pentakis(2,5-dimethylphenyl)cyclopenta-1,3-diene (6b)
A mixture of 66 mg (1.0 mmol) cyclopentadiene (**3**), 1.11 g (6.00 mmol) 1-bromo-2,5-dimethylbenzene, 1.96 g (6.00 mmol) cesium carbonate, 22.4 mg (50 µmol) palladium acetate, 10 mL dimethylformamide, and 52 mg (200 µmol) triphenylphosphine was heated under argon in a screw-capped tube for 23 h at 140 °C. After cooling to room temperature 50 mL CH_2Cl_2 and 12 mmol *para*-toluene sulfonic acid were added, with stirring, and after 10 min the mixture was filtered through 5 g silica with 25 mL of CH_2Cl_2 as eluent. The solvent was evaporated at 50 °C/15 mbar. The residue was dried in vacuo at 120 °C/0.5 mbar and finally fractionated by flash chromatography on silica gel (petroleum ether–ethyl acetate, 50:1; R_f = 0.30, 0.14, 0.03, 0.00). The fraction with R_f = 0.14 is collected, giving 174 mg (62 %) of slightly colored **6b** as a mixture of rotamers with m.p. 194–196 °C; ^{1}H NMR: δ = 1.72–2.46 (m, 30H), 4.86–5.45 (m, 1H; actually singlets at 4.86, 5.07, 5.10, 5.25, 5.29, and 5.45), 6.46–7.10 (m, 15H); MS m/z = 586 (M^+).

1.4.2
Palladium-catalyzed Arylation Reactions via Palladacycles

1.4.2.1 Intramolecular Biaryl Bond Formation – Exemplified by the Synthesis of Carbazoles
Robin B. Bedford, Michael Betham, and Catherine S. J. Cazin

1.4.2.1.1 Introduction and Fundamental Examples
Palladium catalyzed C–H activation is a powerful tool for synthesis of biaryls from "tethered" aryl halide and triflate substrates of type **1** (Scheme 1). This technique

has proved useful in the synthesis of six-membered heterocycles [1] such as quinolinones [2, 3], pyrans, thiazine dioxides, and pyranones [2]. Scheme 2 outlines the synthesis of the pyranone structure with selected illustrative examples. These are the benzylated forms of defucogilocarcins M and E, **2** [4], and compounds **3** and **4**, intermediates in syntheses of dioncophylline A and mastigophorene B respectively [5, 6].

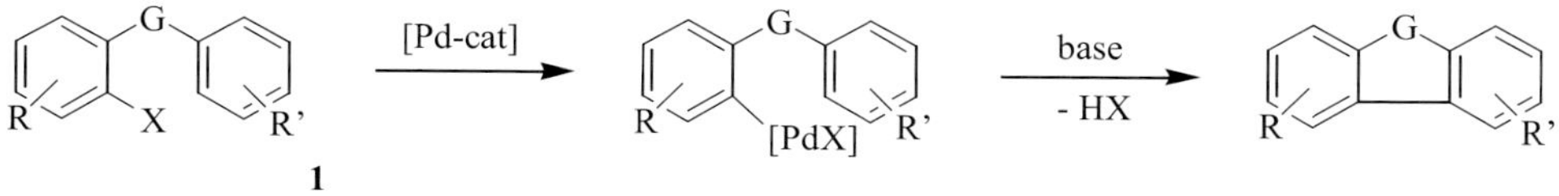

Scheme 1. Intramolecular biaryl bond formation.

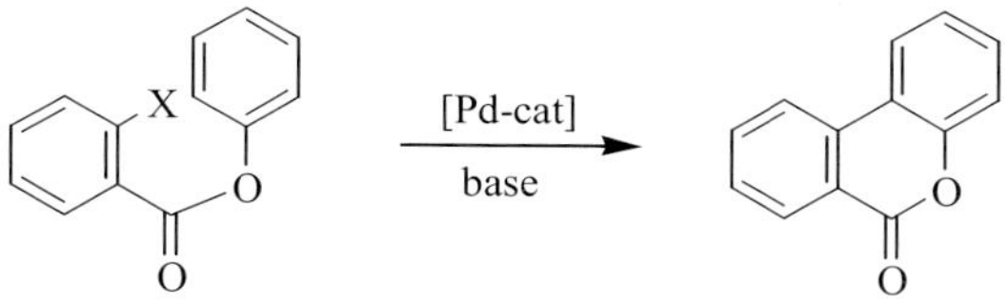

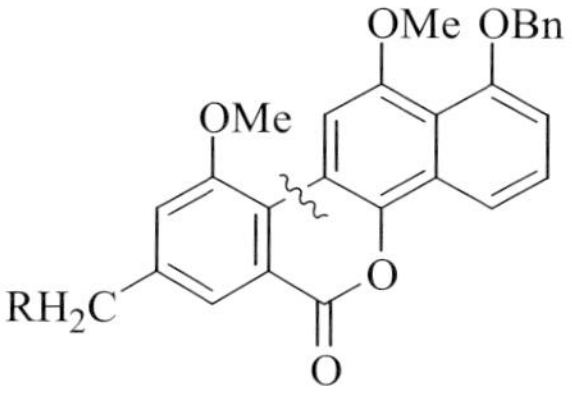

2: R = H, Me

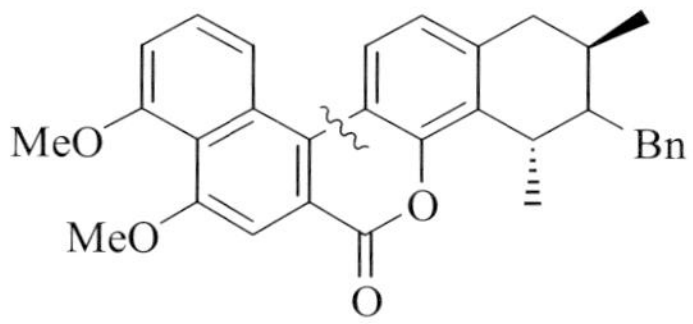

Scheme 2. Intramolecular biaryl bond formation in the synthesis of pyranones and illustrative examples (the new bonds formed are indicated).

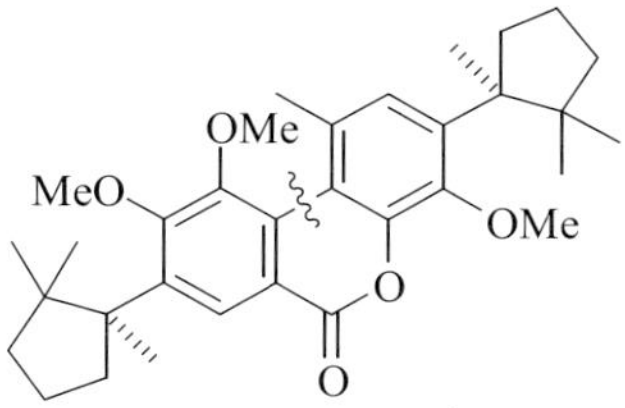

Five-membered rings are also readily synthesized using this coupling methodology, enabling the production of dibenzo[b,d]fused heterocycles such as dibenzofurans [2, 7], carbazoles [7, 8] (Scheme 1, G = O, NR respectively), and related compounds [8, 9]. Carbazoles are important structural motifs found in pharmaceuticals, natural products, agrochemicals, and dyes [10, 11]. For example, carvedilol, (Fig. 1) is a beta-blocker used in the treatment of hypertension and angina and carazostatin is one of a family of carbazoles produced by bacteria that are potent free-radical scavengers and act as anti-oxidants. Ellipticine and its derivatives, in particular 9-hydroxyellipticine, have potent anti-tumor activity.

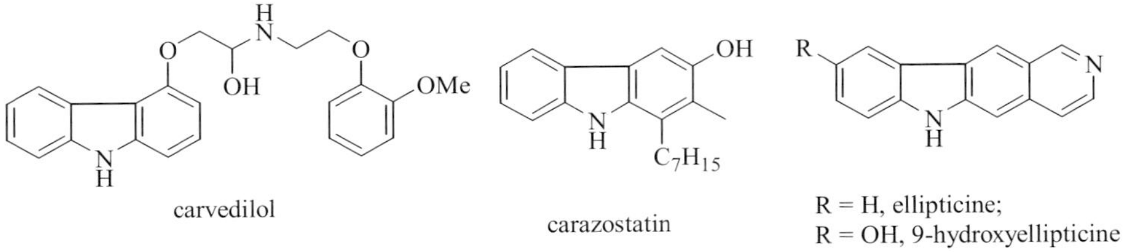

Figure 1. Examples of biologically active carbazoles.

C–H activation can be used to generate carbazoles from N-aryl anilines by palladium catalyzed oxidative coupling (Scheme 3) [12]. Although this is a powerful method, it is ultimately limited by the fact that more heavily substituted N-aryl anilines may not couple selectively, leading to the formation of more than one product.

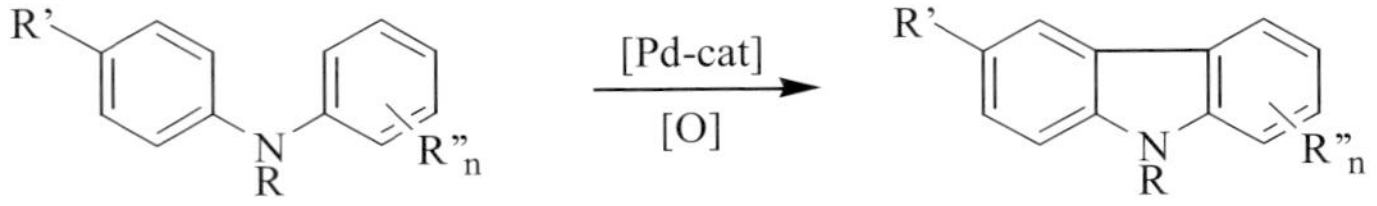

Scheme 3. Formation of carbazoles via palladium catalyzed oxidative coupling.

The dehydrohalogenation approach outlined in Scheme 1 reduces problems of selectivity in the synthesis of carbazoles. The method can be improved further by producing the N-aryl-2-haloaniline starting materials in the same pot as the subsequent carbazole product. This is achieved by Buchwald–Hartwig amination of 2-chloroanilines with aryl bromides [13]. An illustrative example of this one-pot procedure is shown in Scheme 4.

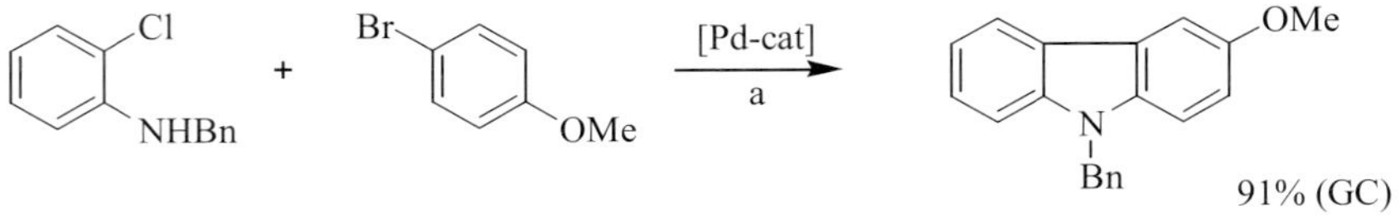

Scheme 4. One-pot synthesis of a carbazole via consecutive Buchwald–Hartwig amination and ring-closing reactions:
(a) Pd(OAc)$_2$ (5 mol%), P^{t}Bu$_3$ (7 mol%), NaO^{t}Bu, toluene, reflux, 24h.

1.4.2.1.2 Mechanism

A plausible mechanism for the one-pot synthesis of carbazoles is shown in Scheme 5. It consists of two interlinked catalytic cycles. In the first cycle a classical Buchwald–Hartwig amination reaction occurs to generate an intermediate **5** which then enters the second cycle by oxidative addition to Pd(0). The resulting Pd(II) complex then undergoes intramolecular C–H activation to give a six-membered palladacycle which subsequently yields the carbazole by reductive elimination.

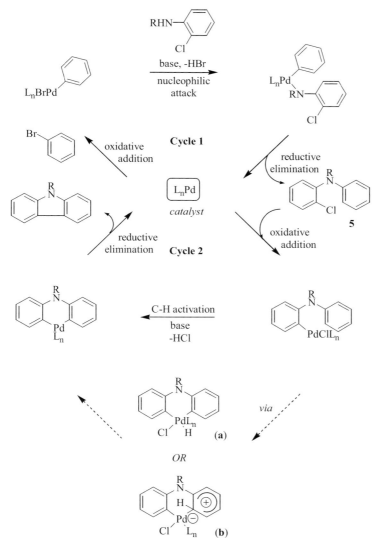

Scheme 5. Possible mechanism for the synthesis of carbazoles from 2-chloroanilines and aryl bromides (ring substituents omitted for clarity).

The C–H activation step could, in principle, occur either by oxidative addition of the C–H bond – pathway (**a**) – or by electrophilic displacement – pathway (**b**). The oxidative addition pathway would proceed via the formation of a palladium(IV) species. Although such intermediates have been postulated in some coupling reactions catalyzed by palladacycles, as yet no conclusive experimental evidence has been presented [14]. It is perhaps more likely that C–H activation results from electrophilic displacement of the ortho proton [15].

To confirm that the compounds **5** are indeed the intermediates that enter the second coupling step, a representative example, **5a**, was subjected to the standard reaction conditions and shown to generate the desired carbazole in good yield (Scheme 6).

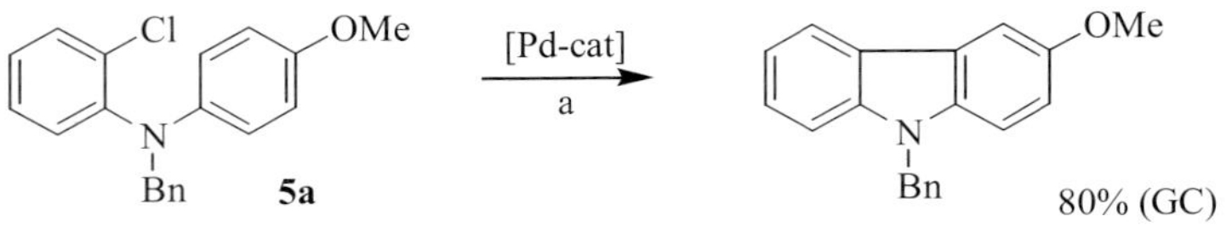

Scheme 6. Ring closing reaction of a pre-formed intermediate: (a) Pd(OAc)$_2$ (5 mol%), P^{t}Bu$_3$ (7 mol%), NaO^{t}Bu, toluene, reflux, 24 h.

1.4.2.1.3 Scope and Limitations

Examples of the one-pot method are given in Table 1 [13]. The coupling works not only with aryl bromides that are electronically non-activated with regard to oxidative addition, for example bromobenzene and 3-bromotoluene, but also for the deactivated substrate 4-bromoanisole. When 3-bromotoluene is used as the substrate, in principle, two isomers can form. This indeed occurs and both the 2- and 4-methyl isomers are produced in a 5:1 ratio. In contrast, when *N*-benzyl-2-chloro-5-trifluoromethylaniline, **6**, is used as substrate there is no evidence for formation of the second isomer.

In all successful double coupling reactions small amounts of the hydrodehalogenated products **7** (R = Me, Bn) are seen and in all the reactions, except with **6**, small amounts of the intermediates **5** are also observed. The lack of any intermediates **5** when **6** is used as a substrate is presumably because the CF$_3$ group activates the chloride for oxidative addition.

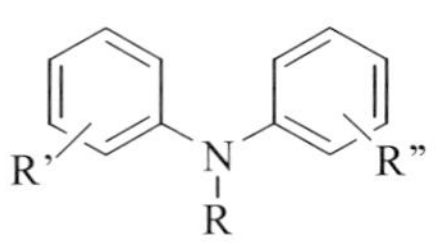

Because 2-haloanilines are strongly electronically deactivated with regard to oxidative addition, it may be expected that there is sufficient disparity in reactivity between *N*-benzyl-2-chloroaniline and 4-chloroanisole for them to undergo the sequential coupling reaction. Very little carbazole product is formed, however, the major product being the intermediate **5a**. In this case it seems that the rate of the

Table 1. Examples of the one-pot, double-coupling method.
Conditions: Pd(OAc)$_2$ (4–5 mol%), P^{t}Bu$_3$ (5–7 mol%), NaO^{t}Bu, toluene, reflux, 24h.

Starting materials	Products	Isolated yield[a] (GC yield)
2-chloro-N-methylaniline + 4-bromoanisole	3-methoxy-9-methylcarbazole	47(73)
2-chloro-N-methylaniline + 3-bromotoluene	3-methyl-9-methylcarbazole	57(75)
	1-methyl-9-methylcarbazole	(~15)
N-benzyl-2-chloro-5-(trifluoromethyl)aniline + 4-bromoanisole	3-methoxy-9-benzyl-6-(trifluoromethyl)carbazole	27(50)
N-benzyl-2-chloro-5-(trifluoromethyl)aniline + 3-bromotoluene	3-methyl-9-benzyl-6-(trifluoromethyl)carbazole	51(69)
2-chloro-N-methylaniline + bromobenzene	9-methylcarbazole	61(85)

[a] Non-optimized yields.

amination is too slow for substantial product formation and conversion of the intermediate **5a** to the product is probably limited by catalyst longevity. The coupling of benzyl-2-bromoaniline and 4-bromoanisole also fails. In this case little of the intermediate is formed, indicating that the problem lies in the amination step. This may be a result of steric hindrance by the bromide in the 2-position of the aniline.

In contrast, with the N-alkylated 2-chloroanilines, simple unsubstituted 2-chloroanilines give no or, under substantially more forcing conditions (10 mol% catalyst, 60 h), trace amounts of the desired carbazoles. In these reactions the main products are the intermediates **5** (R = H) [13]. To circumvent this problem it is necessary to split the double-coupling into its constituent steps [16]. Optimization studies show that the amination proceeds when the solvent is toluene but not 1,4-dioxane, whereas ring-closure is facile in dioxane but does not work in toluene.

Splitting of the steps enables facile, compact synthesis of the previously unsynthesized natural carbazole alkaloid, Clausine P (Scheme 7) [16, 17].

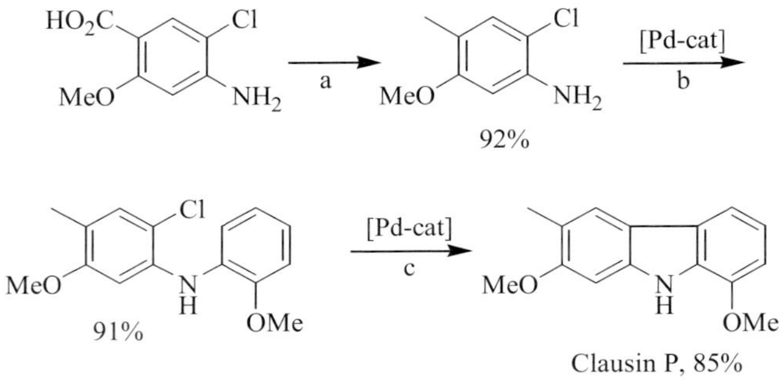

Scheme 7. Synthesis of Clausine P. (a) BH$_3$·SMe$_2$, C$_6$H$_5$Cl, 0 °C for 15 min then 115 °C for 18 h. (b) Pd(OAc)$_2$ (4 mol%), P^{t}Bu$_3$ (5 mol%), NaO^{t}Bu, toluene, reflux, 18 h (high yield, not purified) (c) Pd(OAc)$_2$ (4 mol%), P^{t}Bu$_3$ (5 mol%), NaO^{t}Bu, 1,4-dioxane, reflux, 18 h.

A large drawback of this "double-coupling" method is the need to handle and use the highly air-sensitive, low-melting tri-*tert*-butylphosphine as the ligand [13]. Fu and co-workers demonstrated that the phosphonium salt [HP^{t}Bu$_3$][BF$_4$] can be used as a solid, air-stable replacement for P^{t}Bu$_3$ in a range of coupling reactions [18]. The ligand is generated in situ by deprotonation with the base required in the coupling reaction. The same is found to be true here, with the phosponium salt giving equal or better results than the free phosphine.

Experimental

9-Benzyl-3-methoxycarbazole

A mixture of 4-bromoanisole (0.225 g, 1.22 mmol), N-benzyl-2-chloroaniline (0.283 g, 1.30 mmol), sodium *t*-butoxide (0.586 g, 6.10 mmol), palladium acetate (0.0137 g, 0.061 mmol), tri-*t*-butylphosphine (0.0173 g, 0.086 mmol) in toluene (13 mL) was heated at reflux temperature for 24 h. The reaction was quenched with a solution of hydrochloric acid (3.5 M, 10 mL) and toluene (30 mL) was added. The phases were separated and the aqueous phase was extracted with toluene (1 × 30 mL), dichloromethane (4 × 30 mL) and toluene (1 × 30 mL). The organic extracts were combined, dried over magnesium sulfate, filtered though Celite, and the solvents were removed on a rotary evaporator. The product was purified by flash chromatography (SiO$_2$, petroleum ether–EtOAc, 9:1, R_F = 0.3) to give a pale yellow solid in 40% yield (non-optimized, GC analysis of the reaction shows 91% conversion to the carbazole). ^{1}H-NMR (300 MHz, CDCl$_3$, 298 K): δ 3.96 (s, 3H, OCH_3), 5.52 (s, 2H, CH_2), 7.10 (dd, $^4J(H^2H^4)$ = 2.5 Hz, $^3J(H^2H^1)$ = 8.8 Hz, 1H, Ar-H^2), 7.15 (m, 2H, Ar-H), 7.21–7.34 (m, 5H, Ar-H^1, Ar-H^6 and 3 × Ar-H), 7.37 (br d, $^3J(H^8H^7)$ = 8.2 Hz, 1H, Ar-H^8), 7.45 (ddd, $^4J(H^7H^5)$ = 1.1 Hz, $^3J(H^7H^8)$ = 8.2 Hz,

$^3J(H^7H^6) = 7.0$ Hz, 1H, Ar-H^7), 7.65 (d, $^4J(H^4H^2) = 2.5$ Hz, 1H, Ar-H^4), 8.12 (ddd, $J = 0.6$ Hz, $^4J(H^5H^7) = 1.1$ Hz, $^3J(H^5H^6) = 7.9$ Hz, 1H, Ar-H^5). ^{13}C-{^{1}H}-NMR (300 MHz, CDCl$_3$, 298K): δ 46.5 (s, CH$_2$), 56.0 (s, CH$_3$), 103.3 (s, CH, C^4), 108.9 (s, CH, C^8), 109.6 (s, CH), 114.8 (s, CH, C^2), 118.7 (s, CH), 120.3 (s, CH, C^5), 122.7 (s, C), 123.3 (s, C), 125.8 (s, CH, C^7), 126.3 (s, CH), 127.3 (s, CH), 128.7 (s, CH), 135.6 (s, C), 137.3 (s, C), 141.1 (s, C), 153.7 (s, C). HRMS (CI) [M + H]$^+$ Calcd for C$_{20}$H$_{17}$NO: 288.1388. Found: 288.1386. Anal. Calcd for C$_{20}$H$_{17}$NO: C, 83.6 %; H, 6.0 %; N, 4.9 %. Found: C, 82.2 %; H, 5.9 %; N, 4.65 %.

Clausine P

2-Chloro-5-methoxy-N-(2-methoxyphenyl)-4-methylaniline (1.12 g, 4.0 mmol) was dissolved in 1,4-dioxane (3 mL) and sodium *tert*-butoxide (1.93 g, 20 mmol) was added. Palladium acetate (0.036 g, 0.16 mmol) and tri-*t*-butylphosphine (0.041 g, 0.20 mmol) were dissolved in dioxane (3 mL) and added to the reaction mixture. Finally, dioxane (10 mL) was added and the reaction was heated at reflux for 18 h. The reaction was quenched by addition of hydrochloric acid (2 M, 5 mL). The product was extracted with dichloromethane (3 × 50 mL), the extract was dried over magnesium sulfate and filtered, and the solvent was removed under reduced pressure to give a thick brown oil. The crude product was purified by column chromatography (SiO$_2$, petroleum ether–EtOAc, 5:1, $R_F = 0.2$) to give Clausine P as a yellow solid (0.82 g, 85 %). ^{1}H NMR (300 MHz, CDCl$_3$, 298 K): δ 1.57 (br. s, 2H, H_2O), 2.29 (s, 3H, CH$_3$), 3.84 (s, 3H, OCH$_3$), 3.93 (s, 3H, OCH$_3$), 6.75 (d, $^3J(H^2H^3) = 7.8$ Hz, 1H, Ar-H^2), 6.81 (s, 1H, Ar-H^8), 7.04 (dd, $^3J(H^3H^4) = 7.8$ Hz, $^3J(H^3H^2) = 7.8$ Hz, 1H, Ar-H^3), 7.49 (d, $^3J(H^4H^3) = 7.8$ Hz, 1H, Ar-H^4), 7.69 (s, 1H, Ar-H^5), 8.04 (br. s, 1H, NH). ^{13}C-{^{1}H}-NMR (300 MHz, CDCl$_3$, 298K): δ 17.2 (s, Ar-CH$_3$), 55.9 (s, 2x OCH$_3$), 93.0 (s, CH, C^8), 105.0 (s, CH, C^5), 112.5 (s, CH, C^2), 117.0 (s, C^6), 119.7 (s, C), 120.0 (s, CH, C^3), 122.0 (s, CH, C^4), 125.0 (s, C), 129.7 (s, C), 139.3 (s, C), 145.9 (s, C^1), 157.8 (s, C^7). HRMS (CI) [M + H]$^+$ Calcd for C$_{15}$H$_{15}$NO$_2$: 241.1103. Found 241.1144. Anal. Calcd for C$_{15}$H$_{15}$NO$_2$.H$_2$O: C, 74.7 %; H, 6.3 %; N, 5.8 %. Found C, 73.8 %; H, 6.2 %; N, 5.5 %.

1.4.2.2 Carbopalladation–Cyclopalladation Sequences
Marta Catellani and Elena Motti

1.4.2.2.1 Introduction

We report an sp^2-C–H transformation method based on reactions involving palladacycles. General Eqs (1)–(3) represent the processes described; RX is an alkyl halide and R'Y is an olefin, a terminal alkyne or an arylboronic acid (Y = H or B(OH)$_2$). Asterisks mark the transformed C atoms.

C–H transformation is achieved by cyclometallation by use of a unique catalytic system which involves the in-situ formation of a palladacycle [1]. Our work in this field takes advantage of the stability toward β-hydrogen elimination of *cis,exo*-arylnorbornylpalladium complexes formed by a sequence of oxidative addition of an aryl halide to palladium(0) and stereoselective insertion of norbornene into the resulting arylpalladium bond (Scheme 1). Because of the C–H and C–Pd bond anti configuration these species do not readily undergo β-H elimination but, being rather reactive, are subject to a great variety of transformations. In the presence of a base such as potassium carbonate they readily give palladacycles by activation of a usually inert aromatic C–H bond. The cyclopalladation reaction, which is likely facilitated by coordination of the aromatic ring with palladium, occurs under mild conditions and proceeds through an electrophilic aromatic substitution. In fact the reaction is favored by electron-donating substituents in the aromatic ring of complex **4**, as shown for following the cyclization process by NMR spectroscopy.

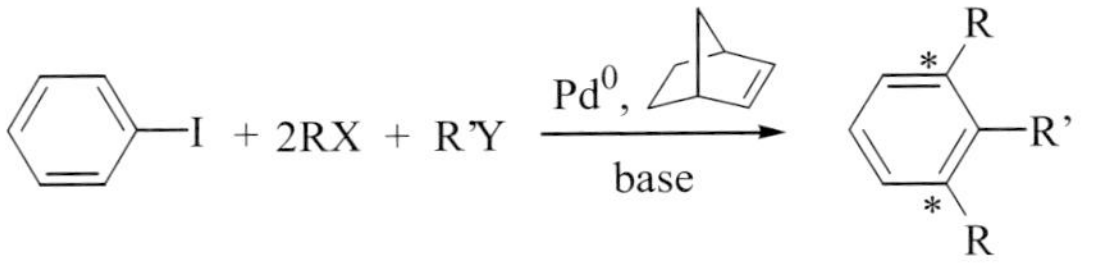

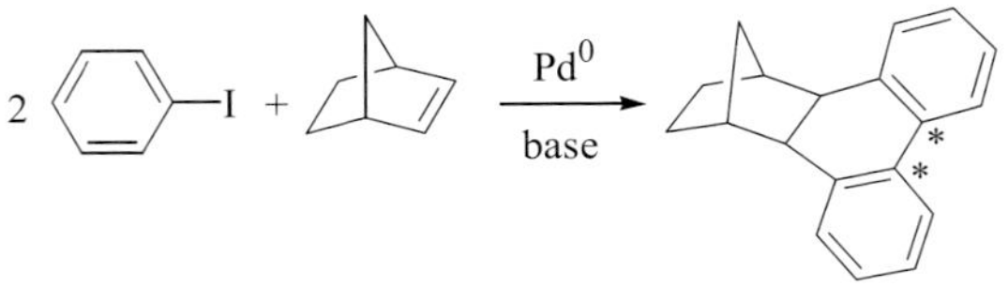

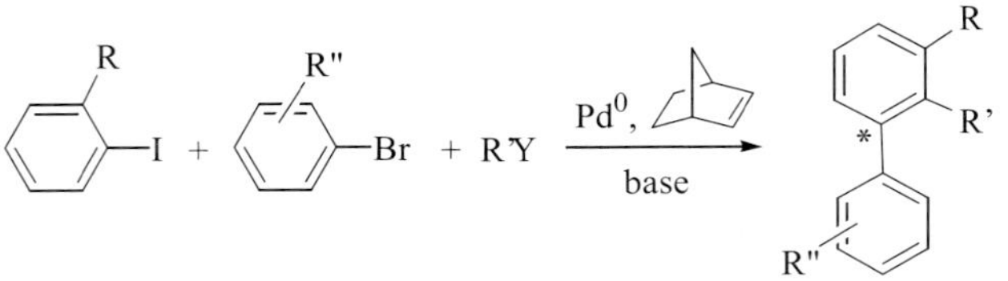

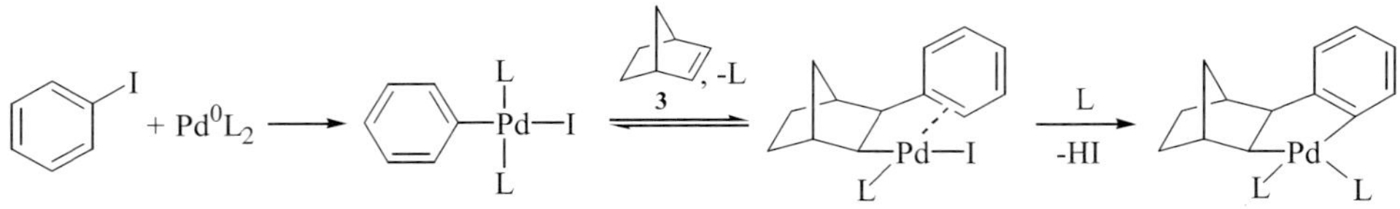

Scheme 1. Palladacycle formation through a sequence of oxidative addition, insertion, and electrophilic aromatic substitution. L = phosphorous or nitrogen ligands, solvent, or coordinating species.

1.4.2.2.2 Palladacycle Reactivity

Palladacycle **5** reacts in various ways depending on ligands and reaction conditions. In particular it readily undergoes oxidative addition of alkyl halides to form a palladium(IV) complex **6**, which has been isolated and characterized with stabilizing nitrogen ligands such as phenanthroline. This palladium(IV) metallacycle has a strong tendency to reductively eliminate to palladium(II) by selective alkyl–aryl coupling, giving species **7** (Scheme 2).

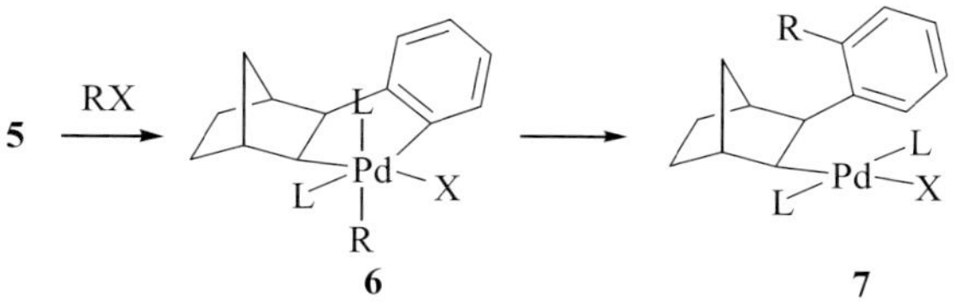

Scheme 2. Palladium(IV) metallacycle formation and subsequent reductive elimination. Isolated species with L–L= phen; R = Me, CH_2=$CHCH_2$, 4-$NO_2C_6H_4$; X = Br, Cl.

At this point the situation becomes analogous to that enabling the formation of palladacycle **5** from **4** and so a new o-substituted palladacycle **8** forms as a result of a second intramolecular C–H activation. Complex **8** again undergoes RX oxidative addition to afford the new palladium(IV) metallacycle **9** which reductively eliminates by selective migration of the R group on to the aromatic site of the same palladacycle to give the palladium(II) species **10**, a disubstituted homolog of complex **7** (Scheme 3).

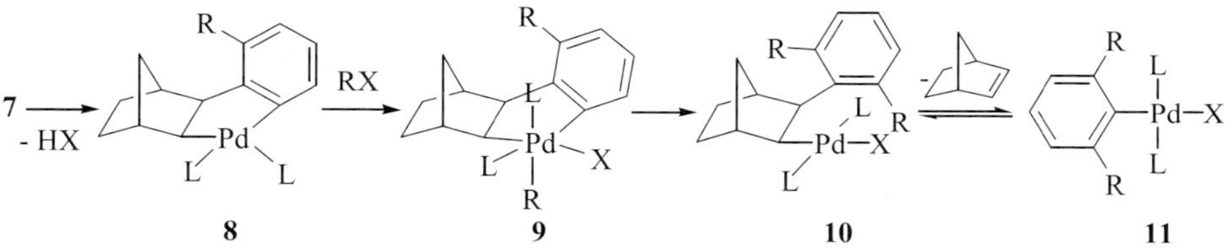

Scheme 3. Intramolecular C–H activation toward mono- and di-o-alkylation of palladacycles and norbornene deinsertion process.

The resulting complex **10** can be viewed as the product of an insertion equilibrium analogous to that leading from **2** to **4** (Scheme 1). This time, however, the presence of the two ortho substituents shifts the equilibrium to the left and norbornene is expelled with formation of the o-dialkylated arylpalladium halide species **11**. This lends itself to a variety of reactions enabling the formation of organic products and palladium(0), which can be represented schematically as follows (Scheme 4).

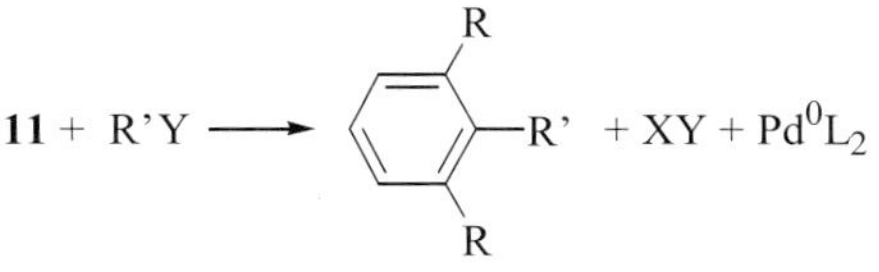

Scheme 4. Termination process regenerating palladium(0).

Complex **5** (Scheme 1) follows a different pathway in reactions with aryl halides (R = aryl in RX). This time the aryl group reacts preferentially with the norbornyl site of the palladacycle rather than with the aryl site (probably via a palladium(IV) metallacycle of type **6**). As a consequence, norbornene is no longer expelled and is incorporated into a condensed ring instead. The minor aryl portion that migrates on to the aryl site (complex **14**) also retains norbornene and gives the same product **13**. A C–H transformation process is thus operating in ring formation from **12** or **14** to **13** (Scheme 5) [2].

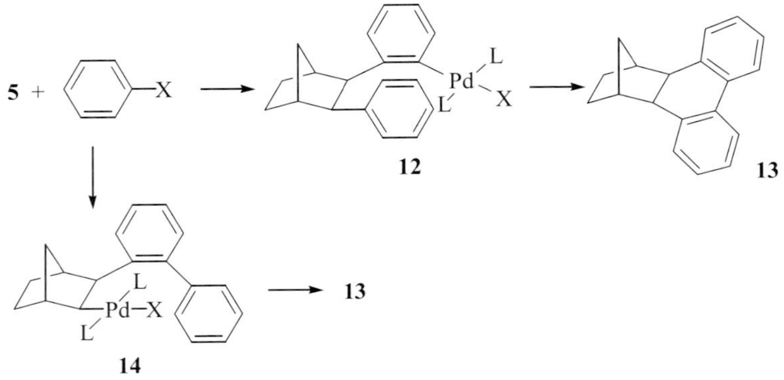

Scheme 5. Reaction pathways for ring closure on aromatic sites.

A recent advance on the method outlined above (Eq. (3), schemes 1–5), is based on the discovery of the ortho effect – when the palladacycle contains an ortho substituent, as in **8**, arylation occurs entirely at the aryl and not at the norbornyl site [3]. This is probably because of the steric effect of the ortho substituent which renders attack at the aryl site much more favorable by preferentially weakening the palladium–carbon bond of the precursor of **15** (Scheme 6).

The consequence is the formation of a sterically hindered complex which readily expels norbornene thus giving rise to the biphenylyl complex **16**. The latter can be caused to react according to Eq. (3).

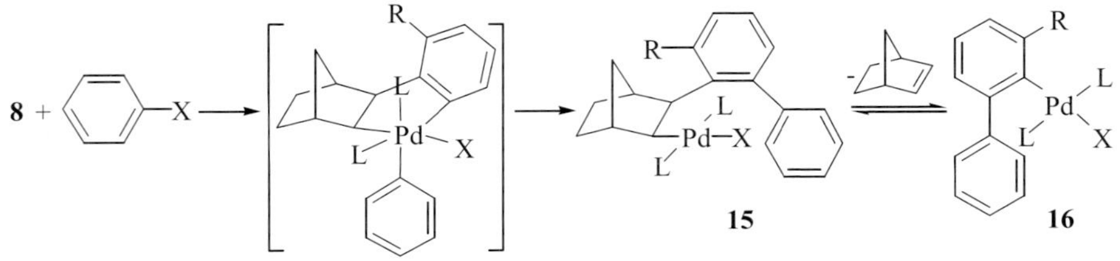

Scheme 6. The ortho effect leading to aromatic arylation.

1.4.2.2.3 Scope and Limitations

The C–H transformations reported here are tolerant of a variety of functional groups but rather sensitive to ortho substituents which strongly interact with palladium or exert significant steric hindrance. The following examples illustrate some specific cases covered by general Eqs (1) and (3).

$$\text{Ph-I} + 2\,n\text{-PrBu} + \text{Ph-B(OH)}_2 \xrightarrow[\text{K}_2\text{CO}_3,\,\text{DMF}]{\text{Pd(OAc)}_2,\,\text{norbornene}} \text{2,6-di-}n\text{-propylbiphenyl} \quad \textbf{17}\ 90\% \tag{4}$$

$$\text{Ph-I} + 2\,n\text{-BuBr} + \text{CH}_2=\text{CHCO}_2\text{Me} \xrightarrow[\text{K}_2\text{CO}_3,\,\text{DMA}]{\text{Pd cat.},\,\text{norbornene}} \text{2,6-di-}n\text{-butyl cinnamate} \quad \textbf{18}\ 93\% \tag{5}$$

Pd cat. = phenylnorbornylpalladium chloride dimer

$$2\ (o\text{-MeO}_2\text{C-C}_6\text{H}_4\text{-Br}) + \text{Ph-B(OH)}_2 \xrightarrow[\text{K}_2\text{CO}_3,\,\text{DMF}]{\text{Pd(OAc)}_2,\,\text{norbornene}} \textbf{19}\ 89\% \tag{6}$$

$$(o\text{-Me-C}_6\text{H}_4\text{-I}) + (p\text{-O}_2\text{N-C}_6\text{H}_4\text{-Br}) + \text{CH}_2=\text{CHCO}_2\text{Me} \xrightarrow[\text{K}_2\text{CO}_3,\,\text{DMF}]{\text{Pd(OAc)}_2,\,\text{norbornene}} \textbf{20}\ 83\% \tag{7}$$

Eqs (4)–(7) show examples of classes of compounds which have been obtained by transformation of one or two sp^2 C–H bonds in the presence of a catalytic system resulting from the cooperation of organic (norbornene) and inorganic (palladium) species. In particular the first example (Eq. 4) entails di-*n*-propylation of the two ortho positions of the phenyl ring of iodobenzene followed by coupling with phenylboronic acid to give 2,6-di-*n*-propyl-1,1′-biphenyl [4]. The second example (Eq. 5) shows the di-*n*-butylation of the same ortho positions, this time followed by a Heck-type reaction with methyl acrylate, leading to 2,6-di-*n*-butyl cinnamic acid methyl ester [5]. The third case (Eq. 6) takes advantage of the ortho effect to form a palladium-bonded biphenylyl species which readily couples with the phenyl group of phenylboronic acid to give a diacid derivative of 2-terphenyl (2,3′-dimethoxycarbonyl-1,1′;2′,1″-terphenyl) [6]. In the last example (Eq. 7) the palladium-bonded biphenylyl group, formed by coupling of two differently substituted aryl moieties, inserts a terminal olefin to form 2-(4-nitrophenyl)-6-methyl cinnamic acid methyl ester [7].

Further extension of the methods shown above has led to new reactions involving condensed ring formation as reported in Eqs (8)–(10).

The synthesis of selectively substituted benzoxepines from ortho-substituted aryl iodides and bromoenoates has been achieved by Lautens and coworkers by palladacycle alkylation followed by an intramolecular Heck reaction under the modified conditions reported in Eq. (8) for synthesis of 1-(1-ethoxycarbonylmethylene)-9-methyl-4,5-dihydro-3-benzoxepine [8]. In the second example (Eq. 9) the palladium-bonded biphenylyl inserts diphenylacetylene to form 1,5-di-i-propyl-9,10-diphenylphenanthrene [9]. In the last case the synthesis of 4-methyl-5H-phenanthridin-6-one is achieved by palladium-catalyzed sequential C–C and C–N bond formation starting from o-iodotoluene and o-bromobenzamide [10].

Experimental

2,6-Di-*n*-propyl-1,1'-biphenyl (17) [4]

A mixture of Pd(OAc)$_2$ (9.0 mg, 0.04 mmol), K$_2$CO$_3$ (332 mg, 2.4 mmol), *o*-iodobenzene (82 mg, 0.4 mmol), *n*-propyl bromide (220 mg, 1.6 mmol), norbornene (38 mg, 0.4 mmol), and phenylboronic acid (59 mg, 0.48 mmol) in dimethylformamide (DMF, 5 mL) was stirred at r.t. under nitrogen for 72 h in a Schlenk-type flask. The mixture was diluted with CH$_2$Cl$_2$ (15 mL) and extracted with 5% H$_2$SO$_4$ (2 × 15 mL). The organic phase was dried over Na$_2$SO$_4$. After removal of the solvent under reduced pressure compound **17** was isolated as a colorless oil (86 mg, 90% yield) by flash chromatography on silica gel using hexane as eluent ($R_F = 0.4$). ^{1}H NMR (CDCl$_3$, 300 MHz): $\delta = 0.75$ (t, $J = 7.3$ Hz, 6H, 2CH$_3$), 1.41 (m, 4H, 2CH_2CH$_3$), 2.27 (m, 4H, 2CH$_2$Ar), 7.15 (m, 2H, H2', H6'), 7.23, 7.11 (AB$_2$ system, $J = 7.3$ Hz, 3H, H4, H3, H5), 7.33 (m, 1H, H4'), 7.40 (m, 2H, H3', H5').

E-3-[6'-Methyl-2'-(4''-nitrophenyl)phenyl]propenoic Acid Methyl Ester (21) [7]

A mixture of Pd(OAc)$_2$ (2.5 mg, 0.011 mmol), K$_2$CO$_3$ (185 mg, 1.34 mmol), *o*-iodotoluene (122 mg, 0.56 mmol), *p*-nitrobromobenzene (114 mg, 0.56 mmol), norbornene (53 mg, 0.56 mmol), and methyl acrylate (78 mg, 0.90 mmol) in DMF (5 mL) was stirred at 105 °C under nitrogen for 24 h in a Schlenk-type flask. After cooling to r.t. the mixture was diluted with dichloromethane (CH$_2$Cl$_2$, 30 mL) and extracted with 5% H$_2$SO$_4$ (2 × 15 mL). The organic phase was washed with water (20 mL) and dried over Na$_2$SO$_4$. After removal of the solvent under reduced pressure compound **21** was isolated as a pale yellow solid (138 mg, 83% yield), m.p. (hexane) 98–99 °C by flash chromatography on silica gel using a mixture of hexane–ethyl acetate, 95:5, as eluent ($R_F = 0.1$). ^{1}H NMR (CDCl$_3$, 300 MHz): $\delta = 2.45$ (s, 3H, CH$_3$), 3.71 (s, 3H, CO$_2$CH$_3$), 5.75 (d, $J = 16.3$ Hz, 1H, C2), 7.17 (dd, $J = 6.5$, 2.5 Hz, 1H, H3'), 7.37–7.28 (m, 1H, H4', H5'), 7.48–7.42 (m, 2H, H2'', H6''), 7.67 (d, $J = 16.4$ Hz, 1H, C3), 8.26–8.21 (m, 2H, H3'', H5''). IR (KBr, cm^{-1}): ν 1718, 1642.

1.4.3
Oxidative Arylation Reactions

Siegfried R. Waldvogel and Daniela Mirk

1.4.3.1 Introduction and Fundamental Examples

On oxidation, electron-rich aryls can undergo oxidative coupling reactions forming aryl–aryl bonds. The dehydrodimerization is a twofold C–H transformation which can proceed by several reaction pathways depending on the nature of the substrate and reagent mixture. Structural requirements include electron-rich arene moieties that incorporate activating substituents. Common donor functionalities are alkyl, alkoxy, amino, and hydroxy groups. If the donor has an acidic proton, e.g. phenols or primary anilines, the oxidative transformation is named a phenol-coupling reaction. Usually, the newly formed bond is ortho to the donor functions, because of the templating effect with the reagent. O-Protected phenols follow a different reaction mechanism and result preferentially in the biaryl coupled para to the donor substituents. When the transformation outlined in Scheme 1 is performed intermolecularly, the well established homo coupling (R = R') occurs, whereas few examples of the oxidative cross-coupling reaction (R ≠ R') are known. In contrast, intramolecular oxidative cyclization can be efficiently achieved when the coupling partners are sufficiently electron rich. Otherwise, dimer formation is observed. The moieties involved in this particular transformation require activation of at least some alkyl substituents on the arene.

Donor—⟨R⟩—H + Donor—⟨R'⟩—H $\xrightarrow[-2HX]{+ MX_n}$ Donor—⟨R⟩—⟨R'⟩—Donor

Scheme 1. Dehydrodimerization of electron-rich aryls (general pictogram).

Many reagents, especially metal salts in high oxidation states, can effect electron transfer and induce the oxidative coupling process. The transformation might be also accomplished catalytically. These procedures are, however, strongly restricted to some substrates. Therefore, a stoichiometric procedure is the most useful for laboratory purposes [1]. Because the reaction rate has to be high to suppress several known side reactions to the desired biaryl formation (dealkylation, Fries-type rearrangements, proto-deiodination, overoxidation, quinol ether formation, etc.), only reagents which promote a rapid transformation are of synthetic value. Furthermore, many powerful reagents for this conversion involve highly toxic metal species, for example Tl(III), Hg(II), Pb(IV), and V(V). Beside the hazardous waste, the products might be contaminated by these toxic metals, which prohibits the synthesis of biologically active compounds. In addition, procedures for this particular transformation should be easy to perform, including removal of the biocompatible by-products. Only a few reagents fulfill these requirements – $FeCl_3$, phenyliodine(III) bis(trifluoroacetate), and $MoCl_5$.

The phenolic coupling can be mediated by different Fe(III) systems. For success of the transformation, the solubility of the reagents is crucial. BINOL (**2**) is efficiently formed from **1** when using the hexahydrate salt. This procedure will result in somewhat lower yields than many reports, but is easily scaled-up to a 5-molar scale [2]. Working in dichloromethane results in more electrophilic conditions which can be applied to more complex structures, for example **3** [3].

Kita et al. demonstrated that reagents based on hypervalent iodine also serve as efficient oxidative agents. The phenolic coupling and the non-phenolic transformation are performed in high yields [4]. Dichloromethane seems to be beneficial for the cationic conversion and enable rapid access to the coupling product **6**. In combination with Lewis acids, PIFA can be employed at low temperatures and accomplishes the transformation without affecting sensitive moieties which are usually not tolerated in the oxidative coupling process. PIFA was found to be the superior reagent for the dehydrodimerization of iodo arenes **7**, providing the corresponding multiple iodinated biaryls **8** [5].

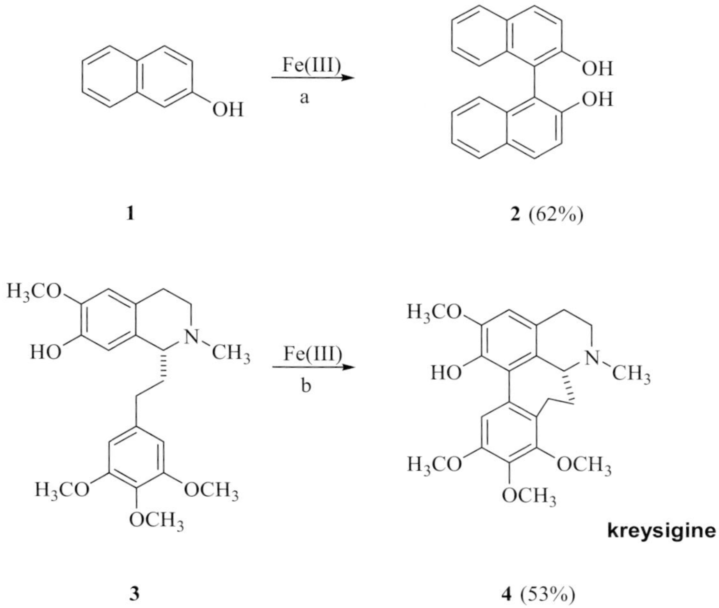

Scheme 2. Oxidative phenol coupling reactions:
(a) FeCl$_3$·(H$_2$O)$_6$, THF, 40 °C, 3 h; (b) FeCl$_3$, CH$_2$Cl$_2$, 25 °C, 9 h.

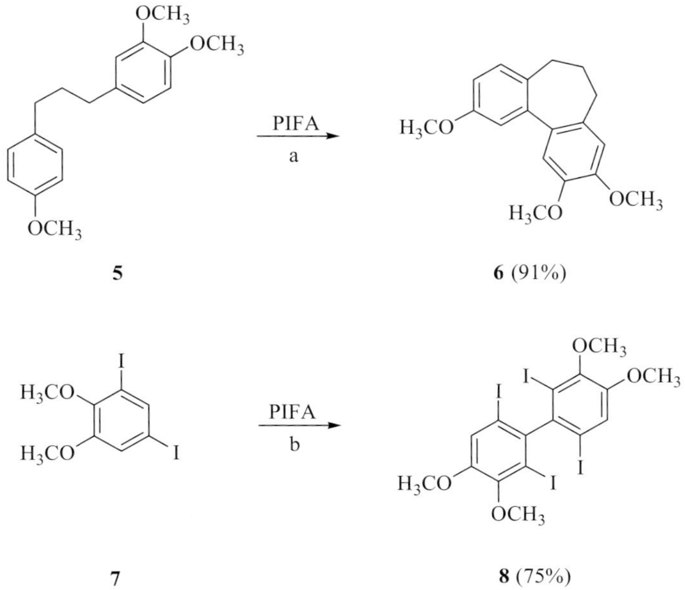

Scheme 3. Dehydrodimerization by phenyliodine(III) bis(trifluoroacetate): (a) PIFA, BF$_3$·OEt$_2$, CH$_2$Cl$_2$, –40 °C, 1.5 h; (b) PIFA, BF$_3$·OEt$_2$, CH$_2$Cl$_2$, 25 °C, 30 min.

The high reaction rate of molybdenum pentachloride-mediated oxidative coupling reactions is compatible with alkyl groups that usually do not survive in strongly electrophilic media and give rise to t-butylated products (**10**) [6]. The oxidative cyclization reaction of **11** proceeds very smoothly when a MoCl$_5$/TiCl$_4$ mixture is applied, providing the seven-membered ring system **12** in excellent yields [7].

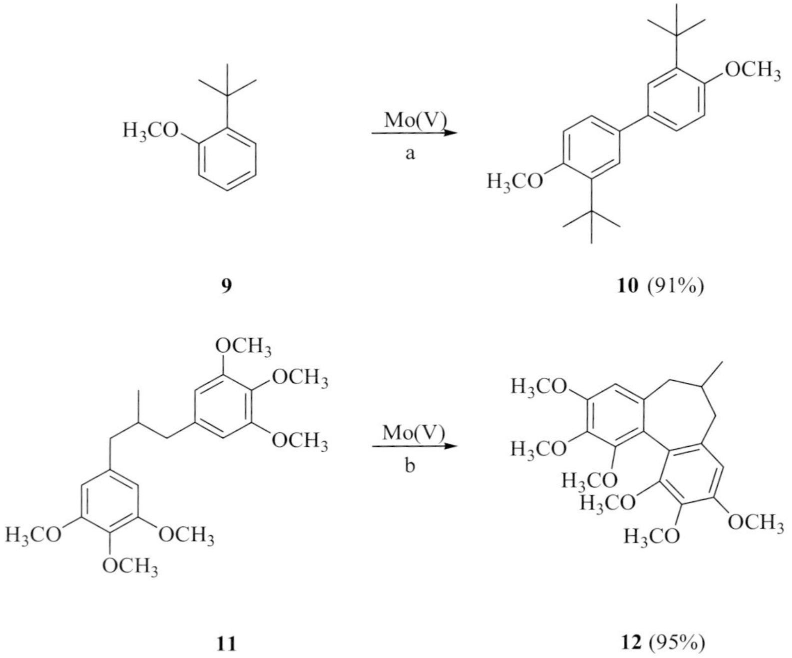

Scheme 4. Oxidative coupling reaction employing MoCl$_5$ systems: (a) MoCl$_5$, CH$_2$Cl$_2$, 0 °C, 30 min; (b) MoCl$_5$, TiCl$_4$, CH$_2$Cl$_2$, 0 °C, 50 min.

1.4.3.2 Mechanism

Depending on the media employed, phenols **13** can be directly oxidized to the corresponding radical cations **14**. The high acidity of this species enables the spontaneous loss of a proton providing the phenoxy radical **15**. Alternatively, **15** can be formed directly from the phenolate **16** by oxidation. This route to the phenoxy radical **15** is easier to perform and therefore preferred. The phenoxy radicals then undergo radical recombination resulting in the dimeric product **17**. Tautomeric processes furnish the biaryl system **18**. Because the intermediate **15** is mesomerically stabilized and has several potential reaction pathways, many coupling products are often obtained. The reaction conditions have a significant effect on product distribution.

1.4 Aryl–Aryl Coupling Reactions | 255

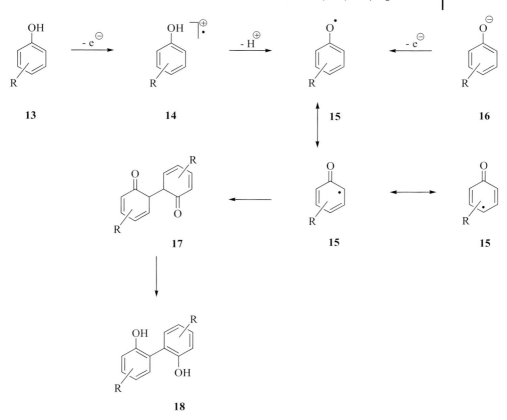

Scheme 5. Mechanism of the phenolic oxidative coupling reaction.

In the non-phenolic oxidative coupling reaction the electron-rich arene **19** undergoes electron transfer yielding the radical cation **20**, which is preferably treated in chlorinated solvents or strongly acidic media. Attack of **20** on the electron-rich reaction partner **21** will proceed in the same way as an electrophilic aromatic substitution involving adduct **22** which extrudes a proton. The intermediate radical **23** is subsequently oxidized to the cationic species **24** which forms the biaryl **25** by rearomatization. In contrast with the mechanism outlined in Scheme 5, two different oxidation steps are required.

For both modern reagents PIFA and MoCl$_5$ an inner-sphere radical transfer is expected, as depicted in Scheme 7. The Lewis acidic additives involved in the PIFA-mediated transformation create an iodonium species that forms a π-complex **26** with the substrate; this subsequently leads to an electron transfer. In contrast, the electrophilic molybdenum chloride most probably coordinates to the oxygen atoms of the donor functions which will then start the transformation. Smooth conversions are obtained if the substituent R″ adjacent to the donor (**27**) is another methoxy group or a bulky moiety.

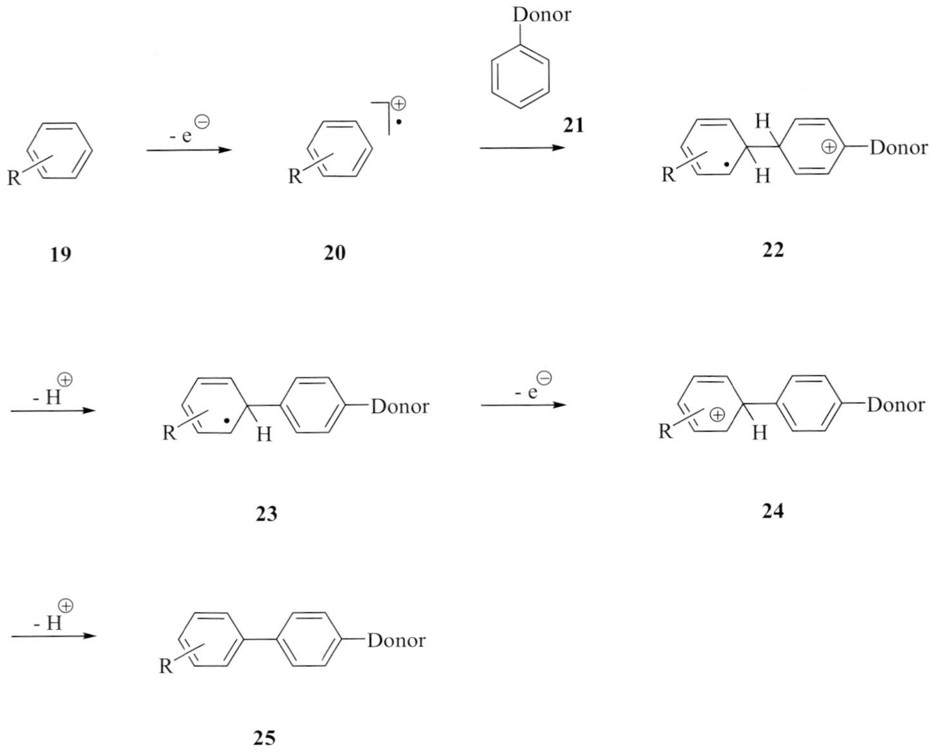

Scheme 6. Dehydrodimerization of non-phenolic substrates.

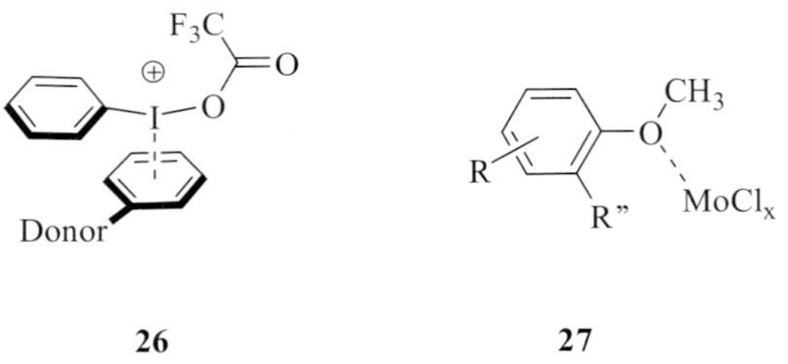

Scheme 7. Expected intermediates enabling inner-sphere radical transfer.

1.4.3.3 Scope and Limitations

The iron(III)-mediated oxidative coupling reaction is convenient to perform, because most of the reagents employed are inexpensive and readily available. In general, the yields obtained are sufficient but can be ameliorated when the conversion is performed heterogeneously. Unfortunately, labile moieties represented by secondary or tertiary alkyl, silyl, or iodo substituents will be shifted or lost during the transformation. For the phenolic coupling transformation, hydrated iron(III) salts or complex species, for example $K_3[Fe(CN)_6]$, can be used. The power of this method is demonstrated in the large-scale route for the intermediate **29** in the galanthamine synthesis, which involves Michael-type addition after formation of a spirodienone moiety (Table 1) [8]. Oxidative treatment of non-phenolic substrates with iron based reagents requires more electrophilic conditions. The conversion of brominated veratrole **30** with anhydrous $FeCl_3$ results in the corresponding biaryl **31** [9]. The combination of ferrous chloride with sulfuric or acetic acid will increase the oxidative power and occasionally lead to partial dealkylation of donor functionalities. Hexamethoxytriphenylene **33** is obtained directly from the parent veratrole **32** when sulfuric acid is added [10].

Table 1. Examples of oxidative coupling with different Fe(III) systems.

Starting material	Product	Yield (%)	Ref.
28	**29**	45–50	8
30	**31**	85	9
32	**33**	98	10

The hypervalent iodo compound PIFA is commercially available and is currently the most versatile reagent for oxidative arylations on a laboratory scale [11]. As the sole reagent, it can be used for the phenolic coupling reaction, which serves as a key transformation for natural product synthesis. Using fluorinated alcohols as solvents in the conversion of **34** turned out to be beneficial (Table 2) [12]. The combination of PIFA with a variety of Lewis acids creates a reliable reagent mixture that enables oxidative arylation of less activated systems **36** [13]. Although PIFA is known to be the best reagent for the dehydrodimerization of iodinated arenes, the treatment of very electron-rich iodo aryls **38** leads to halogen transfer by partial destruction of the aromatic substrate [5].

Table 2. Examples of PIFA-mediated transformations.

Starting material	Product	Yield (%)	Ref.
34	**35**	64	12
36	**37**	81	13
38	**39**	42	5

MoCl$_5$ has a short history as a reagent for the oxidative arylation reaction. It combines strong Lewis-acidic character with high oxidative power. It can, consequently, be successfully employed as the sole reagent in the key step for synthesis of the natural product **41** (Table 3) [14]. Enhancement of the reactivity is achieved

Table 3. Examples of the oxidative coupling reaction using MoCl$_5$.

Starting material	Product	Yield (%)	Ref.
40	**41**	50	14
42	**43**	91	16
44	**45**	80	17
46	**47**	72	20

by combination with other Lewis acids, which essentially bind the by-product, hydrogen chloride [15]. The best results were obtained by employing Lewis acids such as TiCl$_4$, because the electrophilic behavior of the MoCl$_5$ is not reduced. A variety of protective groups for the phenolic oxygen atoms, including silyl, substituted alkyl (**42**), and ketal moieties, were found to be compatible with the oxidative coupling process [16]. The use of alkoxycarbonylmethyl moieties accelerates the oxidative treatment with MoCl$_5$ and gives rise to chlorination reactions. Therefore, the first selective chlorination of iodobenzene **44** was realized [17]. Investigations of biaryl formation from electron-rich benzenes revealed that a 1,2-dialkoxy-substitution pattern was beneficial; two equivalents of $MoCl_5$ are required for each C–C bond formation [18]. A variety of sensitive substituents are tolerated in this transformation [19]. The potential of the reagent is similar to that of PIFA. The oxidative trimerization of catechol ketals **46** to the corresponding triphenylene derivatives **47** was first accomplished by use of $MoCl_5$ [20].

Experimental

1,1′-Binaphthalene-2,2′-diol (BINOL, 2)

Iron(III) chloride hexahydrate (21.6 g, 80 mmol) was dissolved in 13 mL tetrahydrofuran. After 20 min, 2-naphthol **1** (7.2 g, 50 mmol) was added and the reaction mixture was stirred at 80–85 °C for 3 h. The hot mixture was added dropwise to a solution of 20 g citric acid in 300 mL water, with vigorous stirring. The residue in the reaction flask was washed with 50 mL ethyl acetate. After 40 min the suspension was acidified with 30 mL conc. hydrochloric acid and extracted with ethyl acetate (2 × 70 mL). The combined organic layers were washed with 20 mL aqueous hydrochloric acid (10 %), water (2 × 40 mL), and the solvent was evaporated under reduced pressure. Toluene (20 mL) was added and again the solvent was evaporated. The crude product was recrystallized from 40 mL toluene. Standing at room temperature overnight yielded 4.0 g colorless crystals which were removed by filtration, washed with pentane, and dried under high vacuum. After evaporation of the filtrate and subsequent recrystallization from 10 mL toluene, another 500 mg pure product was obtained. Yield: 4.5 g (62 %); m.p. 216–218 °C. ^{1}H NMR ($CDCl_3$, 300 MHz): δ = 5.21 (s, 2H, OH), 7.11–7.52 (m, 8H), 7.91 (d, 2H, J = 7.5 Hz), 7.99 (d, 2H, J = 8.8 Hz).

2,2′,6,6′-Tetraiodo-3,3′,4,4′-tetramethoxybiphenyl (8)

1,5-Diiodo-2,3-dimethoxybenzene (**7**; 585 mg, 1.50 mmol) was dissolved in 20 mL anhydrous dichloromethane. A solution of PIFA (323 mg, 0.75 mmol) and $BF_3 \cdot Et_2O$ (0.19 mL, 1.50 mmol) in 5 mL dichloromethane was added in small portions over a period of 15 min at 25 °C. After stirring for an additional 15 min, the solvent was evaporated under vacuum and without further work-up the crude mixture was directly subjected to column chromatography. TLC (silica, cyclohexane–ethyl acetate, 90:10): R_F = 0.27. On evaporation of the solvents, 439 mg (75 %) biaryl **8** was obtained as a colorless crystalline solid, m.p. >300 °C. ^{1}H NMR ($CDCl_3$, 300 MHz): δ = 3.71 (s, 6H, OCH_3), 3.88 (s, 6H, OCH_3), 7.57 (s, 2H, CH).

3,3′-Di-t-butyl-4,4′-dimethoxybiphenyl (10)

$MoCl_5$ (1.09 g, 4.0 mmol) was dissolved in 20 mL anhydrous dichloromethane under an inert atmosphere and cooled to 0 °C. 2-t-Butyl anisole (**9**; 328 mg, 2.00 mmol) in 5 mL dichloromethane was added and the reaction mixture was stirred for 30 min at 0 °C. After quenching with 25 mL saturated $NaHCO_3$ solution, the aqueous layer was extracted with ethyl acetate (3 × 50 mL). Subsequent treatment with brine, anhydrous $MgSO_4$, and concentration of the organic fraction provided the crude product, which was purified by column chromatography on silica with cyclohexane–ethyl acetate, 95:5; $R_F = 0.58$. Biphenyl **10** (298 mg, 91%), m.p. 124 °C (EtOH), was obtained as a light yellow solid on evaporation of the solvents. ^{1}H NMR (CDCl$_3$, 300 MHz): $\delta = 1.43$ (s, 18H, CH$_3$), 3.86 (s, 6H, OCH$_3$), 6.92 (d, $J = 8.6$ Hz, 2H, 5-H), 7.35 (dd, $J = 8.6$, 2.4 Hz, 2H, 6-H), 7.46 (d, $J = 2.4$ Hz, 2H, 2-H).

11H-9,10-Dihydro-10-methyl-2,3,4,5,6,7-hexamethoxydibenzo[a,c]cycloheptene (12)

Under anhydrous, inert conditions, **11** (1.00 g, 2.56 mmol) was dissolved in 30 mL dichloromethane. The chilled mixture (0 °C) was rapidly treated with TiCl$_4$ (0.53 mL, 4.87 mmol) and MoCl$_5$ (1.33 g, 4.87 mmol). The reaction mixture was then stirred for 50 min at 0 °C and subsequently fractionated between 150 mL ethyl acetate and sat. NaHCO$_3$ solution (2 × 100 mL). The aqueous phase was re-extracted with ethyl acetate. The combined organic phases were treated with brine (2 × 50 mL) and anhydrous MgSO$_4$ providing, on concentration, the crude product, which was purified by column chromatography on silica with cyclohexane–ethyl acetate, 80:20; $R_F = 0.33$. **12** (0.95 g, 95%) was obtained as slightly colored crystals, m.p. 139 °C. ^{1}H NMR (CDCl$_3$, 400 MHz): $\delta = 0.95$ (d, $J = 9.2$ Hz, 3H, 12-H), 1.80–1.87 (m, 1H, 9-/11-H_{eq}), 2.08–2.13 (m, 1H, 9-/11-H_{ax}), 2.25–2.34 (m, 1H, 10-H), 2.35–2.41 (m, 1H, 9-/11-H_{eq}), 2.47–2.54 (m, 1H; 9-/11-H_{ax}), 3.65, 3.66 (s, 6H, 4-H-, 5-H-OCH$_3$), 3.88, 3.89, 3.90, 3.91 (s, 12H, 2-, 3-, 6-, 7-OCH$_3$), 6.50, 6.55 (s, 2H, 1-, 8-H).

References to Chapter 1 – C–H Transformation at Arenes

References and Notes to Section 1.1

1 C. F. Cross, E. J. Bevan, T. Herberg, *Ber.* **1900**, *33*, 2015.
2 C. Walling, R. A. Johnson, *J. Am. Chem. Soc.* **1975**, *97*, 363.
3 J. H. Merz, W. A. Waters, *J. Chem. Soc.* **1949**, 2427.
4 C. Walling, D. M. Camaioni, S. S. Kim, *J. Am. Chem. Soc.* **1978**, *100*, 4814.
5 J. Wellmann, E. Steckhan, *Chem. Ber.* **1977**, *110*, 3561.
6 S. Udenfriend, C. T. Clark, J. Axelrod, B. B. Brodie, *J. Biol. Chem.* **1954**, *208*, 731.
7 B. B. Brodie, J. Axelrod, P. A. Shore, S. Udenfriend, *J. Biol. Chem.* **1954**, *208*, 741.
8 S. Ito, T. Yamasaki, H. Okada, S. Okino, K. Sasaki, *J. Chem. Soc., Perkin Trans. 2* **1988**, 285.
9 S. Ito, A. Kunai, H. Okada, K. Sasaki, *J. Org. Chem.* **1988**, *53*, 296.
10 Y. Mao, A. Bakac, *Inorg. Chem.* **1996**, *35*, 3925.
11 C. Walling, *Acc. Chem. Res.* **1998**, *31*, 155.
12 J. O. Edwards, R. Curci, in *Catalytic Oxidations with Hydrogen Peroxide as Oxidant* (Ed.: G. Strukul), Kluwer Academic Publishers, **1992**, p. 97.
13 G. Sosnovsky, D. J. Rawlinson, in *Organic Peroxides*, Vol. 2 (Ed.: D. Swern), Wiley-Interscience, New York, **1971**, p. 269.
14 W. T. Dixon, R. O. C. Norman, *J. Chem. Soc.* **1964**, 4857.
15 E. A. Karakhanov, S. Y. Narin, T. Y. Filippova, A. G. Dedov, *Dokl. Akad. Nauk SSSR* **1987**, *292*, 1387.
16 D. Bianchi, M. Bertoli, R. Tassinari, M. Ricci, R. Vignolo, *J. Mol. Catal. A: Chem.* **2003**, *204–205*, 419.

17 D. Bianchi, R. Bortolo, R. Tassinari, M. Ricci, R. Vignolo, *Ang. Chem. Int. Ed.* **2000**, *39*, 4321.

18 D. Bianchi, M. Bertoli, R. Tassinari, M. Ricci, R. Vignolo, *J. Mol. Catal. A: Chem.* **2003**, *200*, 111.

19 J. R. L. Smith, B. A. J. Shaw, D. M. Foulkes, A. M. Jeffrey, D. M. Jerina, *J. Chem. Soc., Perkin Trans. 2* **1977**, 1583.

20 H. Mimoun, L. Saussine, E. Daire, M. Postel, J. Fischer, R. Weiss, *J. Am. Chem. Soc.* **1983**, *105*, 3101.

21 G. B. Shul'pin, D. Attanasio, L. Suber, *J. Catal.* **1993**, *142*, 147.

22 M. Bianchi, M. Bonchio, V. Conte, F. Coppa, F. Di Furia, G. Modena, S. Moro, S. Standen, *J. Mol. Catal.* **1993**, *83*, 107.

23 M. Bonchio, V. Conte, F. Di Furia, G. Modena, S. Moro, *J. Org. Chem.* **1994**, *59*, 6262.

24 E. Hata, T. Takai, T. Yamada, T. Mukaiyama, *Chem. Lett.* **1994**, 1849.

25 E. Battistel, R. Tassinari, M. Fornaroli, L. Bonoldi, *J. Mol. Catal. A: Chem.* **2003**, *202*, 107.

26 G. B. Shul'pin, E. R. Lachter, *J. Mol. Catal. A: Chem.* **2003**, *197*, 65.

27 G. A. Olah, R. Ohnishi, *J. Org. Chem.* **1978**, *43*, 865.

28 M. E. Kurz, G. J. Johnson, *J. Org. Chem.* **1971**, *36*, 3184.

29 E. J. Behrman, in *Organic Reactions, Vol. 35*, John Wiley and Sons, New York, **1988**, p. 421.

30 E. J. Behrman, *Biochem. J.* **1982**, *201*, 677.

31 A. Marsella, S. Agapakis, F. Pinna, G. Strukul, *Organometallics* **1992**, *11*, 3578.

32 T. Jintoku, H. Taniguchi, Y. Fujiwara, *Chem. Lett.* **1987**, 1865.

33 T. Jintoku, K. Nishimura, K. Takaki, Y. Fujiwara, *Chem. Lett.* **1990**, 1687.

34 A. Kunai, T. Wani, Y. Uehara, F. Iwasaki, Y. Kuroda, S. Ito, K. Sasaki, *Bull. Chem. Soc. Jpn.* **1989**, *62*, 2613.

35 L. C. Passoni, A. T. Cruz, R. Buffon, U. Schuchardt, *J. Mol. Catal. A: Chem.* **1997**, *120*, 117.

36 F. Shibahara, S. Kinoshita, K. Nozaki, *Org. Lett.* **2004**, *6*, 237.

37 T. Jintoku, K. Takaki, Y. Fujiwara, Y. Fuchita, K. Hiraki, *Bull. Chem. Soc. Jpn.* **1990**, *63*, 438.

38 Y. Fuchita, K. Hiraki, Y. Kamogawa, M. Suenaga, K. Tohgoh, Y. Fujiwara, *Bull. Chem. Soc. Jpn.* **1989**, *62*, 1081.

39 M. Makosza, K. Sienkiewicz, *J. Org. Chem.* **1998**, *63*, 4199.

40 K. Fujimoto, Y. Tokuda, H. Maekawa, Y. Matsubara, T. Mizuno, I. Nishiguchi, *Tetrahedron* **1996**, *52*, 3889.

41 R. H. Thomson, in *The Chemistry of Quinonoid Compounds, Vol. 1* (Ed.: S. Patai), John Wiley and Sons, **1974**, p. 111.

42 M. Periasamy, M. V. Bhatt, *Synthesis* **1977**, 330.

43 W. Adam, W. A. Herrmann, J. Lin, C. R. Saha-Moeller, R. W. Fischer, J. D. G. Correia, *Ang. Chem. Int. Ed. Engl.* **1994**, *33*, 2475.

44 W. A. Herrmann, J. J. Haider, R. W. Fischer, *J. Mol. Catal. A: Chem.* **1999**, *138*, 115.

45 R. Song, A. Sorokin, J. Bernadou, B. Meunier, *J. Org. Chem.* **1997**, *62*, 673.

46 W. Adam, W. A. Herrmann, J. Lin, C. R. Saha-Moeller, *J. Org. Chem.* **1994**, *59*, 8281.

47 W. Adam, W. A. Herrmann, C. R. Saha-Moeller, M. Shimizu, *J. Mol. Catal. A: Chem.* **1995**, *97*, 15.

48 R. Saladino, V. Neri, E. Mincione, S. Marini, M. Coletta, C. Fiorucci, P. Filippone, *J. Chem. Soc., Perkin Trans. 1* **2000**, 581.

49 H. Sun, K. Harms, J. Sundermeyer, *J. Am. Chem. Soc.* **2004**, *126*, 9550.

References and Notes to Section 1.2.1

1 (a) P. M. Esteves, J. Walkimar de Carneiro, S. P. Cardoso, A. G. H. Barbosa, K. K. Laali, G. Rasul, G. K. S. Prakash, G. A. Olah, *J. Am. Chem. Soc.* **2003**, *125*, 4836–4849; (b) R. Taylor, *Electrophilic Aromatic Substitution*, John Wiley & Sons, USA, **1990**.

2 (a) M. R. Crampton, *Organic Reaction Mechanisms*, A. C. Knipe, W. E. Watts (Eds.), **2004**, John Wiley & Sons, UK, pp. 283–294; (b) M. Makosza,

W. Wojciechowski, *Heterocycles* **2001**, *54*, 445–474.
3   R. A. Rossi, A. B. Pierini, A. B. Penenory, *Chem. Rev.* **2003**, *103*, 71–167.
4   (a) M. Makosza, A. Kwast, *Eur. J. Org. Chem.* **2004**, *10*, 2125–2130; (b) M. Makosza, T. Lemek, A. Kwast, F. Terrier, *J. Org. Chem.* **2002**, *67*, 394–400.
5   (a) J. Clayden, *The Chemistry of Organolithium Compounds*, Z. Rappoport, I. Marek (Eds.), John Wiley & Sons, UK, **2004**, pp. 495–646; (b) C. G. Hartung and V. Snieckus, *Modern Arene Chemistry*, D. Astruc (Ed.), Wiley–VCH, USA, **2002**, pp. 330–367; (c) V. Snieckus, *Chem. Rev.* **1990**, *90*, 879–933.
6   (a) S. Murai, Ed. *Activation of Unreactive Bonds and Organic Synthesis*, Vol 3, Springer, Germany, **1999**; (b) F. Kakiuchi, S. Murai, *Acc. Chem. Res.* **2002**, *35*, 826–834.
7   New developments: A. G. M. Barrett, N. Bouloc, D. C. Braddock, D. Chadwick, D. A. Henderson, *Synlett* **2002**, 1653–1656 and refs cited therein.
8   R. J. Mills, V. Snieckus, *Tetrahedron Lett.* **1984**, *25*, 483–486.
9   B. A. Chauder, A. V. Kalinin, V. Snieckus, *Synthesis* **2001**, 140–144.
10 R. J. Mills, N. J. Taylor, V. Snieckus, *J. Org. Chem.* **1989**, *54*, 4372–4385.
11 M. P. Sibi, V. Snieckus, *J. Org. Chem.* **1983**, *48*, 1935–1937.
12 M. P. Sibi, S. Chattopadhyay, J. W. Dankwardt, V. Snieckus, *J. Am. Chem. Soc.* **1985**, *107*, 6312–6315.
13 A. V. Kalinin, A. J. M. Da Silva, C. C. Lopes, R. S. C. Lopes, V. Snieckus, *Tetrahedron Lett.* **1998**, *39*, 4995–4998.
14 B. A. Chauder, A. V. Kalinin, N. J. Taylor, V. Snieckus, *Angew. Chem. Int. Ed.* **1999**, *38*, 1435–1438.
15 M. C. Whisler, S. MacNeil, V. Snieckus, P. Beak, *Angew. Chem. Int. Ed.* **2004**, *43*, 2206–2225.
16 (a) E. J.-G. Anctil, V. Snieckus, *J. Organomet. Chem.* **2002**, *653*, 150–160; (b) E. J.-G. Anctil, V. Snieckus, *Metal-Catalyzed Cross-Coupling Reactions*, 2nd Ed., F. Diederich, A. de Meijere (Eds.), Wiley–VCH, Germany, **2004**, pp. 761–813.

17 J. A, McCubbin, X. Tong, W. Ruiyao, Y. Zhao, V. Snieckus, R. P. Lemieux, *J. Am. Chem. Soc.* **2004**, *126*, 1161–1167.
18 W. Wang, V. Snieckus, *J. Org. Chem.* **1992**, *57*, 424–426.
19 M. Watanabe, M. Sahara, M. Kubo, S. Furukawa, R. J. Billedeau, V. Snieckus, *J. Org. Chem.* **1984**, *49*, 742–747.
20 R. J. Mills, V. Snieckus, *J. Org. Chem.* **1983**, *48*, 1565–1568.
21 J. A. Turner, *J. Org. Chem.* **1983**, *48*, 3401–3408.
22 (a) S. Yu, C. Rabalakos, W. D. Mitchell, W. D. Wulff, *Org. Lett.* **2005**, *7*, 367–369; (b) M. P. Sibi, J. W. Dankwardt, V. Snieckus, *J. Org. Chem.* **1986**, *51*, 271–273.
23 Z. Zhao, V. Snieckus, *Org. Lett.* **2005**, submitted.
24 T. Kremer, M. Junge, P. v. R. Schleyer, *Organomet.* **1996**, *15*, 3345–3359.
25 R. A. Gossage, T. B. H. Jastrzebski, G. van Koten, *Angew. Chem. Int. Ed.* **2005**, *44*, 1448–1454
26 G. Queguiner, F. Marsais, V. Snieckus, J. Epsztajn, *Adv. Heterocycl. Chem.* **1991**, *52*, 187–304.
27 R. R. Fraser, M. Bresse, T. S. Mansour, *J. Am. Chem. Soc.* **1983**, *105*, 7790–7791.
28 (a) M. Uchiyama, H. Naka, Y. Matsumoto, T. Ohwada, *J. Am. Chem. Soc.* **2004**, *126*, 10526–10527; (b) M. Uchiyama, T. Miyoshi, Y. Kajihara, T. Sakamoto, Y. Otani, T. Ohwada, Y. Kondo, *J. Am. Chem. Soc.* **2002**, *124*, 8514–8515; (c) P. B. Kisanga, J. G. Verkade, R. Schwessinger, *J. Org. Chem.* **2000**, *65*, 5431–5432.
29 P. Beak, T. J. Musick, C. Liu, T. Cooper, D. J. Gallagher, *J. Org. Chem.* **1993**, *58*, 7330–7335.
30 G. E. Keck, T. T. Wager, J. F. D. Rodriquez, *J. Am. Chem. Soc.* **1999**, *121*, 5176–5190.
31 M. P. Sibi, K. Shankaran, B. I. Alo, W. R. Hahn, V. Snieckus, *Tetrahedron Lett.* **1987**, *28*, 2933–2936.
32 S. O. De Silva, I. Ahmad, V. Snieckus, *Can. J. Chem.* **1979**, *57*, 1598–1605.
33 S. O. De Silva, J. N. Reed, R. J. Billedeau, X. Wang, D. J. Norris, V. Snieckus, *Tetrahedron* **1992**, *48*, 4863–4878.

34 T. K. Macklin, V. Snieckus, unpublished results.
35 P. Stanetty, B. Krumpak, *J. Org. Chem.* **1996**, *61*, 5130–5133.
36 H. Butenschoen, M. Winkler, K. P. C. Vollhardt, *J. Chem. Soc., Chem. Comm.* **1986**, 388–390.
37 S. Danishefsky, J. Y. Lee, *J. Am. Chem. Soc.* **1989**, *111*, 4829–4837.
38 (a) Y. S. Lo, L. T. Rossano, D. J. Meloni, J. R. Moore, Y. –C. Lee, J. F. Arneett, *J. Heterocycl. Chem.* **1995**, *32*, 355–357; (b) R. D. Larsen, A. O. King, C. Y. Chen, E. G. Corley, B. S. Foster, F. E. Roberts, C. Yang, D. R. Lieberman, R. A. Reamer, D. M. Tschaen, T. R. Verhoeven, P. J. Reider, *J. Org. Chem.* **1994**, *59*, 6391–6394.

References and Notes to Section 1.2.2

1 J. Volhard, *Liebigs Ann. Chem.*, **1892**, *267*, 172.
2 L. Pesci, *Gazz. Chim. Ital.*, **1892**, *22*, 373; **1893**, *23*, 521; *Z. Anorg. Allg. Chem.*, **1897**, *15*, 208.
3 O. Dimroth, *Ber. Dtsch. Chem. Ges.*, **1898**, *31*, 2154; **1899**, *32*, 758; **1902**, *35*, 2032, 2853.
4 A. E. Shilov, G. B. Shul'pin, *Activation and Catalytic Reactions of Saturated Hydrocarbons in the Presence of Metal Complexes*, Kluwer Academic Publishers, Dordrecht, **2000**.
5 M. Wills, *Contemp. Org. Synth.*, **1996**, *3*, 201.
6 R. Taylor, *Electrophilic Aromatic Substitution*; Wiley, New York, **1990**.
7 L. G. Makarova, A. N. Nesmeyanov, *Organic Compounds of Mercury*, North-Holland, Amsterdam, **1967**.
8 W. Kitching, *Organometal. Chem. Rev.*, **1968**, *3*, 35.
9 A. J. Bloodworth, *The Organic Chemistry of Mercury*, C. A. McAuliffe, Ed., Macmillan, London, **1977**. R. C. Larock, *Organomercury Compounds in Organic Synthesis*, Springer-Verlag, Berlin, **1985**.
10 V. I. Boev, A. I. Moskalenko, A. M. Boev, *Russ. Chem. Rev.*, **1997**, *66*, 789.
11 D. K. Breitinger, *Synth. Methods Organomet. Inorg. Chem.*, **1999**, *5*, 193; *Compr. Coord. Chem. II*, **2004**, *6*, 1253.
12 H. C. Brown, R. A. Wirkkala, *J. Am. Chem. Soc.*, **1966**, *88*, 1447.
13 G. B. Deacon, G. J. Farquharson, *J. Organomet. Chem.*, **1974**, *67*, C1.
14 V. I. Bregadze, V. Ts. Kampel, N. N. Godovikov, *J. Organomet. Chem.*, **1976**, *112*, 249.
15 G. A. Olah, S. H. Yu, D. G. Parker, *J. Org. Chem.*, **1976**, *41*, 1983 and references cited therein.
16 W. Lau, J. K. Kochi, *J. Am. Chem. Soc.*, **1986**, *108*, 6720 and reference cited therein.
17 W. Lau, J. C. Huffman, J. K. Kochi, *J. Am. Chem. Soc.*, **1982**, *104*, 5515.
18 L. C. Damude, P. A. W. Dean, *J. Organomet. Chem.*, **1979**, *181*, 1. L. C. Damude, P. A. W. Dean, M. D. Sefcik, J. Schaeffer, *J. Organomet. Chem.*, **1982**, *226*, 105.
19 A. S. Borovik, S. G. Bott, A. R. Barron, *Angew. Chem. Int. Ed.*, **2000**, *39*, 4117; *J. Am. Chem. Soc.*, **2001**, *123*, 11219. A. S. Borovik, A. R. Barron, *J. Am. Chem. Soc.*, **2002**, *124*, 3743. C. S. Branch, A. R. Barron, *J. Am. Chem. Soc.*, **2002**, *124*, 14156.
20 M. R. Haneline, M. Tsunoda, F. P. Gabbai, *J. Am. Chem. Soc.*, **2002**, *124*, 3737. M. R. Haneline, J. B. King, F. P. Gabbai, *Dalton*, **2003**, 2686.
21 J. K. Kochi, *Angew. Chem. Int. Ed. Engl.*, **1988**, *27*, 1227.
22 A. G. Davies, D. C. McGuchan, *Organometallics*, **1991**, *10*, 329. A. G. Davies, K. M. Ng, *J. Chem. Soc., Perkin Trans. 2*, **1998**, 2599.
23 L. Buzhansky, B.-A. Feit, *J. Org. Chem.*, **2002**, *67*, 7523.
24 H. Gilman, R. K. Abbott, Jr., *J. Am. Chem. Soc.*, **1943**, *65*, 122.
25 V. P. Glushkova, K. A. Kocheshkov, *Dokl. Akad. Nauk SSSR*, **1957**, *116*, 233.
26 A. McKillop, J. D. Hunt, M. J. Zelesko, J. S. Fowler, E. C. Taylor, G. McGillivray, F. Kienzle, *J. Am. Chem. Soc.*, **1971**, *93*, 4841. E. C. Taylor, F. Kienzle, R. L. Robey, A. McKillop, J. D. Hunt, *J. Am. Chem. Soc.*, **1971**, *93*, 4845.
27 A. McKillop, E. C. Taylor, *Adv. Organometal. Chem.*, **1973**, *11*, 147.
28 S. Uemura, *Synth. Reagents*, **1983**, *5*, 165.
29 A. Ya. Usyatinskii, V. I. Bregadze, *Russ. Chem. Rev.*, **1988**, *57*, 1054.

30 H. M. C. Ferraz, L. F. Silva Jr., T. de O. Vieira, *Synthesis*, **1999**, 2001.
31 W. Lau, J. K. Kochi, *J. Am. Chem. Soc.*, **1984**, *106*, 7100.
32 E. C. Taylor, E. C. Bigham, D. K. Johnson, A. Mckillop, *J. Org. Chem.*, **1977**, *42*, 362.
33 V. V. Grushin, *Chem. Eur. J.*, **2002**, *8*, 1006.
34 J. T. Pinhey, *Aust. J. Chem.*, **1991**, *44*, 1353; *Lead. In: Comprehensive Organometallic Chemistry II*, E. W. Abel, F. G. A. Stone, G. Wilkinson, Eds., **1995**, Vol. 11, 461; *Pure Appl. Chem.* **1996**, *68*, 819.
35 M. G. Moloney, *Main Group Metal Chem.*, **2001**, *24*, 653.
36 G. I. Elliott, J. P. Konopelski, *Tetrahedron*, **2001**, *57*, 5683.
37 P. J. Guiry, P. J. McCormack, *Sci. Synth.*, **2003**, *5*, 673.
38 E. M. Panov, K. A. Kocheshkov, *Dokl. Akad. Nauk SSSR*, **1958**, *123*, 295. F. R. Preuss, R. Menzel, *Arch. Pharm.*, **1958**, *291*, 350.
39 L. C. Willemsens, D. De Vos, J. Spierenburg, J. Wolters, *J. Organometal. Chem.*, **1972**, *39*, C61.
40 L. M. Stock, T. L. Wright, *J. Org. Chem.*, **1980**, *45*, 4645.
41 J. R. Kalman, J. T. Pinhey, S. Sternhell, *Tetrahedron Lett.* **1972**, 5369.
42 D. M. X. Donnelly, J.-P. Finet, J. M. Kielty, *Tetrahedron Lett.*, **1991**, *31*, 3835.
43 V. V. Grushin, W. J. Marshall, D. L. Thorn, *Adv. Synth. Catal.*, **2001**, *343*, 433.
44 D. Seyferth, M. O. Nestle, *J. Am. Chem. Soc.*, **1981**, *103*, 3320.
45 B. K. Nicholson, S. K. Whitley, *J. Organomet. Chem.*, **2004**, *689*, 515.

References and Notes to Section 1.2.3

1 T. Ishiyama, N. Miyaura, *J. Organomet. Chem.* **2003**, *680*, 3–11.
2 T. Ishiyama, N. Miyaura, *Chem. Rec.* **2004**, *3*, 271–280.
3 C. N. Iverson, M. R. Smith, III., *J. Am. Chem. Soc.* **1999**, *121*, 7696–7697.
4 J.-Y. Cho, C. N. Iverson, M. R. Smith, III., *J. Am. Chem. Soc.* **2000**, *122*, 12868–12869.
5 J.-Y. Cho, M. K. Tse, D. Holmes, R. E. Maleczka, Jr., M. R. Smith, III., *Science* **2002**, *295*, 305–308.
6 T. Ishiyama, J. Takagi, K. Ishida, N. Miyaura, N. R. Anastasi, J. F. Hartwig, *J. Am. Chem. Soc.* **2002**, *124*, 390–391.
7 J. Takagi, K. Sato, J. F. Hartwig, T. Ishiyama, N. Miyaura, *Tetrahedron Lett.* **2002**, *43*, 5649–5651.
8 T. Ishiyama, J. Takagi, J. F. Hartwig, N. Miyaura, *Angew. Chem., Int. Ed.* **2002**, *41*, 3056–3058.
9 T. Ishiyama, J. Takagi, Y. Yonekawa, J. F. Hartwig, N. Miyaura, *Adv. Synth. Catal.* **2003**, *345*, 1103–1106.
10 T. Ishiyama, Y. Nobuta, J. F. Hartwig, N. Miyaura, *Chem. Commun.* **2003**, 2924–2925.
11 P. Nguyen, H. P. Blom, S. A. Westcott, N. J. Taylor, T. B. Marder, *J. Am. Chem. Soc.* **1993**, *115*, 9329–9330.
12 H. Tamura, H. Yamazaki, H. Sato, S. Sakaki, *J. Am. Chem. Soc.* **2003**, *125*, 16114–16126.
13 R. Uson, L. A. Oro, J. A. Cabeza, *Inorg. Synth.* **1985**, *23*, 126–130.

References to Section 1.2.4

1 F. Kakiuchi, N. Chatani, *Adv. Synth. Catal.* **2003**, *345*, 1077–1101.
2 W. A. Gustavson, P. S. Epstein, M. D. Curtis, *Organometallics* **1982**, *1*, 884–885.
3 P. I. Djurovich, A. R. Dolich, D. H. Berry, *J. Chem. Soc., Chem. Commun.* **1994**, 1897–1898.
4 K. Ezbiansky, P. I. Djurovich, M. LaForest, D. J. Sinning, R. Zayes, D. H. Berry, *Organometallics* **1998**, *17*, 1455–1457.
5 Y. Uchimaru, A. M. M. El Sayed, M. Tanaka, *Organometallics* **1993**, *12*, 2065–2069.
6 F. Kakiuchi, K. Igi, M. Matsumoto, N. Chatani, S. Murai, *Chem. Lett.* **2001**, 422–423
7 F. Kakiuchi, K. Igi, M. Matsumoto, T. Hayamizu, N. Chatani, S. Murai, *Chem. Lett.* **2002**, 396–397.

8 F. Kakiuchi, M. Matsumoto, K. Tsuchiya, K. Igi, T. Hayamizu, N. Chatani, S. Murai, *J. Organomet. Chem.* **2003**, *686*, 134–144.
9 F. Kakiuchi, K. Tsuchiya, M. Matsumoto, E. Mizushima, N. Chatani, *J. Am. Chem. Soc.* in press.
10 M. Ishikawa, S. Okazaki, A. Naka, H. Sakamoto, *Organometallics* **1992**, *11*, 4135–4139; M. Shikawa, A. Naka, *Synlett* **1995**, 795–802; A. Naka, K. K. Lee, K. Yoshizawa, T. Yamabe, M. Ishikawa, *Organometallics* **1999**, *18*, 4524.
11 N. A. Williams, Y. Uchimaru, M. Tanaka, *Dalton Trans.* **2003**, 236–243.
12 F. Kakiuchi, M. Matsumoto, M. Sonoda, T. Fukuyama, N. Chatani, S. Murai, N. Furukawa, Y. Seki, *Chem. Lett.* **2000**, 750–751.

References and Notes to Section 1.3.1.1

1 G. Dyker, E. Muth, A. S. K. Hashmi, L. Ding, *Adv. Synth. Catal.* **2003**, *345*, 1247–1252.
2 M. M. Alam, M. Mujahid, R. Varala, S. R. Adapa, *Tetrahedron Lett.* **2003**, *44*, 5115–5119.
3 For other catalysts such as CeCl$_3$ and InCl$_3$, see: G. Bartoli, M. Bartolacci, M. Bosco, G. Foglia, A. Giuliani, E. Marcantoni, L. Sambri, E. Torregiani, *J. Org. Chem.* **2003**, *68*, 4594–4597; J. S. Yadav, S. Abraham, B. V. S. Reddy, G. Sabitha, *Synthesis* **2001**, 2165–2169; K. Manabe, N. Aoyama, S. Kobayashi, *Adv. Synth. Catal.* **2001**, *343*, 174–176; N. Srivastava, B. K. Banik, *J. Org. Chem.* **2003**, *68*, 2109–2114; G. Dujardin, J.-M. Poirier, Bull. Soc. Chim. France **1994**, *131*, 900–909.
4 Y. Nishibayashi, Y. Inada, M. Yoshikawa, M. Hidai, S. Uemura, *Angew. Chem. Int. Ed.* **2003**, *42*, 1495–1498.
5 G. Dyker, E. Muth, unpublished results.
6 J. J. Kennedy-Smith, L. A. Young, F. D. Toste, *Org. Lett.* **2004**, *6*, 1325–1327.

References and Notes to Section 1.3.1.2

1 (a) G. A. Olah, "Friedel–Crafts and Related Reactions" Wiley-Interscience, New York, (1964). Vol. 1; (b) H. Heaney, In "Comprehensive Organic Synthesis," ed by B. M. Trost, I. Fleming, Pergamon Press, Oxford (1991), p. 733.
2 Examples of early trials, (a) M. Hino, K. Arata, *Chem. Lett.* **1978**, 325; b) K. Nomita, Y. Sugaya, S. Sasa, M. Miwa *Bull. Chem. Soc. Jpn.* **1980**, *53*, 2089; (c) T. Yamaguchi, A. Mitoh, K. Tanabe, *Chem. Lett.* **1982**, 1229; (d) F. Effenberger, G. Epple, *Angew. Chem. Int. Ed.* **1972**, *11*, 300; (e) T. Mukaiyama, H. Nagaoka, M. Ohshima, M. Murakami, *Chem. Lett.* **1986**, 165; (f) F. Effenberger, D. Steegmiller, *Chem. Ber.* **1988**, *121*, 117; (g) T. Mukaiyama, T. Ohno, T. Nishimura, S. J. Han, S. Kobayashi, *Chem. Lett.* **1991**, 1059.
3 For example, (a) J. O. Morley, *J. Chem. Soc. Perkin Trans. 2* **1977**, 601; (b) G. Harvey, G. Mader, *Collect. Czech. Chem. Commun.* **1992**, *57*, 862; (c) D. E. Akporiaye, K. Daasvatn, J. Solberg, M. Stocker, *Stud. Surf. Sci. Catal.* **1993**, *78*, 521; (d) H. van Bekkum, A. J. Hoefnagel, M. A. van Koten, E. A. Gunnewegh, A. H. G. Vogt, H. W. Kouwenhoven, *Stud. Surf. Sci. Catal.* **1994**, *83*, 379; (e) F. Richard, H. Carreyre, G. Perot, *J. Catal.* **1996**, *159*, 427; (f) A. K. Pandy, A. P. Singh, *Catal. Lett.* **1997**, *44*, 129; (f) K. Smith, Z. Zhenhua, P. K. G. Hodgson, *J. Mol. Catal. A: Chem.* **1998**, *134*, 121; (g) B. Jacob, S. Sugunan, A. P. Singh, *J. Mol. Catal. A: Chem.* **1999**, *139*, 43; (h) S. Fujiyama and T. Kasahara, *Hydrocarbon Process* **1978**, *57*, 147.
4 (a) S. Kobayashi, *Synlett* **1994**, 689; (b) S. Kobayashi, M. Sugiura, H. Kitagawa, W. W.-L. Lam, *Chem. Rev.* **2002**, *102*, 2227; (c) S. Kobayashi, *Chem. Lett.* **1991**, 2187; (d) S. Kobayashi, I. Hachiya, *J. Org. Chem.* **1994**, *59*, 3590.
5 (a) A. Kawada, S. Mitamura, J.-I. Matsuo, T. Tsuchiya, S. Kobayashi, *Bull. Chem. Soc. Jpn.* **2000**, *73*, 2325; (b) A. Kawada, S. Mitamura,

S. Kobayashi, *J. Chem. Soc. Chem. Commun.* **1993**, 1157; (c) A. Kawada, S. Mitamura, S. Kobayashi, *Synlett* **1994**, 545; (d) A. Kawada, S. Mitamura, S. Kobayashi, *J. Chem. Soc. Chem. Commun.* **1996**, 183.

6 (a) S. Kobayashi, *Eur. J. Org. Chem.* **1999**, 15; (b) S. Kobayashi, I. Hachiya, M. Araki, H. Ishitani, *Tetrahedron Lett.* **1993**, *34*, 3755.

7 F. A. Cotton, G. Wilkinson, "*Advanced Inorganic Chemistry, Fifth Edition,*" John Wiley and Sons, New York, 1988, p. 973.

8 (a) G. A. Olah, S. J. Kuhn, W. S. Tolgyesi, E. B. Becker, *J. Am. Chem. Soc.* **1962**, *84*, 2733; (b) G. A. Olah, S. J. Kuhn, S. H. Flood, B. A. Hardie, *J. Am. Chem. Soc.* **1964**, *86*, 2203; (c) G. A. Olah, M. B. Comisarow, *J. Am. Chem. Soc.* **1968**, *88*, 3313.

9 (a) H. Burton, P. F. G. Praill, *J. Chem. Soc.* **1950**, 1203; (b) H. Burton, P. F. G. Praill, *J. Chem. Soc.* **1950**, 2034; (c) H. Burton, P. F. G. Praill, *J. Chem. Soc.* **1951**, 522; (d) H. Burton, P. F. G. Praill, *J. Chem. Soc.* **1951**, 529; (e) A. M. Avedikian, A. Kergomard, J. C. Tardivat, and J. P. Vullerme, *Bull. Soc. Chim. Fr.* **1974**, *11*, 2652; (f) K. Baum, C. D. Beard, *J. Org. Chem.* **1975**, *40*, 81.

10 (a) T. Mukaiyama, K. Suzuki, S. J. Han, S. Kobayashi, *Chem. Lett.* **1992**, 435; (b) I. Hachiya, M. Moriwaki, S. Kobayashi, *Bull. Chem. Soc. Jpn.* **1995**, *68*, 2053.

11 For example, (a) S. Kobayashi, S. Iwamoto, *Tetrahedron Lett.* **1998**, *39*, 4697; (b) J. Matsuo, K. Odashima, S. Kobayashi, *Synlett* **2000**, 403; (c) S. Kobayashi, I. Komoto, J.-I. Matsuo, *Adv. Synth. Catal.* **2001**, *343*, 71; (d) J. R. Desmurs, M. Labrouillere, C. Le Roux, H. Gaspard, A. Laporterie, J. Dubac, *Tetrahedron Lett.* **1997**, *38*, 8871.

References and Notes to Section 1.3.1.3

1 R. M. Roberts, A. A. Khalaf, In *Friedel–Crafts Alkylation Chemistry. A Century of Discovery*, Marcel Dekker: New York, **1984**.

2 M. Bandini, A. Melloni, A. Umani-Ronchi, *Angew. Chem. Int. Ed.* **2004**, *43*, 550–556.

3 N. A. Paras; D. W. C. MacMillan, *J. Am. Chem. Soc.* **2002**, *124*, 7894–7895.

4 M. Johannsen, *Chem. Commun.* **1999**, 2233–2234.

5 H. Yamamoto, In *Lewis Acids in Organic Synthesis*, Wiley–VCH: Weinheim, **2000**.

6 N.A. Paras; D.W.A. MacMillan, *J. Am. Chem. Soc.* **2001**, *123*, 4370–4371.

7 (a) A. Ishii; V.A. Soloshonok; K. Mikami, *J. Org. Chem.* **2000**, *65*, 1597–1599; (b) N. Gathergood; W. Zhuang; K. A. Jørgensen, *J. Am. Chem. Soc.* **2000**, *122*, 12517–12522; (c) Y. Yuan; X. Wang; W. Li; K. Ding, *J. Org. Chem.* **2004**, *69*, 146–149; (d) W. Zhuang; N. Gathergood; R. G. Hazell; K.A. Jørgensen *J. Org. Chem.* **2001**, *66*, 1009–1013; (e) A. Corma; H. García; A. Moussaif; M. J. Sabater; R. Zniber; A. Redouane, *Chem. Commun.* **2002**, 1058–1059; (f) S. Saaby, P. Bayón, P. S. Aburel, K. A. Jørgensen, *J. Org. Chem.* **2002**, *67*, 4352–4361.

8 (a) W. Zhuang, T. Hansen, K. A. Jørgensen, *Chem. Commun.* **2001**, 347–348; (b) J. Zhou, Y. Tang, *J. Am. Chem. Soc.* **2002**, *124*, 9030–9031; (c) K. B. Jensen, J. Thorhauge, R. G. Hazell, K. A. Jørgensen, *Angew. Chem. Int. Ed.* **2001**, *40*, 160–163; (d) M. Bandini, M. Fagioli, A. Melloni, A. Umani-Ronchi, *Tetrahedron Lett.* **2003**, *44*, 5843–5846; (e) J. A. Austin, D. W. C. MacMillan, *J. Am. Chem. Soc.* **2002**, *124*, 1172–1173; (f) D. A. Evans, K. A. Scheidt, K. R. Fandrick, H. W. Lam, J. Wu, *J. Am. Chem. Soc.* **2003**, *125*, 10780–10781; (g) M. Bandini, A. Melloni, S. Tommasi, A. Umani-Ronchi, *Helv. Chim. Acta*, **2003**, *86*, 3753–3763.

9 D. W. C. MacMillan, K.A. Ahrendt, Patent, No.: US 6,515,137 B2.

10 M. Bandini, P. G. Cozzi, P. Melchiorre, A. Umani-Ronchi, *Angew. Chem. Int. Ed.* **2004**, *43*, 84–87.

References and Notes to Section 1.3.1.4

1 (a) J. F. Normant, A. Alexakis, *Synthesis* **1981**, 841–870; (b) E. Negishi, T. Takahashi, *Synthesis* **1988**, 1–19.
2 M. T. Reetz, K. Sommer, *Eur. J. Org. Chem.* **2003**, 3485–3496.
3 (a) C. Jia, W. Lu, J. Oyamada, T. Kitamura, K. Matsuda, M. Irie, Y. Fujiwara, *J. Am. Chem. Soc.* **2000**, *122*, 7252–7263; (b) C. Jia, D. Piao, J. Oyamada, W. Lu, T. Kitamura, Y. Fujiwara, *Science* **2000**, *287*, 1992–1995; (c) C. Jia, D. Piao, T. Kitamura, Y. Fujiwara, *J. Org. Chem.* **2000**, *65*, 7516–7522; (d) T. Tsuchimoto, T. Maeda, E. Shirakawa, Y. Kawakami, *Chem. Commun. (Cambridge, U. K.)* **2000**, 1573–1574; (e) G. Sartori, F. Bigi, A. Pastorio, C. Porta, A. Arienti, R. Maggi, N. Moretti, G. Gnappi, *Tetrahedron Lett.* **1995**, *36*, 9177–9180.
4 K. Sommer, Dissertation, Ruhr-Universität Bochum, Germany, **2003**.
5 Gold(III)-catalyzed addition of oxygen nucleophiles to alkynes: (a) R. O. C. Norman, W. J. E. Parr, C. B. Thomas, *J. Chem. Soc, Perkin Trans. I* **1976**, 1983–1987; (b) Y. Fukuda, K. Utimoto, *J. Org. Chem.* **1991**, *56*, 3729–3731; (c) Y. Fukuda, K. Utimoto, *Bull. Chem. Soc. Jpn.* **1991**, *64*, 2013–2015; (d) Y. Fukuda, K. Utimoto, *Synthesis* **1991**, 975–978; (e) F. Gasparrini, M. Giovannoli, D. Misiti, G. Natile, G. Palmieri, L. Maresca, *J. Am. Chem. Soc.* **1993**, *115*, 4401–4402; Nitrogen nucleophiles: (f) M. Deetlefs, H. G. Raubenheimer, M. W. Esterhuysen, *Catal. Today* **2002**, *72*, 29–41; (g) A. Arcadi, S. D. Giuseppe, F. Marinelli, E. Rossi, *Adv. Synth. Catal.* **2001**, *343*, 443–446; Gold(I)-catalyzed addition of oxygen nucleophiles to alkynes: (h) J. H. Teles, S. Brode, M. Chabanas, *Angew. Chem.* **1998**, *110*, 1475–1478; *Angew. Chem. Int. Ed.* **1998**, *37*, 1415–1418; (i) J. H. Teles, M. Schulz, BASF AG, WO-A 97/21648, **1997** [*Chem. Abstr.* **1997**, *127*, 121499u]; (j) E. Mizushima, K. Sato, T. Hayashi, M. Tanaka, *Angew. Chem.* **2002**, *114*, 4745–4747; *Angew. Chem. Int. Ed.* **2002**, *41*, 4563–4565; Gold(III)-catalyzed C–C, C–O and C–N bond coupling reactions: (k) A. S. K. Hashmi, L. Schwarz, J.-H. Choi, T. M. Frost, *Angew. Chem.* **2000**, *112*, 2382–2385; *Angew. Chem. Int. Ed.* **2000**, *39*, 2285–2288; (l) A. S. K. Hashmi, T. M. Frost, J. W. Bats, *Catal. Today* **2002**, *72*, 19–27; (m) A. S. K. Hashmi, T. M. Frost, J. W. Bats, *Org. Lett.* **2001**, *3*, 3769–3771; (n) A. S. K. Hashmi, T. M. Frost, J. W. Bats, *J. Am. Chem. Soc.* **2000**, *122*, 11553–11554; (o) A. Hoffmann-Röder, N. Krause, *Org. Lett.* **2001**, *3*, 2537–2538; (p) N. Asao, K. Takahashi, S. Lee, T. Kasahara, Y. Yamamoto, *J. Am. Chem. Soc.* **2002**, *124*, 12650–12651; (q) S. Kobayashi, K. Kakumoto, M. Sugiura, *Org. Lett.* **2002**, *4*, 1319–1322; (r) A. Arcadi, G. Bianchi, S. Di Giuseppe, F. Marinelli, *Green Chem.* **2003**, *5*, 64–67; (s) Z. Shi, C. He, *J. Am. Chem. Soc.* **2004**, *126*, 5964–5965; (t) J. J. Kennedy-Smith, S. T. Staben, F. D. Toste, *J. Am. Chem. Soc.* **2004**, *126*, 4526–4527; (u) A. Arcadi, G. Bianchi, M. Chiarini, G. D'Anniballe, F. Marinelli, *Synlett* **2004**, 944–950; (v) C. Nieto-Oberhuber, M. P. Muñoz, E. Buñuel, C. Nevado, D. J. Cárdenas, A. M. Echavarren, *Angew. Chem.* **2004**, *116*, 2456–2460; *Angew. Chem. Int. Ed.* **2004**, *43*, 2402–2406; (w) V. Mamane, T. Gress, H. Krause, A. Fürstner, *J. Am. Chem. Soc.* **2004**, *126*, 8654–8655; Recent reviews on gold catalysis: (x) G. Dyker, *Angew. Chem.* **2000**, *112*, 4407–4409; *Angew. Chem. Int. Ed.* **2000**, *39*, 4237–4239; (y) G. C. Bond, *Catal. Today* **2002**, *72*, 5–9.
6 (a) M. S. Kharasch, H. S. Isbell, *J. Am. Chem. Soc.* **1931**, *53*, 3053–3059; (b) K. S. Liddle, C. Parkin, *J. Chem. Soc., Chem. Commun.* **1972**, 26; (c) P. W. J. de Graaf, J. Boersma, G. J. M. van der Kerk, *J. Organomet. Chem.* **1976**, *105*, 399–406; (d) Y. Fuchita, Y. Utsunomiya, M. Yasutake, *J. Chem. Soc., Dalton Trans.* **2001**, 2330–2334.
7 M. Preisenberger, A. Schier, H. Schmidbaur, *J. Chem. Soc., Dalton Trans.* **1999**, 1645–1650.
8 (a) A. Fürstner, V. Mamane, *J. Org. Chem.* **2002**, *67*, 6264–6267; see also:

(b) S. J. Pastine, S. W. Youn, D. Sames, *Org. Lett.* **2003**, *5*, 1055–1058.

9 Z. Shi, C. He, *J. Org. Chem.* **2004**, *69*, 3669–3671.

References to Section 1.3.2.1

1 R. H. Crabtree, *Chem. Rev.* **1985**, *85*, 245–269.

2 F. Kakiuchi, S. Murai, *Top. Organomet. Chem.* **1999**, *3*, 47–79; F. Kakiuchi, N. Chatani, N. *Adv. Synth. Catal.* **2003**, *345*, 1077–1101.

3 S. Murai, F. Kakiuchi, S. Sekine, Y. Tanaka, A. Kamatani, M. Sonoda, N. Chatani, *Nature* **1993**, *366*, 529–531.

4 L. N. Lewis, J. F. Smith, *J. Am. Chem. Soc.* **1986**, *108*, 2728–2735.

5 R. F. Jordan, Taylor, D. F. *J. Am. Chem. Soc.* **1989**, *111*, 778–779; S. Rodewald, R. F. Jordan, *J. Am. Chem. Soc.* **1994**, *116*, 4491–4492.

6 F. Kakiuchi, Y. Yamamoto, N. Chatani, S. Murai, *Chem Lett* **1995**, 681–682.

7 F. Kakiuchi, S. Murai, *Acc. Chem. Res.* **2002**, *35*, 826–834.

8 F. Kakiuchi, S. Sekine, Y. Tanaka, A. Kamatani, M. Sonoda, N. Chatani, S. Murai, *Bull. Chem. Soc. Jpn.* **1995**, *68*, 62–83; M. Sonoda, F. Kakiuchi, N. Chatani, M. Murai, *Bull. Chem. Soc. Jpn.* **1997**, *70*, 3117–3128.

9 T. Matsubara, N. Koga, D. G. Musaev, K. Morokuma, *J. Am. Chem. Soc.* **1998**, *120*, 12692–12693; T. Matsubara, N. Koga, D. G. Musaev, K. Morokuma, *Organometallics* **2000**, *19*, 2318–2329.

10 C. P. Lenges, M. Brookhart, *J. Am. Chem. Soc.* **1999**, *121*, 6616–6623

11 P. W. R. Harris, P. D. Woodgate, *J. Organomet. Chem.* **1997**, *530*, 211–223.

12 Y. Guari, S. Sabo-Etienne, B. Chaudret, *J. Am. Chem. Soc.* **1998**, *120*, 4228–4229.

13 R. Grigg, V. Savic, *Tetrahedron Lett.* **1997**, *38*, 5737–5740.

14 S. K. Gupta; W. P. Weber, *Macromolecules* **2002**, *35*, 3369–3373.

15 M. Sonoda, F. Kakiuchi, A. Kamatani, N. Chatani, S. Murai, *Chem. Lett.* **1996**, 109–110; F. Kakiuchi, H. Ohtaki, M. Sonoda, N. Chatani, S. Murai, *Chem. Lett.* **2001**, 918–919.

16 F. Kakiuchi, T. Sato, K. Igi, N. Chatani, S. Murai, *Chem. Lett.* **2001**, 386–387.

17 R. Aufdenblatten, S. Diezi, A. Togni, *Monatsh. Chem.* **2000**, *131*, 1345–1350.

18 P. W. R. Harris, C. E. F. Rickard, P. D. Woodgate, *J. Organomet. Chem.* **1999**, *589*, 168–179.

19 T. M. Londergan, Y. You, M. E. Thompson, W. P. Weber, *Macromolecules* **1998**, *31*, 2784–2788.

20 F. Kakiuchi, S. Murai, *Org. Synth.* **2003**, *80*, 104–110.

References and Notes to Section 1.3.2.2

1 F. Kakiuchi, N. Chatani, *Adv. Synth. Catal.* **2003**, *345*, 1077–1101.

2 E. J. Moore, W. R. Pretzer, T. J. O'Connell, J. Harris, L. LaBounty, L. Chou, S. S. Grimmer, *J. Am. Chem. Soc.* **1992**, *114*, 5888.

3 N. Chatani, T. Fukuyama, F. Kakiuchi, S. Murai, *J. Am. Chem. Soc.* **1996**, *118*, 493–494. N. Chatani, T. Fukuyama, H. Tatamidani, F. Kakiuchi, S. Murai, *J. Org. Chem.* **2000**, *65*, 4039–4047.

4 T. Fukuyama, N. Chatani, J. Tatsumi, F. Kakiuchi, S. Murai, *J. Am. Chem. Soc.* **1998**, *120*, 11522–11523.

5 N. Chatani, Y. Ie, F. Kakiuchi, S. Murai, *J. Org. Chem.* **1997**, *62*, 2604–2610.

6 Y. Ie, N. Chatani, T. Ogo, D. R. Marshall, T. Fukuyama, F. Kakiuchi, S. Murai, *J. Org. Chem.* **2000**, *65*, 1475–1488.

7 T. Asaumi, N. Chatani, T. Matsuo, F. Kakiuchi, S. Murai, *J. Org. Chem.* **2003**, *68*, 7538–7340. T. Asaumi, T. Matsuo, T. Fukuyama, Y. Ie, F. Kakiuchi, N. Chatani, *J. Org. Chem.* **2004**, *69*, 4433–4440.

8 N. Chatani, S. Yorimitsu, T. Asaumi, F. Kakiuchi, S. Murai, *J. Org. Chem.* **2002**, *67*, 7557–7560.

9 Y. Ishii, N. Chatani, F. Kakiuchi, S. Murai, *Organometallics* **1997**, *16*, 3615–3622.

10 J. W. Szewczyk, R. L. Zuckerman, R. G. Bergman, J. A. Ellman, *Angew. Chem. Int. Ed.* **2001**, *40*, 216–219.

11 N. Chatani, T. Asaumi, S. Yorimitsu, T. Ikeda, F. Kakiuchi, S. Murai, *J. Am. Chem. Soc.* **2001**, *123*, 10935–10941.

12 N. Chatani, T. Asaumi, T. Ikeda, S. Yorimitsu, Y. Ishii, F. Kakiuchi, S. Murai, *J. Am. Chem. Soc.* **2000**, *122*, 12882–12883.

References and Notes to Section 1.3.2.3

1 M. Lail, B. N. Arrowood, T. B. Gunnoe, *J. Am. Chem. Soc.* **2003**, *125*, 7506.
2 M. Lail, C. M. Bell, D. Conner, T. R. Cundari, T. B. Gunnoe, J. L. Petersen, *Organometallics* **2004**, *23*, 5007–5020.
3 K. A. Pittard, J. P. Lee, T. R. Cundari, T. B. Gunnoe, J. L. Petersen, *Organometallics* **2004**, *23*, 5514–5523.
4 B. N. Arrowood, M. Lail, T. B. Gunnoe, P. D. Boyle, *Organometallics* **2003**, *22*, 4692.
5 R. A. Periana, X. Y. Liu, G. Bhalla, *Chem. Commun.* **2002**, *24*, 3000.
6 T. Matsumoto, D. J. Taube, R. A. Periana, H. Taube, H. Yoshida, *J. Am. Chem. Soc.* **2000**, *122*, 7414.
7 T. Matsumoto, R. A. Periana, D. J. Taube, H. Yoshida, *J. Mol. Catal. A.: Chemical* **2002**, *180*, 1.
8 J. Oxgaard, W. A. Goddard III, *J. Am. Chem. Soc.* **2004**, *126*, 442.
9 J. Oxgaard, R. P. Muller, W. A. Goddard III, R. A. Periana, *J. Am. Chem. Soc.* **2004**, *126*, 352.
10 I. P. Beletskaya, A. V. Cheprakov, *Chem. Rev.* **2000**, *100*, 3009.
11 R. F. Heck, in *Comp. Org. Syn.*, Vol. 4 (Eds.: B. M. Trost, I. Fleming, M. F. Semmelhack), Pergamon Press, Oxford, **1999**, pp. 833.
12 J. K. Stille, *Angew. Chem. Int. Ed. Engl.* **1986**, *25*, 508.
13 N. Miyaura, A. Suzuki, *Chem. Rev.* **1995**, *95*, 2457.
14 E.-I. Negishi, L. Anastasia, *Chem. Rev.* **2003**, *103*, 1979.
15 J. Oxgaard, R. A. Periana, W. A. Goddard III, *J. Am. Chem. Soc.* **2004**, *126*, 11658–11665.

References and Notes to Section 1.3.2.4

1 Jordan, R., Taylor, D. *J. Am. Chem. Soc.* **1989**, *111*, 778–779.
2 Moore, E. J.; Pretzer, W. R.; O'Connell, T. J.; Harris, J.; Labounty, L.; Chou, L.; Grimmer, S. S. *J. Am. Chem. Soc.* **1992**, *114*, 5888–5890.
3 Szewczyk, J. W.; Zuckerman, R. L.; Bergman, R. G.; Ellman, J. A. *Angew. Chem., Int. Ed.* **2001**, *40*, 216–219. Chatani, N.; Fukuyama, T.; Kakiuchi, F.; Murai, S. *J. Am. Chem. Soc.* **1996**, *118*, 493–494. Fukuyama, T.; Chatani, N.; Tatsumi, J.; Kakiuchi, F.; Murai, S. *J. Am. Chem. Soc.* **1998**, *120*, 11522–11523. Chatani, N.; Fukuyama, T.; Tatamidani, H.; Kakiuchi, F.; Murai, S. *J. Org. Chem.* **2000**, *65*, 4039–4047.
4 Tan, K. L.; Bergman, R. G.; Ellman, J. A. *J. Am. Chem. Soc.* **2001**, *123*, 2685–2686.
5 Thalji, R. K.; Ahrendt, K. A.; Bergman, R. G.; Ellman, J. A. *J. Am. Chem. Soc.* **2001**, *123*, 9692–9693. Jun, C. H.; Hong, J. B.; Kim, Y. H.; Chung, K. Y. *Angew. Chem., Int. Ed.* **2000**, *39*, 3440–3442. Lim, Y. G.; Kang, J. B.; Kim, Y. H. *J. Chem. Soc., Perkin Trans. 1* **1996**, 2201–2206.
6 Tan, K. L.; Bergman, R. G.; Ellman, J. A. *J. Am. Chem. Soc.* **2002**, *124*, 3202–3203.
7 Herrmann, W. A.; Kocher, C. *Angew. Chem., Int. Ed. Engl.* **1997**, *36*, 2163–2187.
8 Tan, K. L.; Bergman, R. G.; Ellman, J. A. *J. Am. Chem. Soc.* **2002**, *124*, 13964–13965.
9 Tan, K. L.; Vasudevan, A.; Bergman, R. G.; Ellman, J. A.; Souers, A. J. *Org. Lett.* **2003**, *5*, 2131–2134.
10 Wiedemann, S. H.; Bergman, R. G.; Ellman, J. A. *Org. Lett.* **2004**, *6*, 1685–1687.
11 Lewis, J. C.; Wiedemann, S. H.; Bergman, R. G.; Ellman, J. A. *Org. Lett.* **2004**, *6*, 35–38.
12 This notion is supported by the observation that, when the nitrogen lone pair and the reactive C–H bond are not proximal (as in the thermodynamically favored acyclic isomer of an acyclic formamidine), no alkylation takes place.
13 This prediction is born out by the observed zero-order dependence of the reaction on substrate concentration.
14 Larhed, M.; Moberg, C.; Hallberg, A. *Acc. Chem. Res.* **2002**, *35*, 717–727.

15 Tan, K. L.; Park, S.; Ellman, J. A.; Bergman, R. G., submitted for publication.
16 Gant, T. G.; Meyers, A. I. *Tetrahedron* **1994**, *50*, 2297–2360.
17 Mori, A.; Sekiguchi, A.; Masui, K.; Shimada, T.; Horie, M.; Osakada, K.; Kawamoto, M.; Ikeda, T. *J. Am. Chem. Soc.* **2003**, *125*, 1700–1701. Sezen, B.; Sames, D. *J. Am. Chem. Soc.* **2003**, *125*, 5274–5275.
18 Wiedemann, S. H.; Lewis, J. C.; Bergman, R. G.; Ellman, J. A. manuscript in preparation.
19 $PCy_3 \cdot HCl$ is prepared by precipitation from a mixture of HCl and PCy_3 solutions in ether [9].

References and Notes to Section 1.3.2.5

1 (a) I. Moritani, Y. Fujiwara, *Tetrahedron Lett.* **1967**, 1119–1122. (b) I. Moritani, Y. Fujiwara, *Synthesis* **1973**, 524–533. (c) C. Jia, T. Kitamura, Y. Fujiwara, *Acc. Chem. Res.* **2001**, *34*, 633–639. (d) Y. Fujiwara, *Palladium-Promoted Alkene–Arene Coupling via C–H Activation*, in: *Handbook of Organopalladium Chemistry for Organic Synthesis*, E.-i. Negishi (Ed.), John Wiley and Sons, 2002, pp. 2863–2871.
2 F. Kakiuchi, S. Murai, *Acc. Chem. Res.* **2002**, *35*, 826–834.
3 (a) C. Jia, D. Piao, J. Oyamada, W. Lu, T. Kitamura, Y. Fujiwara, *Science* **2000**, *287*, 1992–1995. (b) C. Jia, W. Lu, J. Oyamada, T. Kitamura, K. Matsuda, M. Irie, Y. Fujiwara, *J. Am. Chem. Soc.* **2000**, *122*, 7252–7263. (c) C. Jia, D. Piao, T. Kitamura, Y. Fujiwara, *J. Org. Chem.* **2000**, *65*, 7516–7522.
4 W. Lu, C. Jia, T. Kitamura, Y. Fujiwara, *Org. Lett.* **2000**, *2*, 2927–2930.
5 Y. Fuchita, K. Hiraki, Y. Kamogawa, M. Suenaga, K. Tohgoh, Y. Fujiwara, *Bull. Chem. Soc. Jpn.* **1989**, *62*, 1081–1085.
6 J. Oyamada, W. Lu, C. Jia, T. Kitamura, Y. Fujiwara, *Chem. Lett.* **2002**, 20–21.
7 (a) J. Oyamada, C. Jia, Y. Fujiwara, T. Kitamura, *Chem. Lett.* **2002**, 380–381. (b) T. Kitamura, K. Yamamoto, M. Kotani, J. Oyamada, C. Jia, Y. Fujiwara, *Bull. Chem. Soc. Jpn.* **2003**, *76*, 1889–1895.
8 T. Mitsudome, T. Umetani, T. Mizugaki, K. Ebitani, K. Kaneda, *Catalyst Catalysis* **2004**, *46*, 78–80.

References and Notes to Section 1.3.2.6

1 (a) J. G. de Vries, *Can. J. Chem.* **2001**, *79*, 1086–1092; (b) M. S. Stephan, A. J. J. M. Teunissen, G. K. M. Verzijl, J. G. de Vries, Johannes, *Angew. Chem. Int. Ed.* **1998**, *37*, 662–664; (c) L. J. Goossen, J. Paetzold, *Synlet*, **2002**, 1721–1723.
2 (a) A. E. Shilov, G. B. Shul'pin, *Activation and Catalytic Reactions of Saturated Hydrocarbons in the Presence of Metal Complexes*, **2000**, Catalysis by Metal Complexes, Volume 21, B. R. James, P. W. N. M. van Leeuwen, Eds. Kluwer Academic Publishers; (b) F. Kakiuchi, N. Chatani, *Adv. Synth. Catal.* **2003**, *345*, 1077–1101.
3 (a) I. Moritani, Y. Fujiwara, *Tetrahedron Lett.* **1967**, 1119–1122; (b) I. Moritani, Y. Fujiwara, *Synthesis* **1973**, 524–533; (c) C. Jia, T. Kitamura, Y. Fujiwara, *Acc. Chem. Res.* **2001**, *34*, 633–639; (d) Y. Fujiwara, *Palladium-Promoted Alkene–Arene Coupling via C–H Activation*, in: *Handbook of Organopalladium Chemistry for Organic Synthesis*, E.-i. Negishi (Ed.) John Wiley and Sons, 2002, pp 2863–2871; (e) F. Kakiuchi, S. Murai, *Acc. Chem. Res.* **2002**, *35*, 826–834.
4 (a) Y. Fujiwara, I. Moritani, S. Danno, S. Teranishi, *J. Am. Chem. Soc.* **1969**, *91*, 7166–7169 and references therein; (b) J. Tsuji, H. Nagashima, *Tetrahedron* **1984**, *40*, 2699–702.
5 C. Jia, W. Lu, T. Kitamura, Y. Fujiwara, *Org. Lett.* **1999**, *1*, 2097–2100.
6 A. C. Cope, R. W. Siekman, *J. Am. Chem. Soc.* **1965**, *87*, 3272–3273; (b) V. Ritleng, C. Sirlin, M. Pfeffer, *Chem. Rev.* **2002**, *102*, 1731–1769; (c) M. Pfeffer, *Rec. Trav. Chim. Pays-Bas* **1990**, *109*, 567–76.
7 (a) H. Horino, N. Inoue, *J. Org. Chem.* **1981**, *46*, 4416–4422; (b) S. J. Tremont,

H. Ur Rahman, *J. Am. Chem. Soc.* **1984**, *106*, 5759–60.

8 M. D. K. Boele, G. P. F. van Strijdonck, A. H. M. de Vries, P. C. J. Kamer, J. G. de Vries, P. W. N. M. van Leeuwen, *J. Am. Chem. Soc.* **2002**, *124*, 1586–1587.

9 C. Jia, D. Piao, J. Oyamada, W. Lu, T. Kitamura, Y. Fujiwara, *Science* **2000**, *287*, 1992–1995.

10 (a) Z. Zhang, X. Lu, Z. Xu, Q. Zhang, X. Han, *Organometallics* **2001**, *20*, 3724–3728; (b) Z. Wang, Z. Zhang, X. Lu, *Organometallics* **2000**, *19*, 775–780.

11 H. Grennberg, A. Gogoll, J.-E. Bäckvall, *Organometallics* **1993**, *12*, 1790–1793.

12 (a) J.-E. Bäckvall, *Pure Appl. Chem.* **1992**, *64*, 429–437 and references therein (b) P. Roffia, F. Conti, G. Gregorio, G. F. Pregaglia, R. Ugo, *J. Organomet. Chem.* **1973**, *391*, 391–394 and references therein.

13 A. D. Ryabov, I. K. Sakodinskaya, A. K. Yatsimirsky, *J. Organomet. Chem.* **1991**, *406*, 309–321.

14 (a) T. Yokota, M. Tani, S. Sakaguchi, Y. Ishii, *J. Am. Chem. Soc.* **2003**, *125*, 1476–1477; (b) M. Tani, S. Sakaguchi, Y. Ishii, *J. Org. Chem.* **2004**, *69*, 1221–1226.

15 S. S. Stahl, *Angew. Chem. Int. Ed.* **2004**, *43*, 3400–3420.

16 M. Dams, D. E. de Vos, S. Celen, P. A. Jacobs, *Angew. Chem. Int. Ed.* **2003**, *42*, 3512–3515.

17 E. M. Ferreira, B. M. Stoltz, *J. Am. Chem. Soc.* **2003**, *125*, 9578–9579.

18 H. Zhang, E. M. Ferreira, B. M. Stoltz, *Angew. Chem. Int. Ed.* **2004**, *43*, 6144–6148.

19 S. Ma, S. Yu, *Tetrahedron Lett.* **2004**, *45*, 8419–8422.

20 E. J. Hennessy, S. L. Buchwald, *J. Am. Chem. Soc.* **2003**, *125*, 12084–12085.

21 G. Karig, M.-T. Moon, N. Thasana, T. Gallagher, *Org. Lett.* **2002**, *4*, 3115–3118.

22 C. Zhou, R. C. Larock, *J. Am. Chem. Soc.* **2004**, *126*, 2302–2303.

23 A. D. Ryabov, *Chem. Rev.*, **1990**, *90*, 403/424.

References to Section 1.3.3

1 A. Studer and M. Bossart *Radicals in Organic Synthesis* P. Renaud and M. P. Sibi Eds, Vol. II, pg 62, Wiley–VCH, **2001**.

2 Reviews: (a) F. Minisci, *Synthesis*, **1973**, 1; (b) F. Minisci and O. Porta, *Advances in Heterocyclic Chemistry*, A. R. Katritzky Ed., Vol. 16, pg 123, Academic Press, **1974**; (c) F. Minisci, *Top. Curr. Chem.*, **1976**, 621; (d) F. Minisci and O. Porta, Zh. Yses. Khim.. D-Va **1979**, *24*, 121; (e) F. Minisci and O. Porta, *La Chimica e l'Industria*, **1980**, *62*, 769–775; (f) F. Minisci, *Fundamental Research in Homogeneous Catalysis*, M. Graziani Ed., Vol 4, pg 173, Plenum, **1984**; (g) F. Minisci, *Substituent Effects in radical Chemistry* H. G. Viehe Ed., pg 391, Reidel, **1986**; (h) F. Minisci and E. Vismara, *Organic Synthesis: Modern Trends*, O. Chizhov Ed., pg 229, Blackwell Sci. Publ., **1987**; (i) F. Minisci, F. Fontana and E. Vismara, *Heterocycles*, **1989**, *28*, 489; (l) F. Minisci, F. Fontana and E. Vismara, *J. Heterocyclic Chem.*, **1990**, *27*, 79.

3 (a) A. Citterio, F. Minisci, O. Porta and G. Sesana *J. Am. Chem. Soc.*, **1977**, *99*, 7960; (b) F. Recupero, A. Bravo, H. R. Bjorsvik, F. Fontana , F. Minisci and M. Piredda, *J. Chem. Soc. Perkin Trans II*, **1997**, 2399.

4 A. Clerici, N. Pastori and O. Porta, *Tetrahedron Lett.*, **2004**, *45*, 1825.

5 F. Minisci, F. Recupero, A. Cecchetto, C. Punta, C. Gambarotti, F. Fontana and G. F. Pedulli, *J. Heterocyclic Chem.*, **2003**, *40*, 325.

6 F. Minisci, A. Citterio, E. Vismara e C. Giordano, *Tetrahedron*, **1985**, *41*, 4157.

7 C. Giordano, F. Minisci, E. Vismara and S. Levi, *J. Org. Chem.*, **1986**, *51*, 536.

8 F. Minisci, E. Vismara, F. Fontana, G. Morini, M. Serravalle and C. Giordano, *J. Org. Chem.*, **1986**, *51*, 4411.

9 T. Caronna, A. Citterio, T. Crolla and F. Minisci, *J. Chem. Soc. Perkin Trans II*, **1977**, 865.

10 F. Minisci, A. Citterio, M. Perchinunno and F. Bertini, *Gazz. Chim. Ital.*, **1975**, *105*, 1083.

11 F. Fontana, F. Minisci, M. C. Nogueira Barbosa and E. Vismara, *Tetrahedron*, **1990**, *46*, 2525.

12 F. Fontana, F. Minisci, M. C. Nogueira Barbosa and E. Vismara, *J. Org. Chem.*, **1991**, *56*, 2866.

13 F. Coppa, F. Fontana, E. Lazzarini and F. Minisci, *Heterocycles*, **1993**, *36*, 2687.

14 F. Coppa, F. Fontana, E. Lazzarini, F. Minisci and G. Pianese, *Tetrahedron Lett.*, **1992**, *33*, 3057.

15 F. Minisci, O. Porta, F. Recupero, C. Punta, C. Gambarotti, B. Pruna, M. Pierini and F. Fontana, *Synlett*, **2004**, 874–876.

16 E. Vismara, A. Donna, F. Minisci, A. Naggi, N. Pastori and G. Torri, *J. Org. Chem.*, **1993**, *58*, 959.

17 F. Minisci, F. Fontana and E. Vismara, *Free Radicals in Synthesis and Biology* F. Minisci Ed., Kluwer, Dordrecht, **1989**, p. 53.

18 F. Antonietti, A. Mele, F. Minisci, C. Punta, F. Recupero and F. Fontana, *J. Fluorine Chem.*, **2004**, *125*, 205.

References to Section 1.4.1.1

1 A. de Meijere, F. Diederich (Eds) *Metal-Catalyzed Cross-Coupling Reactions*, 2nd edn, Wiley–VCH, Weinheim, **2004**.

2 J. Tsuji, *Palladium Reagents and Catalysts*, 2nd edn, Wiley, Chichester, **2004**.

3 J. Hassan, M. Sévignon, C. Gozzi, E. Schulz, M. Lemaire, *Chem. Rev.* **2002**, *102*, 1359–1469.

4 M. Miura, M. Nomura, *Top. Curr. Chem.* **2002**, *219*, 211–241.

5 T. Satoh, Y. Kawamura, M. Miura, M. Nomura, *Angew. Chem. Int. Ed. Engl.* **1997**, *36*, 1740–1742.

6 T. Satoh, J. Inoh, Y. Kawamura, Y. Kawamura, M. Miura, M. Nomura, *Bull. Chem. Soc. Jpn.* **1998**, *71*, 2239–2246.

7 G. Dyker, *Chem. Ber./Recueil* **1997**, *130*, 1567–1578.

8 Y. Kawamura, T. Satoh, M. Miura, M. Nomura, *Chem. Lett.* **1999**, 961.

9 D. A. Culkin, J. F. Hartwig, *Acc. Chem. Res.* **2003**, *36*, 234–245.

10 Y. Kawamura, T. Satoh, M. Miura, M. Nomura, *Chem. Lett.* **1998**, 931–932.

11 A. R. Muci, S. L. Buchwald, *Top. Curr. Chem.* **2002**, *219*, 131–209.

12 R. B. Bedford, S. J. Coles, M. B. Hursthouse, M. E. Limmert, *Angew. Chem. Int. Ed.* **2003**, *42*, 112–114.

13 R. B. Bedford, M. E. Limmert, *J. Org. Chem.* **2003**, *68*, 8669–8682.

14 S. Oi, S. Watanabe, S. Fukita, Y. Inoue, *Tetrahedron Lett.* **2003**, *44*, 8665–8668.

15 T. Satoh, Y. Kametani, Y. Terao, M. Miura, M. Nomura *Tetrahedron Lett.* **1999**, *40*, 5345–5348.

16 Y. Kametani, T. Satoh, M. Miura, M. Nomura, *Tetrahedron Lett.* **2000**, *41*, 2655–2658.

17 S. Oi, Y. Ogino, S. Fukita, Y. Inoue, *Org. Lett.* **2002**, *4*, 1873–1785.

18 S. Oi, S. Fukita, N. Hirata, N. Watanuki, S. Miyano, Y. Inoue, *Org. Lett.* **2001**, *3*, 2579–2581.

19 B. Sezen, D. Sames, *J. Am. Chem. Soc.* **2003**, *125*, 10580–10585.

20 F. Kakiuchi, S. Kan, K. Igi, N. Chatani, S. Murai, *J. Am. Chem. Soc.* **2003**, *125*, 1698–1699.

21 S. Oi, S. Fukita, Y. Inoue, *Chem. Commun.* **1998**, 2439–2440.

References to Section 1.4.1.2

1 M. Miura, M. Nomura, *Top. Curr. Chem.* **2002**, *219*, 211–241.

2 J. Hassan, M. Sévignon, C. Gozzi, E. Schulz, M. Lemaire, *Chem. Rev.* **2002**, *102*, 1359–1469.

3 S. Pivsa-Art, T. Satoh, Y. Kawamura, M. Miura, M. Nomura, *Bull. Chem. Soc. Jpn.* **1998**, *71*, 467–473.

4 J. F. D. Chabert, L. Joucla, E. David, M. Lemaire, *Tetrahedron* **2004**, *60*, 3221.

5 Y. Akita, Y. Itagaki, S. Takizawa, A. Ohta, *Chem. Pharm. Bull.* **1989**, *37*, 1477–1480.

6 Y. Aoyagi, A. Inoue, I. Koizumi, R. Hashimoto, K. Tokunaga, K. Gohma, J. Komatsu, K. Sekine, A. Miyafuji, J. Kunoh, R. Honma, Y. Akita, A. Ohta, *Heterocycles* **1992**, *33*, 257–272.

7 B. Sezen, D. Sames, *J. Am. Chem. Soc.* **2003**, *125*, 5274–5275.
8 B. Sezen, D. Sames, *J. Am. Chem. Soc.* **2003**, *125*, 10580–10585.
9 C.-H. Park, V. Ryabova, I. V. Seregin, A. W. Sromek, V. Gevorgyan, *Org. Lett.* **2004**, *6*, 1159–1162.
10 A. Ohta A, Y. Akita, T. Ohkuwa, M. Chiba, R. Fukunaga, A. Miyafuji, T. Nakata, N. Tani, Y. Aoyagi, *Heterocycles*, **1990**, *31*, 1951–1958.
11 M. S. McClure, B. Glover, E. McSorley, A. Millar, M. H. Osterhout, F. Roschanger, *Org. Lett.* **2001**, *3*, 1677–1680.
12 B. Glover, K. A. Harvey, B. Liu, M. J. Sharp, M. F. Tymoschenko, *Org. Lett.* **2003**, *5*, 301.
13 K. Masui, A. Mori, K. Okano, K. Takamura, M. Kinoshita, T. Ikeda, *Org. Lett.* **2004**, *6*, 2011–2014.
14 J. Hassan, C. Gozzi, E. Schulz, M. Lemaire, *J. Organomet. Chem.* **2003**, *687*, 280–283.
15 T. Okazawa, T. Satoh, M. Miura, M. Nomura, *J. Am. Chem. Soc.* **2002**, *124*, 5286–5287.
16 A. Yokoorji, T. Satoh, M. Miura, M. Nomura, *Tetrahedron* **2004**, *60*, 6757–6763.
17 M. Sévignon, J. Papillon, E. Schulz, M. Lemaire M, *Tetrahedron Lett.* **1999**, *40*, 5873–5876.
18 *Comprehensive Heterocyclic Chemistry*, vols. 5 and 6 (Ed.: K. T. Potts), Pergamon Press, Oxford, **1984**.
19 J. C. Lewis, S. H. Wiedemann, R. G. Bergman, J. A. Ellman, *Org. Lett.* **2004**, *6*, 35–38.
20 W. Li, D. P. Nelson, M. S. Jensen, R. S. Hoerrner, G. J. Javadi, D. Cai, R. D. Larsen, *Org. Lett.* **2003**, *5*, 4835–4837.
21 B. Sezen, D. Sames, *Org. Lett.* **2003**, *5*, 3607–3610.
22 A. Yokoorji, T. Okazawa, T. Satoh, M. Miura, M. Nomura, *Tetrahedron* **2003**, *69*, 5685–5689.
23 A. Mori, A. Sekiguchi, K. Masui, T. Shimada, M. Horie, K. Osakada, M. Kawamoto, T. Ikeda, *J. Am. Chem. Soc.* **2003**, *125*, 1700–1701.
24 Y. Kondo, T. Komine, T. Sakamoto, *Org. Lett.* **2000**, *2*, 3111–3113.

References and Notes to Section 1.4.1.3

1 (a) C. Janiak, H. Schumann, *Adv. Organomet. Chem.* **1991**, *33*, 291–393; (b) J. Okuda, *Topics Curr. Chem.* **1991**, *160*, 97–145; (c) C. Janiak, R. Weimann, F. Görlitz, *Organometallics* **1997**, *16*, 4933–4936; (d) M. Bond, R. Colton, D. A. Fiedler, L. D. Field, D. Leslie, T. He, P. A. Humphery, C. M. Lindall, F. Marken, A. F. Masters, H. Schumann, K. Suehring, V. Tedesco, *Organometallics* **1997**, *16*, 2787–2797; (e) D. Matt, M. Huhn, M. Bonnet, I. Tkatchenko, U. Englert, W. Kläul, *Inorg. Chem.* **1995**, *34*, 1288–1291; (f) S. Barry, A. Kucht, H. Kucht, M. D. Rausch, *J. Organomet. Chem.* **1995**, *489*, 195–199; (g) C. Ruble, H. A. Latham, G. C. Fu, *J. Am. Chem. Soc.* **1997**, *119*, 1492–1493; (h) A. M. Bond, R. Colton, D. A. Fiedler, L. D. Field, T. He, P. A. Humphrey, C. M. Lindall, F. Marken, A. F. Masters, H. Schumann, K. Sühring, V. Tedesco, *Organometallics* **1997**, *16*, 2787–2797; (i) H. Schumann, A. Lentz, R. Weimann, J. Pickardt, *Angew. Chem.* **1994**, *106*, 1827–1828, *Angew. Chem. Int. Ed. Engl.* **1994**, *33*, 1731–1732; (j) W. M. Harrison, C. Saadeh, S. B. Colbran, D. C. Craig, *J. Chem. Soc., Dalton Trans.* **1997**, 3785–3792.
2 (a) C. Adachi, T. Tsuji and S. Saito, *Appl. Phys. Lett.* **1990**, *56*, 799–801; (b) Y. Ohmori, Y. Hironaka, M. Yoshida, N. Tada, A. Fujii, K. Yoshino, *Synth. Metal* **1997**, *85*, 1241–1242.
3 (a) W. Broser, P. Siegle and H. Kurreck, *Chem. Ber.* **1968**, *101*, 69–83; (b) L. D. Field, K. M. Ho, C. M. Lindall, A. F. Masters, A. G. Webb, *Aust. J. Chem.* **1990**, *43*, 281–291; (c) T. R. Jack, C. J. May, J. Powell, *J. Am. Chem. Soc.* **1977**, *99*, 4707–4716; (d) H. Schumann, H. Kucht, A. Kucht, *Z. Naturforsch.* **1992**, *47b*, 1281–1289; (e) R. H. Lowack, K. P. C. Vollhardt, *J. Organomet. Chem.* **1994**, *476*, 25–32.
4 M. Miura, S. Pivsa-Art, G. Dyker, J. Heiermann, T. Satoh, M. Nomura, *Chem. Commun.* **1998**, 1889–1890.
5 G. Dyker, J. Heiermann, M. Miura, J.-I. Inoh, S. Pivsa-Art, T. Satoh,

M. Nomura, *Chem. Eur. J.* **2000**, *6*, 3426–3433.
6 G. Dyker, J. Heiermann, M. Miura, *Adv. Synth. Catal.* **2003**, *345*, 1127–1132.
7 (a) A. F. Littke, G. C. Fu., *J. Org. Chem.* **1999**, *64*, 10–11; (b) G. Mann, C. Incarvito, A. L. Rheingold, J. F. Hartwig, *J. Am. Chem. Soc.* **1999**, *121*, 3224–3225.
8 W. A. Herrmann, C. Brossmer, C.-P. Reisinger, T. H. Riermeier. K. Öffele, M. Beller, *Chem. Eur. J.* **1997**, *3*, 1357–1364, and references cited therein.
9 (a) N. A. Beare, J. F. Hartwig, *J. Org. Chem.* **2002**, *67*, 541–555; (b) M. Jørgensen, S. Lee, X. Liu, J. P. Wolkowski, J. F. Hartwig, *J. Am. Chem. Soc.* **2002**, *124*, 12557–12565; (c) D. A. Culkin, J. F. Hartwig, *Acc. Chem. Res.* **2003**, *36*, 234–245, and references cited therein.
10 G. Dyker, S. Borowski, J. Heiermann, J. Körning, K. Opwis, G. Henkel, M. Köckerling, *J. Organomet. Chem.* **2000**, *606*, 108–111.

References and Notes to Section 1.4.2.1

1 This method is also applicable to the synthesis of carbocyclic compounds, see: (a) J. E. Rice and Z.-W. Cai, *Tetrahedron Lett.*, **1992**, *33*, 1675; (b) J. E. Rice, Z.-W. Chen, *J. Org. Chem.*, **1993**, *58*, 1415; (c) B. E. Gómez-Lor, O. de Frutos and A. M. Echavarren, *Chem. Commun*, **1999**, 2431; (d) J. E. Rice, Z.-W. Cai, Z.-M. He and E. J. LaVoie, *J. Org. Chem.*, **1995**, *60*, 8101; (e) J. J. Gonzáles, N. García, B. Gómez-Lor and A. M. Echavarren, *J. Org. Chem.*, **1997**, *62*, 1286; (f) O. de Frutos, B. Gómez-Lor, T. Granier. M. Á. Monge, E. Gutiérrez-Puebla and A. M. Echavarren, *Angew. Chem. Int. Ed.*, **1999**, *38*, 205; (g) L. Wang and P. B. Shevlin, *Tetrahedron Lett.*, **2000**, *41*, 285.
2 D. E. Ames and A. Opalko, *Tetrahedron*, **1984**, *40*, 1919.
3 T. Kuroda and F. Suzuki, *Tetrahedron Lett.*, **1991**, *47*, 6915.
4 P. P. Deshpande and O. R. Martin, *Tetrahedron Lett.*, **1990**, *44*, 6313.

5 G. Bringmann, J. R. Jansen, H. Reuscher, M. Rubenacker, K. Peters and H. G. von Schnering, *Tetrahedron Lett.*, **1990**, *31*, 643. See also: G. Bringmann, R. Walter and R. Weirich, *Angew. Chem. Int. Ed.*, **1990**, *29*, 977.
6 G. Bringmann, T. Pabst, P. Henschel, J. Kraus, K. Peters, E.-M. Peters, D. S. Rycroft and J. D. Connolly, *J. m. Chem. Soc.*, **2000**, *122*, 9127.
7 D. E. Ames and A. Opalko, *Synthesis*, **1983**, 235.
8 T. Iwaki, A. Yasuhara and T. Sakamoto, *J. Chem. Soc., Perkin Trans. 1*, **1999**, 1505.
9 D. E. Ames and D. Bull, *Tetrahedron*, **1982**, *38*, 383.
10 A. R. Katritzky, C. W. Rees, and E. F. Scriven (Eds) *Comprehensive Heterocyclic Chemistry II*, Pergamon, Oxford, 1996.
11 For a recent review on the isolation and synthesis of biologically active carbazoles see: H.-J. Knölker and K. R. Reddy, *Chem. Rev.* **2002**, *102*, 4303
12 For examples of this oxidative coupling method and leading references, see Ref. [11].
13 R. B. Bedford and C. S. J. Cazin, *Chem. Commun.*, **2002**, 2310.
14 R. B. Bedford, *Chem. Commun.*, **2003**, 1787.
15 B. Martín-Matute, C. Mateo, D. J. Cárenas and A. M. Echavarren, *Chem. Eur. J.*, **2001**, *7*, 2341.
16 R. B. Bedford and M. Betham, unpublished data
17 Originally isolated from the roots of *Clausina excavate*: T.S. Wu, S. C. Huang, P. L. Wu, C. C. Kuoh, *Phytochemistry*, **1999**, *52*, 523.
18 M. R. Netherton and G. C. Fu, *Org. Lett.*, **2001**, *3*, 4295.

References to Section 1.4.2.2

1 M. Catellani, *Synlett* and references cited therein **2003**, 298–313.
2 M. Catellani, G.P. Chiusoli, *J. Organomet. Chem.* **1985**, *286*, C13–C16.
3 M. Catellani, E. Motti, *New J. Chem.* **1998**, 759–761.

4 M. Catellani, E. Motti, M. Minari, *J. Chem. Soc., Chem. Commun.* **2000**, 157–158.
5 M. Catellani, F. Frignani, A. Rangoni, *Angew. Chem. Int. Ed. Engl.* **1997**, *36*, 119–122.
6 E. Motti, A. Mignozzi, M. Catellani, *J. Mol. Catal. A: Chem.* **2003**, *204–205*, 115–124.
7 F. Faccini, E. Motti, M. Catellani, *J. Am. Chem. Soc.* **2004**, *126*, 78–79 and unpublished results.
8 M. Lautens, J.-F. Paquin, S. Piguel, *J. Org. Chem.* **2002**, *67*, 3972–3974.
9 M. Catellani, E. Motti, S. Baratta, *Org. Lett.* **2001**, *3*, 3611–3614.
10 R. Ferraccioli, D. Carenzi, O. Rombolà, M. Catellani, *Org. Lett.* **2004**, *6*, 4759–4762.

References and Notes to Section 1.4.3

1 G. Lessene, K. S. Feldman in *Modern Arene Chemistry*, Ed. D. Astruc, pages 479–538, Wiley–VCH, Weinheim, **2002**.
2 H.-J. Deuben, P. Frederiksen, T. Bjørmhom, K. Bechgaard, *Org. Prep. Proc. Int.* **1996**, *52*, 484–486.
3 R. B. Herbert, E. E. Kattah, A. J. Murtagh, P. W. Sheldrake, *Tetrahedron Lett.* **1995**, *36*, 5649–5650.
4 Y. Kita, M. Gyoten, M. Ohtsubo, H. Tohma, T. Takada, *Chem. Commun.* **1996**, 1481–1482; T. Takada, M. Arisawa, M. Gyoten, R. Hamada, H. Tohma, Y. Kita, *J. Org. Chem.* **1998**, *63*, 7698–7706; H. Tohma, H. Morioka, S. Takizawa, M. Arisawa, Y. Kita, *Tetrahedron* **2001**, *57*, 345–352; H. Tohma, M. Iwata, T. Maegawa, Y. Kita, *Tetrahedron Lett.* **2002**, *43*, 9241–9244.
5 D. Mirk, A. Willner, R. Fröhlich, S. R. Waldvogel, *Adv. Synth. Catal.* **2004**, *346*, 675–681.
6 D. Mirk, B. Wibbeling, R. Fröhlich, S. R. Waldvogel, *Synlett* **2004**, 1970–1975.
7 B. Kramer, S. R. Waldvogel, *Angew. Chem. Int. Ed.* **2004**, *43*, 2446–2449.
8 L. Czollner, W. Frantsits, B. Küenburg, U. Hedenig, J. Fröhlich, U. Jordis, *Tetrahedron Lett.* **1998**, *39*, 2087–2088.
9 N. Boden, R. J. Bushby, Z. Lu, G. Headdock, *Tetrahedron Lett.* **2000**, *41*, 10117–10120.
10 N. Naarmann, M. Hanack, R. Mattmer, *Synthesis* **1994**, 477–478.
11 G. Pohnert, *J. Prakt. Chem.* **2000**, *342*, 731–734.
12 Y. Kita, M. Arisawa, M. Gyoten, M. Nakajima, R. Hamada, H. Tohma, T. Takada, *J. Org. Chem.* **1998**, *63*, 6625–6633; Y. Kita, T. Takada, M. Gyoten, H. Tohma, M. H. Zenk, J. Eichhorn, *J. Org. Chem.* **1996**, *61*, 5857–5864.
13 R. Olivera, R. San Martin, S. Pascual, M. Herrero, E. Dominguez, *Tetrahedron Lett.* **1999**, *40*, 3479–3480.
14 B. Kramer, A. Averhoff, S. R. Waldvogel, *Angew. Chem. Int. Ed.* **2002**, *41*, 2981–2982.
15 B. Kramer, R. Fröhlich, S. R. Waldvogel, *Eur. J. Org. Chem.* **2003**, 3549–3554.
16 B. Kramer, R. Fröhlich, K. Bergander, S. R. Waldvogel, *Synthesis* **2003**, 91–96.
17 D. Mirk, O. Kataeva, R. Fröhlich, S. R. Waldvogel, *Synthesis* **2003**, 2410–2414.
18 S. R. Waldvogel, *Synlett* **2002**, 622–624.
19 S. R. Waldvogel, E. Aits, C. Holst, R. Fröhlich, *Chem. Commun.* **2002**, 1278–1279.
20 S. R. Waldvogel, R. Fröhlich, C. A. Schalley, *Angew. Chem. Int. Ed.* **2000**, *39*, 2472–2475.

2
C–H Transformation at Alkenes

2.1
The Heck Reaction
Lukas Gooßen and Käthe Baumann

2.1.1
Introduction and Fundamental Examples

Over 35 years ago, Richard F. Heck found that olefins can insert into the metal–carbon bond of arylpalladium species generated from organomercury compounds [1]. The carbopalladation of olefins, stoichiometric at first, was made catalytic by Tsutomu Mizoroki, who coupled aryl iodides with ethylene under high pressure, in the presence of palladium chloride and sodium carbonate to neutralize the hydroiodic acid formed (Scheme 1) [2]. Shortly thereafter, Heck disclosed a more general and practical procedure for this transformation, using palladium acetate as the catalyst and tri-*n*-butylamine as the base [3]. After investigations on stoichiometric reactions by Fitton et al. [4], it was also Heck who introduced palladium phosphine complexes as catalysts, enabling the decisive extension of the olefination reaction to inexpensive aryl bromides [5].

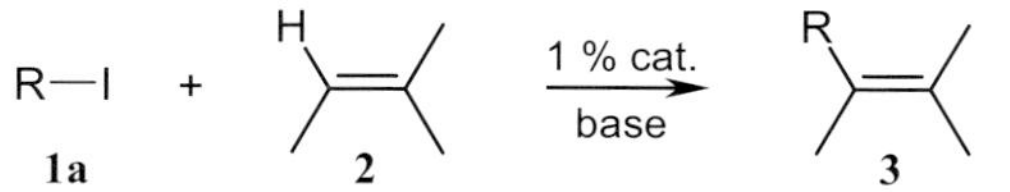

Mizoroki: $PdCl_2$/KOAc, R = Ph
Heck: $Pd(OAc)_2$/N(*n*-Bu)$_3$
R = Ph, styryl, benzyl

Scheme 1. Catalytic olefination reactions by Heck and Mizoroki.

Interestingly, it then took almost 10 years until the chemical community became fully aware of the enormous synthetic utility of this transformation. Since then, the Mizoroki–Heck reaction (or simply Heck reaction) has become one of the most popular C–C bond-forming reactions, accounting for several thousand publications overall.

The original reaction procedures have steadily been extended and improved, and an overwhelming number of catalyst systems are now known. Besides aryl halides, many additional substrates, for example aryl triflates, diazonium salts, sulfonyl and aroyl halides, carboxylic and phosphonic acids, and even arenes have been used as

Handbook of C–H Transformations. Gerald Dyker (Ed.)
Copyright © 2005 WILEY-VCH Verlag GmbH & Co. KGaA, Weinheim
ISBN: 3-527-31074-6

substrates. The combination of carbopalladation (Scheme 2) with other reaction steps into domino processes has further increased the preparative utility of this reaction.

The term "Heck reaction" is associated with such a huge variety of transformations that it is impossible to cover them all within the scope of this chapter. We therefore restricted ourselves to giving an outline of the reaction principle and discussing a number of illustrative developments in this exciting field of research. For a more comprehensive overview, we recommend recent review articles [6–8].

2.1.2
Mechanism

Because the catalytic Heck olefination evolved from stoichiometric organometallic transformations, its key elemental steps are relatively well-understood (Scheme 2) [9], although many mechanistic details are still unclear. The widely accepted mechanism of the Heck reaction starts with an oxidative addition of the organic halide **1** to a coordinatively unsaturated Pd(0) complex *a*, generated in situ from Pd(0) precursors or Pd(II) salts, yielding the σ-arylpalladium(II) complex *b*[10]. As a result of this reaction step, the electrophilicity of the Pd increases, favoring precoordination of the olefin (intermediate *c*). Insertion of the olefin **2** into the Pd–C bond of *c* results in the formation of the σ-alkylpalladium(II) complex *d*. Because β-hydride elimination requires at least one β-hydrogen in synperiplanar orientation relative to the halopalladium residue, elimination towards the original position of the C–C double bond can only proceed after an internal rotation around this bond, giving rise to the hydridopalladium olefin complex *e*. In this case, no new chiral center is formed in the product molecule. If, however, β-hydride elimination is forced in a different direction (intermediate *e'*), the stereogenic carbon atom created into the insertion step will be retained, and enantioselective reactions become possible (Section 2.1.3.4). Dissociation of the product olefins **3** or **3a** leads to the formation of the hydridopalladium species *f*. This can finally be converted back into the catalytically active Pd(0) complex *a* by reductive elimination of HX with the help of a base.

These fundamental steps of the catalytic cycle have been confirmed by stoichiometric reactions starting from isolated stable complexes, and by DFT calculations [11]. Although many aspects of the Heck olefination can be rationalized by this "textbook mechanism", it provides no explanation of the pronounced influence that counter-ions of Pd(II) pre-catalysts or added salts have on catalytic activity [12]. This led Amatore and Jutand to propose a slightly different reaction mechanism [13]. They revealed that the preformation of the catalytically active species from Pd(II) salts does not lead to neutral Pd(0)L$_2$ species *a*; instead, three-coordinate anionic Pd(0)-complexes *g* are formed (Scheme 3, top). They also observed that on the addition of aryl iodides **1a** to such an intermediate *g*, a new species forms quantitatively within seconds and the solution remains free of iodide and acetate anions. It may then take several minutes before the expected stable, four-coordinate trans complex b^1 can be detected. They therefore postulated that the catalytic cycle is dominated by anionic three- and five-coordinate species.

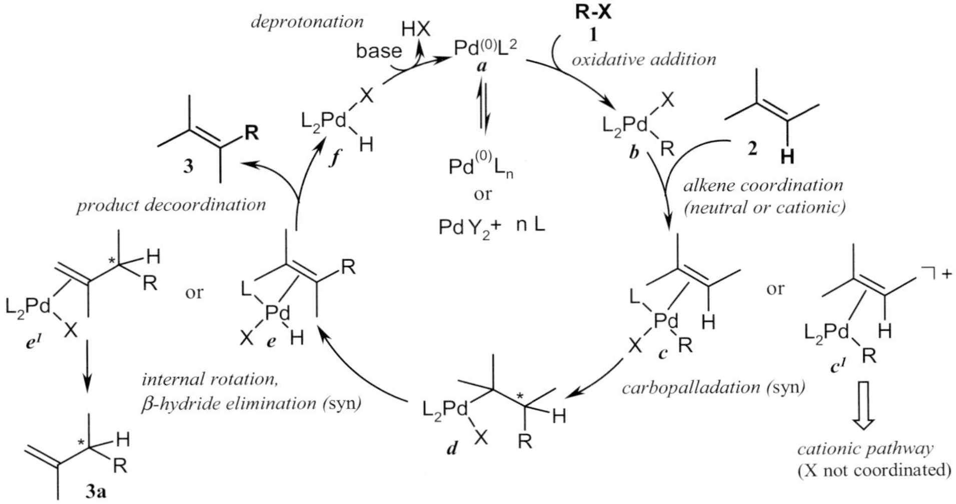

Scheme 2. Catalytic cycle of the Heck reaction.

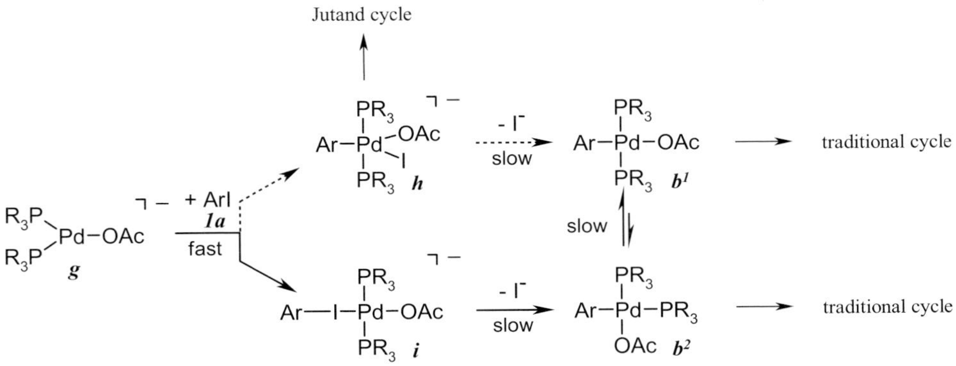

Scheme 3. Alternative mechanistic concepts for the oxidative addition process.

With the help of DFT calculations the stability of the three-coordinate complexes **g** was recently confirmed, but no evidence has yet been found for the five-coordinate intermediates **h** [14, 15]. Instead, Goößen et al. disclosed the linear complex **i** as an alternative structure for the initial intermediate [15]. It offers a valid explanation for Jutand's findings and reinstates the original four-membered intermediates in the catalytic cycle, albeit starting from anionic precursors (Scheme 3, bottom).

2.1.3
Scope and Limitations

2.1.3.1 Substrates

In the Heck reaction, a vinylic hydrogen is replaced by an organic fragment, usually an alkenyl, aryl, allyl, or benzyl group, but there are also examples of analogous olefinations of alkoxycarbonylmethyl, alkynyl, silyl, and even some alkyl groups. Both intra- and intermolecular Heck reactions are known. The regioselectivity of intramolecular reactions is largely controlled by entropic factors, so the exo-trig products are formed preferentially. In intermolecular reactions, the olefin reacts preferentially at the sterically less crowded position, the polarization of the olefin playing a minor role [16]. Selected examples are shown in Fig. 1.

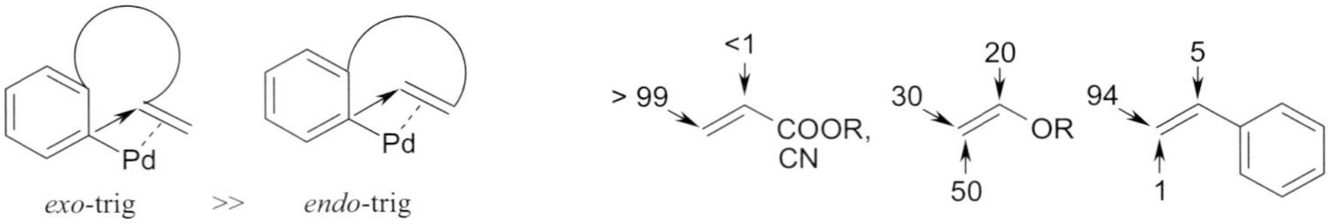

Figure 1. Regiochemical outcomes of the Heck reaction.

Mono- and 1,1-disubstituted olefins are the most reactive, but even tetrasubstituted alkenes have been used [17]. Because double-bond isomerization is a common side-reaction, olefins with a non-isomerizable bond are usually used.

2.1.3.2 Heck Reactions of Aryl Bromides and Iodides

The most widely used version of the Heck reaction is the coupling of aryl bromides or iodides with activated olefins such as styrene or acrylic acid derivatives, using conditions first described by Spencer [18]. The catalysts are generated in situ by combining $Pd(OAc)_2$ with two to four equivalents of triphenyl- or tri-o-tolyl-phosphine. Triethylamine, K_2CO_3, $NaHCO_3$, or NaOAc are often used as the base.

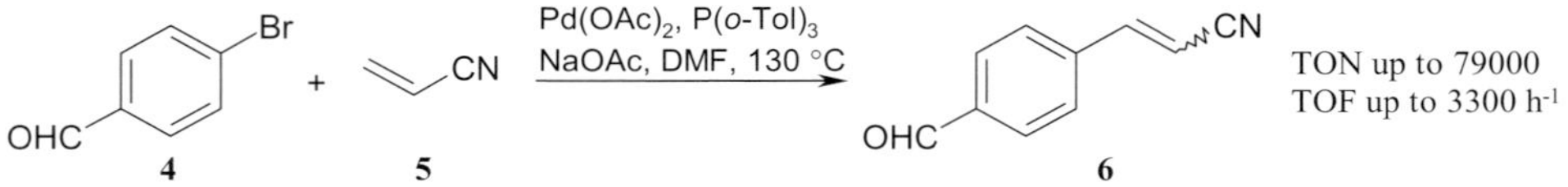

Scheme 4. Typical example of a Heck reaction following Spencer's improved procedure.

The most effective solvents for this reaction are polar aprotic solvents such as acetonitrile, DMF, NMP, or DMSO. Addition of halide salts such as tetra-n-butyl-ammonium bromide strongly facilitates the reaction and enables the conversion of aryl iodides and activated bromides even in the absence of phosphines (Jeffery conditions) [19]. Under these conditions the presence of water was sometimes found to be beneficial and there are even examples for Heck olefinations in aqueous solutions [20].

Because aryl phosphines are not only costly but can also act as aryl sources themselves, giving rise to unwanted byproducts, there has been steady interest in extending ligand-free Heck reactions to aryl bromides and aryl chlorides. Reetz and de Vries recently found that these can be performed with high efficiency using stabilized colloidal Pd catalysts [21]. If the palladium is kept at a low concentration between 0.01 and 0.1 mol%, precipitation of the Pd(0) is avoided and the colloids serve as a reservoir for the catalytically active species (Scheme 5). This economically attractive method has been successfully applied on an industrial scale by DSM [22].

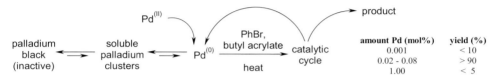

	amount Pd (mol%)	yield (%)
	0.001	< 10
	0.02 - 0.08	> 90
	1.00	< 5

Scheme 5. Heck reaction using colloidal palladium as catalyst.

2.1.3.3 Domino Reactions Involving Carbometallation Steps

For domino Heck reactions the initial carbometallation step must lead to intermediates that do not readily undergo β-hydride elimination and can thus serve as starting points for additional Pd-mediated transformations [7]. Such processes, as exemplified in Scheme 6, have been used extensively by de Meijere, Grigg, Overman and others for assembly of complex carbon skeletons.

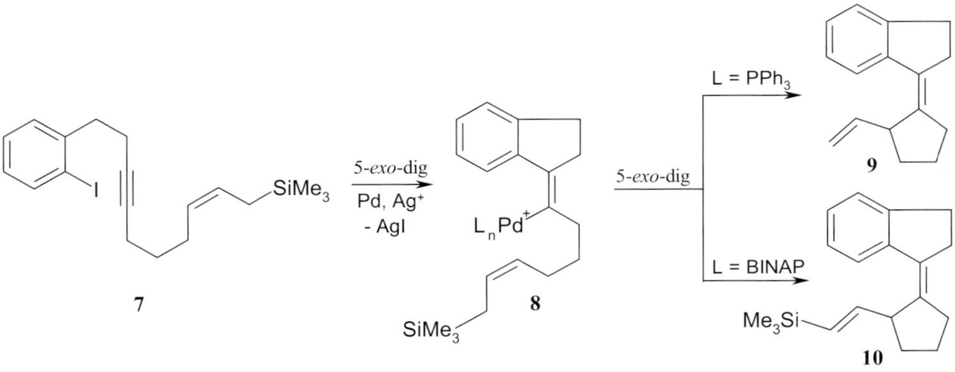

Scheme 6. Allylsilane-terminated cascade Heck reaction by Tietze et al.

In Tietze's example shown in Scheme 6 the palladium is eliminated away from the tertiary carbon center formed after two cyclization steps. Both the silane **10** and the desilylated product **9** are accessible, depending on the ligand employed [23].

The name "zipper reactions" has been coined for cascade reactions in which a whole series of carbometalations proceed in a single reaction step [24]. In the example by Negishi shown in Scheme 7, palladium initially inserts into the vinylic carbon–iodine bond, then progresses along a line of multiple bonds, literally zipping the two chains of the substrate together to form the fused steroidal ring system **12**.

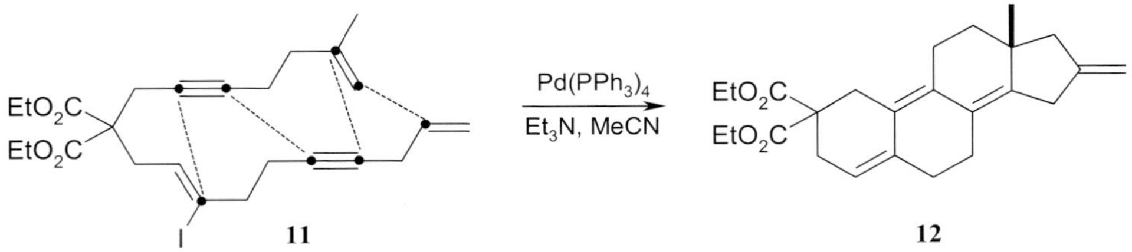

Scheme 7. Cascade Heck reaction.

2.1.3.4 Enantioselective Heck Reactions

In intramolecular Heck reactions, chiral centers can be generated by differentiation of either two enantiotopic leaving groups or two double bonds, or by double-bond migration. In the classical synthesis of a decalin ring system **14** (Scheme 8), two stereocenters were generated with high selectivity using the chiral ligand (*R*)-binaphthyldiphenylphosphine (BINAP) [25].

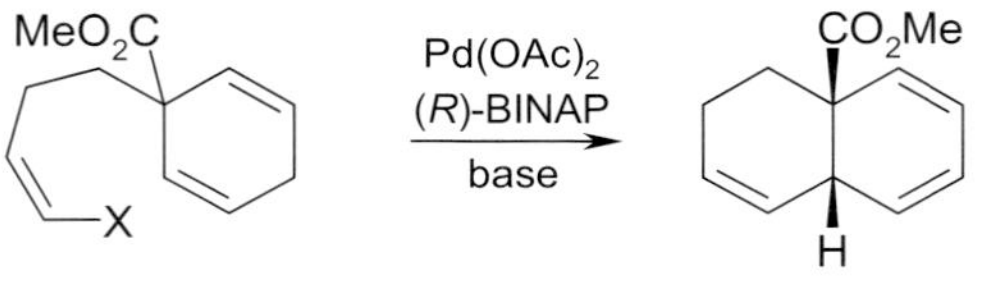

Scheme 8. Two stereocenters generated in an intramolecular Heck reaction.

In this example, significant chiral induction can only be realized if silver salts are added to trap the iodide ions, and thus a chelating coordination mode of the chiral bidentate ligand is facilitated during coordination of the olefin (Scheme 2, cationic pathway). Even better results are obtained from substrates with non-coordinating leaving groups such as vinyl triflates.

These "cationic" conditions are also used for intermolecular Heck olefinations in which the chiral centers are generated via double-bond migration. The synthesis of 2-aryl-2,3-dihydrofurans **17** is a good example (Scheme 9) [26].

A cationic pathway is not compulsory for achieving high enantioselectivity, however – there are also examples in which conditions facilitating the neutral pathway (Scheme 2) are most effective. Interestingly, in an intramolecular Heck coupling performed by Overman et al., both enantiomers of spirocycles could be selectively generated using the same BINAP ligand (Scheme 10), depending on whether silver salts were used to sequester iodide (cationic pathway), or a tertiary amine was employed as an HI scavenger (neutral pathway) [27].

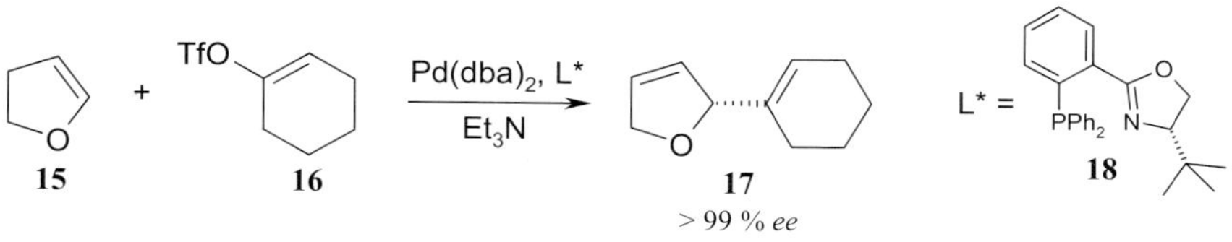

Scheme 9. Enantioselective intermolecular Heck reaction.

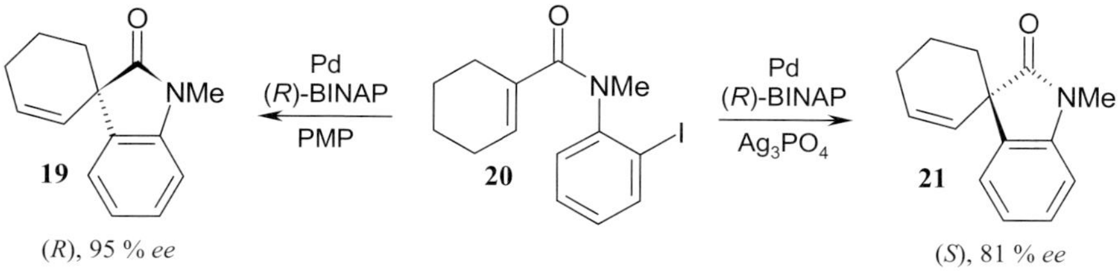

Scheme 10. Reverse enantioselectivity for the neutral and cationic pathways.

2.1.3.5 Heck Reactions of Aryl Chlorides

The profound industrial interest in extending the Heck reaction to the inexpensive but notoriously unreactive chlorohydrocarbons has triggered extensive studies in this field. Based on key pioneering work by Spencer [28], much progress has been achieved in recent years. The main challenge is to facilitate oxidative addition to stable C–Cl bonds (ca. 95 kcal mol^{-1}) while suppressing degradation reactions of the activating phosphine ligands.

Milstein et al. found that Pd complexes with chelating alkylphosphines such as bis(diisopropylphosphino)butane (dippb) efficiently catalyze the olefination of aryl chlorides with styrenes in the presence of elemental zinc [29]. Unfortunately, these electron-rich phosphines are apparently incompatible with electron-poor olefins such as acrylic acid derivatives. The latter were successfully coupled with activated chloroarenes by Herrmann et al., who used palladacycles or Pd-catalysts with heterocyclic carbenes [30].

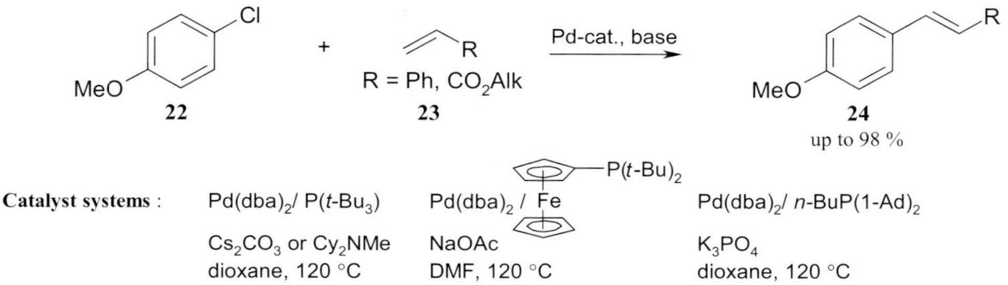

Scheme 11. Heck olefination of deactivated aryl chlorides.

Almost simultaneously, several groups developed efficient procedures for Heck reactions of deactivated chloroarenes **22** involving sterically crowded monodentate phosphines as activating ligand on the palladium (Scheme 11) [31]. Littke and Fu employed commercially available P(t-Bu)$_3$, Hartwig P(t-Bu)$_3$ or bis-t-butyl-ferrocenylphosphine, and Beller di(1-adamantyl)-n-butylphosphine. The use of biscyclohexylmethylamine as the base instead of alkali metal carbonates or phosphates significantly extends the scope of the Fu procedure in respect of the olefin partner.

2.1.3.6 Heck Reactions of Diazonium Salts

Another particularly attractive substrate class for industrial Heck reactions are the arenediazonium salts, because of to their ready availability from anilines. Matsuda et al. found that for these highly reactive compounds, ligand-free Pd(dba)$_2$ suffices as catalyst [32]. Even palladium catalysts on solid supports are highly effective [33]. The diazotization of an aniline **25** and its olefination can be performed separately, or in one pot (Scheme 12).

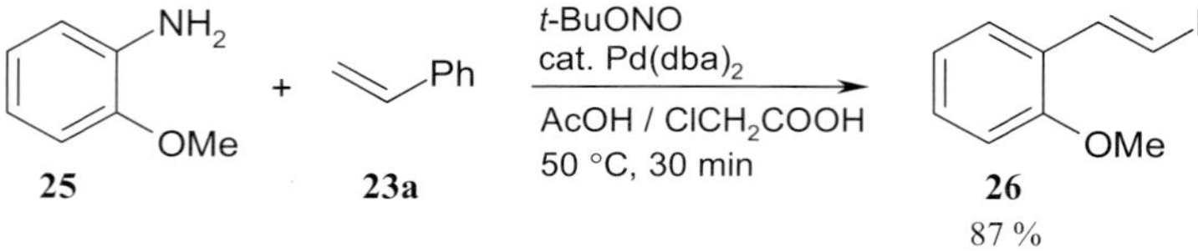

Scheme 12. Example for the one-pot diazotization – Heck reaction of an aniline.

2.1.3.7 Heck Reactions of Carboxylic Acid Derivatives

Blaser and Spencer discovered that when using aroyl halides in the place of aryl halides, the olefination proceeds with decarbonylation, giving rise to the same vinyl arenes [34]. The ligandless catalysts are generated in situ from palladium halides – coordinating phosphines were found to hamper the decarbonylation step. Interestingly, Miura et al. reported that aroyl chlorides can be olefinated efficiently, even in the absence of base, by using a rhodium catalyst; gaseous HCl is produced as byproduct [35].

For all standard Heck olefinations, stoichiometric amounts of base are required, leading to production of the corresponding quantities of waste salts. De Vries et al. first presented a possible solution to this long-standing problem by employing aromatic carboxylic anhydrides as the aryl source [36]. In the presence of a PdCl$_2$/NaBr catalyst system carboxylic anhydrides were cleanly converted into one equivalent of the corresponding vinyl arene and one equivalent of the carboxylic acid, which in principle can be converted back into the starting anhydride.

Using optimized catalyst systems, even esters of electron-deficient phenols **27** or of enols were successfully coupled to olefins **23** to give the vinyl arenes **28**, along with CO and the corresponding phenol **29** or ketone, respectively

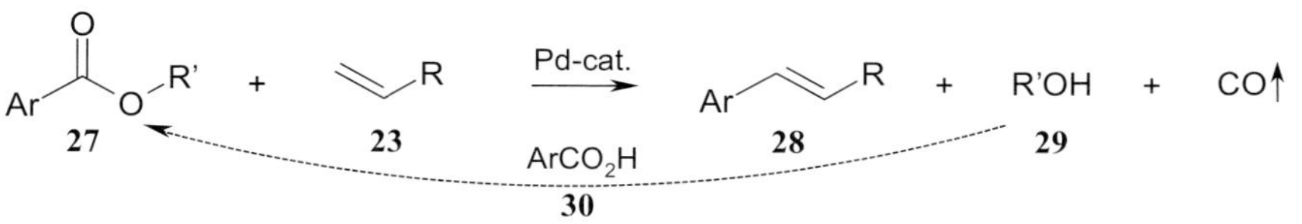

Scheme 13. Salt-free Heck olefination.

(Scheme 13) [37]. The phenols were successfully recycled into the starting material **27** by esterification with fresh carboxylic acid **30** (Scheme 4), thus demonstrating for the first time that the production of waste salts in Heck reactions is avoidable. This approach opens up interesting opportunities for the development of environmentally friendly Pd-catalyzed reactions.

2.1.3.8 Miscellaneous Substrates

In a process developed by Myers et al., aromatic carboxylic acids were directly employed as substrates for Heck olefinations, albeit in the presence of a large excess of silver carbonate [38]. This base both facilitates the decarboxylation step and acts as an oxidant, generating arylpalladium(II) intermediates. In related processes, arylphosphonic [39] and arylboronic acids [40] were used as aryl sources in the presence of an oxidant.

Under similar oxidative conditions, with activation of the aromatic C–H bond, some arenes could be used directly as aryl sources [41]. Unfortunately, by analogy with the Friedel–Crafts acylation, this reaction is regioselective for very few substrates only. High regioselectivity was, however, obtained if coordinating substituents on the arenes facilitate an orthopalladation reaction by a Pd(II) species [42]. After carbometallation and reductive elimination, Pd(0) is released, which has to be converted into the initial Pd(II) species in an extra oxidation step. Usually, quinines are used for this purpose, but in combination with certain heteropolyacids as cocatalysts even molecular oxygen can be employed as the oxidant.

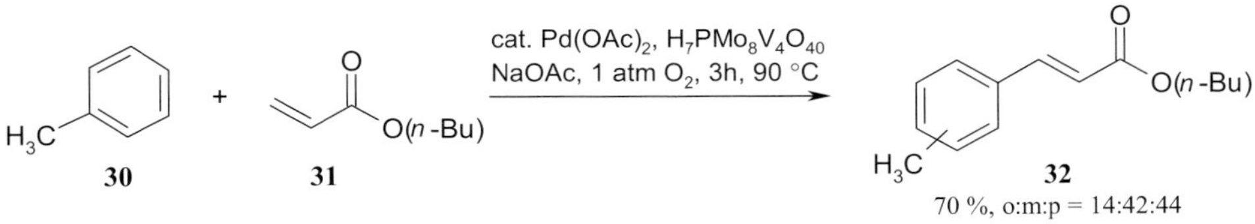

Scheme 14. Example for the oxidative Heck reaction of an arene.

As demonstrated in the preceding sections, the Heck reaction, although already a reliable and versatile tool, is still under development, and one can still expect major improvements in the coming years.

2.1.3.9 Industrial Applications

Several commercial products have been produced via Heck reactions on a scale in excess of one ton year^{-1} [43]. The sunscreen agent 2-ethylhexyl-*p*-methoxycinnamate has been synthesized on a pilot scale by using Pd/C as the catalyst [44]. Albermarle produces Naproxen via a Heck reaction of 2-bromo-6-methoxynaphthalene with ethylene, followed by carbonylation of the product [45]. A key step in the production of Singulair (montelukast sodium), a leukotriene receptor antagonist for treatment of asthma, is Heck reaction of methyl 2-iodobenzoate with an allylic alcohol to give a ketone [43].

Heck olefination of a diazonium salt is a key step in the industrial synthesis of sodium 2-(3,3,3-trifluoropropyl)benzenesulfonate, en route to Novartis' sulfonylurea herbicide Prosulfuron (Scheme 15) [46]. Pd(dba)$_2$ (0.5–1.5 mol%, prepared in situ from PdCl$_2$), is used as catalyst. After completion of the arylation step, charcoal is added, thus directly providing a heterogeneous catalyst for the subsequent hydrogenation step and enabling easy removal of the now supported catalyst by filtration.

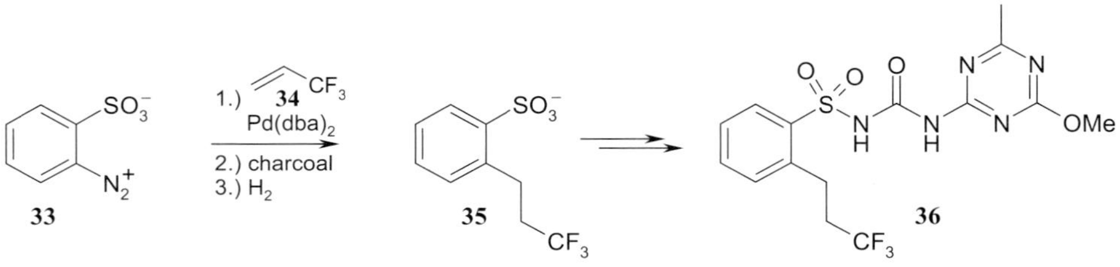

Scheme 15. Heck reaction in the synthesis of Prosulfuron.

The effective removal of residual palladium has become one of the most critical issues in the implementation of homogeneously catalyzed Heck reactions in industrial syntheses, particularly in the production of pharmaceuticals, in which tight specifications (sometimes less than 0.5 ppm Pd) must be met. Only very few scalable, inexpensive removal techniques are known [47], for example, binding of Pd to *N*-acetylcysteine and removal of the adduct by extraction or crystallization [48]. When such practical, inexpensive separation and recycling tools for palladium catalysts have been fully developed and industrially implemented, the Heck reaction will certainly find commercial use to the large extent that would be expected from its enormous synthetic utility.

Experimental

Heck Olefination of Aryl Bromides According to Spencer: Synthesis of Ethyl 4-Cyanocinnamate [18]

A mixture of palladium acetate (22.4 mg, 0.1 mmol) and tri(*o*-tolyl)phosphine (122 mg, 0.4 mmol) in DMF (20 mL) was stirred until a homogeneous solution had formed (5 min). A dry three-necked flask equipped with a magnetic stirrer, thermometer, and reflux condenser with argon bubbler and argon inlet-tube was charged with DMF (49 mL), 4-bromobenzonitrile (9.10 g, 50.0 mmol), anhydrous sodium acetate (4.51 g, 55.0 mmol), and ethyl acrylate (5.96 g, 55.0 mmol). The reaction mixture was purged with argon and heated to 130 °C. Catalyst stock solution (1 mL) was added by means of a syringe and the reaction was stirred until complete conversion was indicated by gas chromatography (ca. 1 h), the reaction mixture was poured into water, extracted with dichloromethane, dried over $MgSO_4$, filtered, and the volatiles were removed in vacuo. The residue was purified by distillation, giving the title compound in 97 % yield.

Heck Olefination of Aryl Chlorides According to Fu: Synthesis of (E)-3-(2-Methylphenyl)-2-methylacrylic Acid Methyl Ester [49]
An oven-dried Schlenk tube equipped with a magnetic stirrer bar was charged with $Pd_2(dba)_3$ (12.9 mg, 0.014 mmol). The Schlenk tube was fitted with a rubber septum, evacuated and then refilled with argon. 2-Chlorotoluene (0.110 mL, 0.941 mmol), Cy_2NMe (0.220 mL, 1.03 mmol), P(t-Bu)$_3$ (0.15 M stock solution in dioxane; 0.38 mL, 0.057 mmol), methyl methacrylate (0.110 mL, 1.03 mmol), and then dioxane (0.56 mL) were added by means of a syringe. The septum was replaced with a Teflon stopper and the reaction mixture was stirred at 110 °C for 24 h. The mixture was diluted with diethyl ether, filtered through a pad of silica gel with copious washing, concentrated, and purified by column chromatography (eluting with 5 % Et_2O–hexanes) to yield 158 mg (88 %) of the title compound as a clear, colorless liquid.

2.2
Wacker Oxidation
Lukas Hintermann

2.2.1
Introduction and Fundamental Examples

The palladium(II)-mediated oxidative coupling of olefins with oxygen-nucleophiles (ROH; water, alcohols, carboxylic acids) is a stoichiometric reaction with respect to Pd(II), resulting in an oxygenated product and Pd(0). To convert Pd(0) back to Pd(II) and start a new reaction cycle, a *reoxidation reaction* (which can itself be stoichiometric or catalytic) using a *terminal oxidant* is required. In this way, the overall process becomes catalytic with respect to the expensive Pd salt.

The first and foremost reaction of this type is the now classical Wacker–Hoechst oxidation [1–4] of ethene to ethanal (acetaldehyde) by means of a $PdCl_2/CuCl_2$ as catalyst and O_2 as terminal oxidant (Scheme 1, a).

This process with its ingenious $Cu(I)/Cu(II)/O_2$ reoxidation cycle for Pd(0) (Scheme 2) was highly influential in the further development of homogeneous catalysis, because it demonstrated that even an "exotic" and expensive metal such as palladium can be used in the large-scale production of cheap base-chemicals.

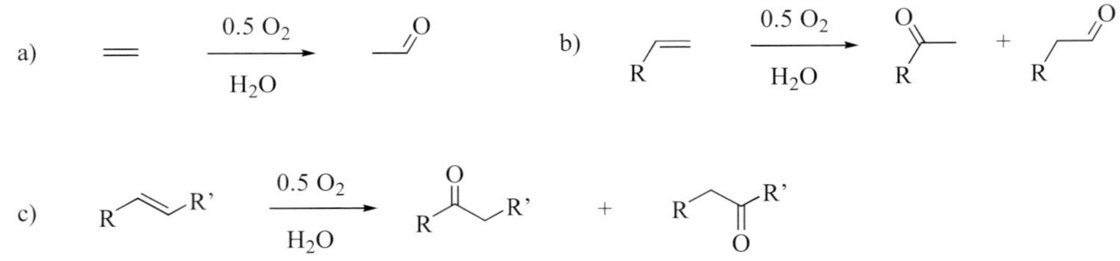

Scheme 1. Palladium-mediated oxidative coupling reactions of olefins and water.

I. $PdCl_2 + R-CH=CH_2 + H_2O = R-C(=O)CH_3 + Pd(0) + 2 HCl$

II. $Pd(0) + 2 CuCl_2 = PdCl_2 + 2 CuCl$

III. $2 CuCl + 0.5 O_2 + 2 HCl = 2 CuCl_2 + H_2O$

Scheme 2. The Cu(I)/Cu(II)-reoxidation system in the Wacker–Hoechst olefin oxidation reaction.

Application of Wacker chemistry to higher olefins can give isomeric products (Scheme 1, b and c), but with terminal olefins, methyl ketones are usually formed (Scheme 1, b). On an industrial scale, propene is converted to acetone [4]. Additional examples of higher olefin oxidation were reported by the Wacker group [1], but the reaction did not find much use in synthetic chemistry before it was adapted for use with organic solvents, and reliable reoxidation systems working in such media were found. In that respect, Moiseev et al. observed that benzoquinone (BQ) is a rapid and homogeneous stoichiometric reoxidant for Pd(0) [5], and practical procedures for synthetic oxidations were presented by Clement and Selwitz in 1964 [6], who also introduced aqueous N,N-dimethylformamide (DMF) as a solvent of choice. Later, influential work came from Tsuji and coworkers, who popularized $CuCl/O_2$ in place of $CuCl_2/O_2$ as a powerful and reliable reoxidation system [7–10] which, apparently, produces none of the chlorinated side-products of the Wacker process and operates under almost neutral conditions. Although water is the oxygen-nucleophile in the original Wacker reaction, mechanistically similar oxidative couplings with carboxylic acids give rise to vinyl esters (Scheme 3, d) and alcohols give acetals [11–13] (Scheme 3, e). In fact, the palladium-catalyzed coupling of acetic acid with ethene and O_2 as direct terminal oxidant (Cu is not necessary) to produce vinyl acetate is now also an important industrial process, based on the discovery by Moiseev et al. in 1960 [14].

Examples of oxidative coupling to vinyl derivatives remain limited in number, however, because the reaction takes a different course whenever allylic hydrogen atoms are present in a substrate (Scheme 3, f and g). Under these conditions allylic sp^3 C–H bonds are activated and such reactions are therefore *allylic oxidations* (Section IV.1.2.4). In recent years, however, several examples of reactions of type g (Scheme 3) have been performed enantioselectively and called *asymmetric*

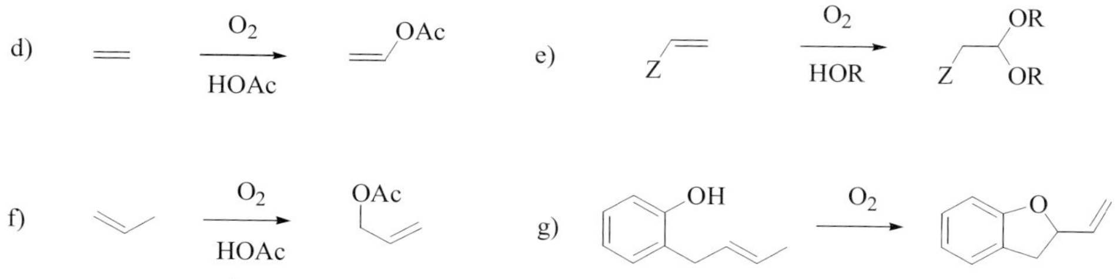

d) $\text{CH}_2=\text{CH}_2 \xrightarrow[\text{HOAc}]{\text{O}_2} \text{CH}_2=\text{CHOAc}$ e) $Z-\text{CH}=\text{CH}_2 \xrightarrow[\text{HOR}]{\text{O}_2} Z-\text{CH(OR)-CH}_2\text{OR}$

f) propene $\xrightarrow[\text{HOAc}]{\text{O}_2}$ allyl acetate   g) 2-allylphenol $\xrightarrow{\text{O}_2}$ 2-vinyl-2,3-dihydrobenzofuran

Scheme 3. Palladium-mediated coupling reactions between olefins and alcohols or carboxylic acids.

Wacker reactions, which is why we refer to them here [12, 15–17]. Finally, the general concept of a Wacker reaction could be regarded as the palladium-catalyzed oxidative coupling of heteronucleophiles and olefins, and this can obviously be extended to nitrogen nucleophiles and others [18]; conversely, the principle of the Cu(I)/Cu(II)/O_2 reoxidation system for Pd(0) can be applied to other oxidation reactions (for example that of CO to CO_2), but the present overview is limited to sp^2-C–H activation in olefins.

2.2.2
Mechanism

The overall mechanistic pathway of the Wacker reaction is easily rationalized, but details of the individual reaction steps are still a matter of discussion and continuing research [19–21]. Considering the manifold reaction conditions and reoxidation systems known from the literature, *a single mechanism is certainly not operating in all cases*. A good starting point is the classical Wacker–Hoechst ethanal synthesis [4], a simplified catalytic cycle of which is depicted in Fig. 1 [19–21]. Starting from tetrachloropalladate(II) (**A**), ligand exchange with ethene leads to a π-complex **B**, in which the olefin is activated toward nucleophilic attack by water (vide infra), resulting in a 2-hydroxyethylpalladium species **C**. Now, β-hydrogen elimination leads to a hydridopalladium species **D**, and reinsertion results in a 1-hydroxyethylpalladium species **E**. There is neither dissociation of vinyl alcohol nor exchange of protons from **D**, as shown by isotope-labeling studies, because ethene in D_2O yields ethanal free from deuterium [2].

Subsequently, **E** releases ethanal to give a Pd(0) species which is reoxidized to Pd(II) by means of Cu(II), possibly via dinuclear species such as **F**, and Cu(II) is regenerated from Cu(I) by reaction with O_2 and HCl. From the kinetics of the Wacker oxidation, which are roughly expressed [4, 22] by the equation:

$$-d[C_2H_4]/dt = k\,[C_2H_4][PdCl_4^{2-}]/[H^+][Cl^-]^2$$

it can be concluded that **B** undergoes at least one additional ligand exchange of Cl^- by H_2O and loss of a proton, occuring *before* the rate-determining step. What exactly happens on a molecular level is not yet clear, but the interpretation is

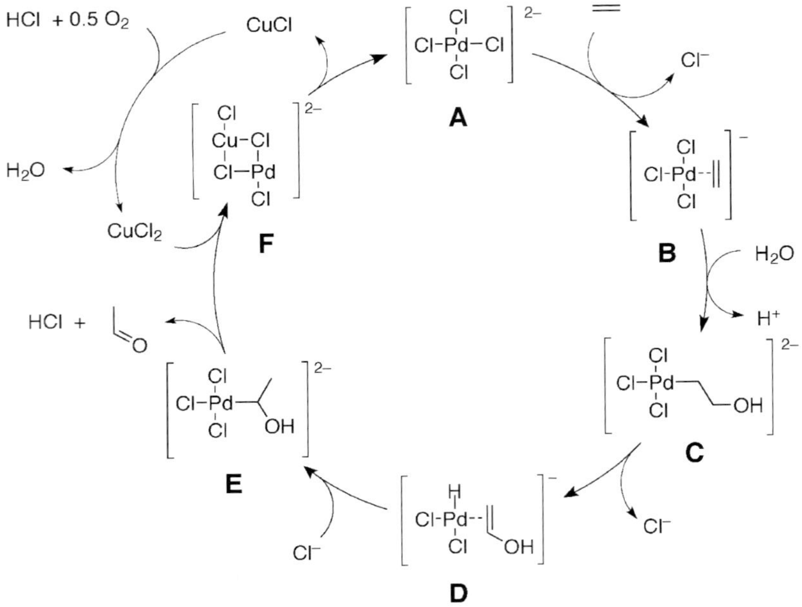

Figure 1. Schematic catalytic cycle of the Wacker–Hoechst ethene oxidation.

directly connected to the debate about the stereochemistry of *hydroxypalladation*, which might be anti, because of external attack of water on coordinated ethene [20, 21], or alternatively syn, with insertion of ethene into a Pd–OH bond [4, 19]. Anti hydroxypalladation has been established in many model reactions with palladium–olefin complexes and even for ethene at high chloride concentrations [20], but syn hydroxypalladation is not excluded for ethene at low chloride concentrations [23]. In fact, a recent investigation of the cyclization of ortho allylphenols nicely illustrates that the stereochemistry of oxypalladation can depend on the presence of chloro ligands on palladium(II) [24]. For higher olefins the mechanism is probably similar, but it is an additional challenge to explain and predict the *regioselectivity* of the Wacker reaction on mechanistic grounds. Deviations from the usual Markovnikov regioselectivity (methyl ketones from terminal olefins) are not uncommon and are usually explained by invoking directing effects of coordinating functional groups within the substrate [8, 25, 26]. In such cases, regioselective hydroxypalladation of the olefin is governed by reaction pathways via kinetically favored cyclopalladated intermediates. An illustration for this comes from an X-ray structure of a dimeric five-membered cyclopalladated complex **2** (Scheme 4) which is formed via oxypalladation from olefin **1** [27]:

Intramolecular coordination is apparently responsible for most examples of regioselective Wacker oxidations of *internal* olefins, but *electronic* effects are also operating [28], specifically in acceptor-substituted olefins. Steric effects are currently not well explored [8]. Recent theoretical studies on the mechanism of the Wacker and related reactions are available elsewhere [29, 30].

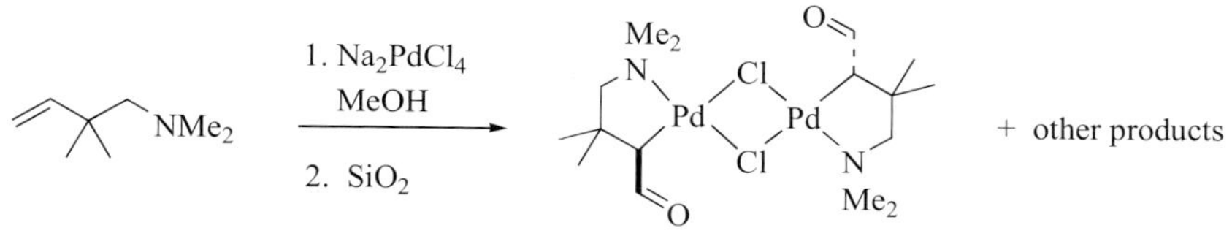

Scheme 4. Palladacycle from the Pd-mediated oxidation of a homoallylic tertiary amine.

2.2.3
Scope and Limitations

The major emphasis in this section is on examples from the recent literature, thus the priority of the cited authors for a certain type of reaction cannot generally be assumed. Many earlier examples will be found in the reviews by Tsuji [8, 10].

2.2.3.1 Reactions Initiated by the Addition of Water to Terminal Alkenes

Terminal non-functionalized alkenes react selectively to give methyl ketones (Table 1), with the corresponding aldehydes as by-products in variable amounts. In terms of synthetic planning, the terminal alkene can be regarded as a masked methyl ketone. The reaction has been performed on a wide range of substrates from simple olefins, for example 1-decene (**3**, gives 2-decanone **4**) [9] to fairly complex intermediates bearing many functional groups (**5** to **6** [31], **9** to **10** [32], **11** to **12** [33]) or even polymers bearing vinyl groups (**7** to **8**) [34]. The "vinyl to acetyl" conversion has been included in various *synthetic sequences,* in which the generation of particular olefins is followed by a Wacker oxidation and then another specific reaction involving the methyl ketone. Examples [8, 10] include: (1) cyclopentenones are generated by α-allylation of ketones via subsequent oxidation and aldol-condensation. (2) Analogously, annelation of cyclohexenones is possible via oxidation of α-3-butenyl ketones, then aldol-condensation. (3) If asymmetric allylation of an aldehyde to a homoallylic alcohol is followed by O-protection and Wacker oxidation, the equivalent of an asymmetric acetone aldol reaction is achieved (for the second part, cf. **11** to **12** [33]), etc. An elegant new one-pot tandem sequence combines the palladium-catalyzed chloroallylation of alkynes (e.g. **13** with **14**) and subsequent Wacker oxidation (simply by adding CuCl and O_2) of the intermediary chlorodiene **15** (not drawn) [35].

Among the most popular reaction conditions for this conversion are those popularized by Tsuji and coworkers (experiment 1, below), using 10 mol% $PdCl_2$ and 1 equiv. CuCl in aqueous DMF under an atmosphere of O_2 at room temperature [9]. Everything can be varied, however. In terms of solvents, alcohols tend to give faster reactions, but are said to speed up olefin isomerization. $CuCl_2 \cdot 2H_2O$ can be used as a reoxidant, but chlorinated side-products are potentially formed. This is precluded in halide-free systems, e.g. with $Pd(OAc)_2$ and $Cu(OAc)_2/O_2$,

Table 1. Wacker oxidation of terminal olefins.

Starting materials	Products	Yield (%)	Ref.
3	4	65–73	9
5	6	65	31
7	8	n. d.	34
9	10	84[a]	32
11	12	73	33
13 + 14	16	66	35

Reaction conditions similar to those in experiment 1, below, plus
(a) Pd(OAc)$_2$, Cu(OAc)$_2$/O$_2$ reoxidant

which are also recommended for acid-sensitive substrates such as **9** [32]. Benzoquinone (BQ) is a very interesting reoxidant for synthetic purposes, resulting in rapid reactions in homogeneous solution and enabling rather low catalyst loadings (e.g. ≤1 mol%; experiment 3, below) [5, 6]. A particularly active system for olefin oxidation is obtained with Pd(OAc)$_2$ and BQ, if HClO$_4$ (0.3 M) is present in the reaction medium [36] (experiment 2). In addition, many variations of reaction medium and reoxidation system have been suggested but, currently, none is widely applied in synthesis. In an interesting line of research, reaction conditions have been developed in which O_2 is used as a *direct* terminal oxidant for Pd(0) [37].

2.2.3.1.1 Specific Substrates

Although allyl-arenes are prone to olefin isomerization, several successful reactions have been performed, for example in the chemoselective oxygenation of **22** to aryl-acetone **23** (Table 2) [38]. Allyl alcohols sometimes react sluggishly, but examples with high ketone selectivity are known, for example the oxidation of tertiary alcohol **24** to α-hydroxyketone **25** [39].

Table 2. Oxidation of specific terminal olefins.

Starting materials	Products	Yield (%)	Ref.
22	**23**	75	38
24	**25**	88[a]	39
26	**27**	77	40

Reaction conditions similar to those in experiment 1, below, plus
(a) BQ as terminal oxidant.

2.2.3.1.2 Styrenes

The oxidation of styrenes under Wacker conditions is relatively slow and not always regioselective, but useful yields of acetophenones have been obtained, for example the highly substituted **27** from olefin **26** [40]. Acetophenones, on the other hand, have consistently been obtained in yields higher than 75 % by using a Pd-diketonate complex with *t*-BuOOH as oxidant in a biphasic $C_8F_{17}Br/C_6H_6$ medium [41]. The oxidation with Pd(II)/t-BuOOH is mechanistically distinct from the usual Wacker oxidation, and is rather general for terminal olefins [42].

2.2.3.1.3 Inversion of Regioselectivity in Terminal Alkene Oxidation

The basic *Markovnikov* selectivity pattern is partially or fully overrun in the presence of neighboring coordinating groups within the olefin substrate (Section 2.2.2). Known functionalities where inversed selectivity can occur include 3-alkenoylamides (e.g. **17** reacts to give a mixture of **18** and **19**, Table 3) [43], homoallyl esters and alcohols, allyl ethers (but not necessarily allyl alcohols) [44], allyl amines, allyl amides, or carbamates (cf. **20** to **21**) [45], allyl sulfides [46] or 1,5-dienes [47]. As a matter of fact, aldehyde by-products are quite normal in Wacker reactions, but tend to be overlooked.

Table 3. Inversion of regioselectivity in terminal alkene oxidation.

Starting materials	Products		Yield (%)	Ref.
17	**18** + **19**		65 (18) 5 (19)	43
20	**21**		76	45

Reaction conditions similar to those in experiment 1

A catalyst system based on the complex $PdCl(NO_2)(MeCN)_2$ with $CuCl_2/O_2$ as reoxidant in t-BuOH as solvent has a tendency to give high selectivity for aldehydes [48], but preparative applications remain to be explored.

2.2.3.2 Reactions Initiated by Addition of Water to Internal Alkenes

Internal olefins react more slowly than terminal olefins, and mixtures of regioisomers are usually formed. Accompanying olefin isomerization can further complicate product selectivity. Despite these words of warning, the reaction can still give satisfactory results (Table 4), for example in the final step of the total synthesis of annuionone A (**29**), which involves oxidation of bicyclic olefin **28** [49]. A small amount of regioisomer **30** is formed, though. The key to high regioselectivity is the presence of certain coordinating or electronically directing groups within the substrate: Allyl ethers and esters yield β-alkoxyketones (**31** to **32**) and β-acyloxyketones, respectively, with rather high selectivity [10, 50]. The α-propenyl-tetrahydropyrane **33** was converted to the building block **34** for total synthesis of leucascandrolide A [51]. Homoallyl esters are regioselectively converted into the 4-acyloxyketones [10].

Table 4. Oxidation of internal olefins.

Starting materials	Products		Yield (%)	Ref.
28	**29**	**30**	54 (**29**) 16 (**30**)	49
31	**32**		67	10
33	**34**		86	51
35	**36**		69[a]	54

Reaction conditions are similar to those in experiment 1, plus (a) Na$_2$PdCl$_4$ (40 mol%)/t-BuOOH in HOAc/H$_2$O

1-Propenylarenes (β-methylstyrenes) are oxidized with varying degrees of regioselectivity, depending on the electronic effects of the substituents in the aromatic ring [28, 52]. Reactions using Na$_2$PdCl$_4$ as catalyst in aqueous acetic acid with either H$_2$O$_2$ or t-BuOOH as oxidant convert 2-alkenones and 2-alkenoates to β-diketones or β-ketoesters, respectively [8, 10, 53], e.g. alkenone **35** was transformed to a tricyclic ether **36** by oxidation, followed by acidic cleavage of the acetal unit and condensation [54]. 3-Alkenones and 3-alkenoates are oxidized to 1,4-difunctionalized compounds while 4-Alkenones give 1,5-dicarbonyl compounds [8, 10]. The regioselectivity of oxidation can be difficult to predict in substrates containing multiple functional groups [44]. Changes of protecting groups can also affect the reaction selectivity [44].

2.2.3.3 Reactions Initiated by the Addition of Alcohols or Carboxylic Acids to Alkenes

If carboxylic acids and alcohols act as nucleophiles in the Wacker reaction, the products are vinyl esters or acetals, respectively. As mentioned, the substrate scope of these oxidations is usually limited to olefins not bearing hydrogen in allylic position, because of competing *allylic oxidation*.

2.2.3.3.1 ROH Addition then Elimination to Vinyl Derivatives

The industrial synthesis of vinyl acetate [14] via palladium-catalyzed oxidative coupling of acetic acid and ethene using direct O_2 reoxidation has already been mentioned (Scheme 3, *d*). Some NaOAc is required in the reaction medium, and catalysis by Pd clusters, as alternative to Pd(II) salts, was proposed to proceed with altered reaction characteristics [14]. Similarly, the alkenyl ester **37** (Table 5) containing an "isolated" vinyl group yields the expected enol acetate **38** [55] whereas allylphenol **39** cyclizes to benzofuran **40** with double bond isomerization [56].

2.2.3.3.2 Acetalization Reactions

Acetals result from oxidative coupling of alcohols with electron-poor terminal olefins followed by a second, redox-neutral addition of alcohol [11–13]. Acrylonitrile (**41**) is converted to 3,3-dimethoxypropionitrile (**42**), an intermediate in the industrial synthesis of thiamin (vitamin B1), by use of an alkyl nitrite oxidant [57]. A stereoselective acetalization was performed with methacrylates **43** to yield **44** with variable de [58]. Rare examples of *intermolecular* acetalization with nonactivated olefins are observed with chelating allyl and homoallyl amines and thioethers (**45**, give acetals **46**) [46]. As opposed to intermolecular acetalizations, the *intramolecular* variety do not require activated olefins, but a suitable spatial relationship of hydroxy groups and the alkene [13]. Thus, Wacker oxidation of enediol **47** gave bicyclic acetal **48** as a precursor of a fluorinated analogue of the pheromone frontalin [59].

Table 5. Carboxylic acids and alcohols as nucleophiles in the Wacker oxidation of sp^2-C–H bonds.

Starting materials	Products	Yield (%)	Ref.
37	**38**	63[a]	55
39	**40**	92[b]	56
41	**42**	n.a.[c]	57
43	**44**	up to 92 %[d] de up to 95 %	58
45	**46** (n = 1, 2; X = -NR$_2$ or -SR)	up to 85 %[e]	46
47	**48**	80 %[f]	59

Reaction conditions: (a) HOAc, O$_2$ (5 MPa), Pd(OAc)$_2$ (2 mol%), 80 °C, 22 h. (b) Cu(OAc)$_2$·H$_2$O, LiCl, PdCl$_2$ (2 mol%), DMF/H$_2$O, r.t. (c) PdCl$_2$, MeOH, MeONO, N$_2$O, O$_2$. (d) PdCl$_2$ (10 mol%), CuCl (100 mol%), DME/MeOH, r.t. (e) CuCl$_2$ (300 mol%), Li$_2$PdCl$_4$ (10 mol%), MeOH, 50 °C. (f) PdCl$_2$ (59 mol%), CuCl$_2$, DME, r.t.

Experimental

Olefin Oxidation Using a CuCl/O_2 Reoxidation System (Tsuji Conditions), General Working Procedure. Adapted From Ref. [9]

A mixture of $PdCl_2$ (10 mol%) and CuCl (100 mol%) in N,N-dimethylformamide/H_2O (7:1) is stirred vigorously for 1 h under an atmosphere of oxygen. The olefin is added and the reaction mixture stirred at r.t. until consumption of the starting material. The reaction is quenched with dilute HCl and extracted with ether. The ratio of reagents and solvents can be varied widely as described in many examples throughout this chapter. The oxygen atmosphere is maintained by use of a simple balloon technique. Reactions will also proceed in air, but at a slower rate.

Fast Olefin Oxidation Using Pd(OAc)$_2$ and Benzoquinone in Acetonitrile/H_2O/$HClO_4$, General Working Procedure. Adapted From Refs [25, 36]

A mixture of 0.2 mmol Pd(OAc)$_2$ (2 mol%), 10 mmol benzoquinone, and 1 mL perchloric acid (72%) in 50 mL acetonitrile–H_2O (7:1) is stirred under Ar at r.t. until complete dissolution. The olefin (10 mmol) is added and the reaction mixture stirred at r.t. (aliphatic olefins, cyclohexene) or 60 °C (styrene, cycloheptene, internal olefins) until completion of the reaction (from 10 min to 90 min), as judged by TLC. The mixture is poured into diethyl ether and washed with aqueous NaOH (30%). The aqueous layer is extracted with ether. The combined organic layers are dried over $MgSO_4$, filtered, and the solvent is evaporated. The crude product is purified by chromatography.

Oxidation of 1-Octene, Using PdCl$_2$/Benzoquinone, at Low Catalyst Loading. Conditions Adapted From Ref. [6]

A solution of 1-octene (5.00 g, 44.55 mmol) in MeOH (5 mL) was slowly added to a solution of PdCl$_2$(MeCN)$_2$ (52 mg, 0.2 mmol, 0.45 mol%), and benzoquinone (5.40 g, 50 mmol) in 25 mL MeOH and 3 mL H_2O, held at 60 °C. After 5 h at 60 °C, the reaction mixture was cooled to r.t. and the solvent removed by rotary evaporation. The residue was dissolved in t-BuOMe and the organic solution washed with 2 M HCl (2×), 2 M NaOH (3×, removes hydroquinone) and water (1×). After drying (Na$_2$SO$_4$) and evaporation, the residual oil was distilled in vacuo (75 °C/ca. 80 mbar) to give 4.01 g (70%) of a fruity smelling clear liquid, consisting of octanones. GC–MS: Peak area (2-octanone:3-octanone:4-octanone:octanal) = 82:11:7:<0.5; substances were identified from their mass spectra; no olefins detected. ^{1}H NMR (CDCl$_3$): 2-octanone (82%): δ = 0.88 (t, J = 7.1 Hz, 3 H, H-C(8)), 1.20–1.08 (m, 6 H, H–C(7,6,5)), 1.63–1.51 (m, 2 H, H–C(4)), 2.13 (s, 3 H, H-C(1)), 2.42 (t, J = 7.4 Hz, 2 H, H-C(3)). 3-octanone, selected signal (12%): 1.05 (t, J = 7.3 Hz, 3 H, H-C(1)); 4-octanone; overlapping signals only; octanal, selected signal (<1%): 9.77 (t, J = 1.9 Hz, 1 H, H-C(1)).

References to Chapter 2 – C–H Transformation at Alkenes

References to Section 2.1

1 R. F. Heck, *J. Am. Chem. Soc.* **1968**, *90*, 5518–5546 (7 papers).
2 T. Mizoroki, K. Mori, A. Ozaki, *Bull. Chem. Soc. Japan* **1971**, *44*, 581.
3 R. F. Heck, J. P. Nolley, Jr., *J. Org. Chem.* **1972**, *37*, 2320–2322.
4 P. Fitton, J. E. McKeon, *J. Chem. Soc., Chem. Commun.* **1968**, 4–6; P. Fitton, M. P. Johnson, J. E. McKeon, *ibid.*, 6–7.
5 H. A. Dieck, R. F. Heck, *J. Am. Chem. Soc.* **1974**, *96*, 1133–1136.
6 A. de Meijere, F. E. Meyer, *Angew. Chem. Int. Ed.* **1994**, *33*, 2379–2411; I. P. Beletskaya, A. V. Cheprakov, *Chem. Rev.* **2000**, *100*, 3009–3066.
7 Domino reactions: A. de Meijere, S. Bräse, *J. Organomet. Chem.* **1999**, *576*, 88–110.
8 Enantioselective Heck reactions: M. Shibasaki, E. M. Vogl, *J. Organomet. Chem.* **1999**, *576*, 1–15; O. Loiseleur, M. Hayashi, M. Keenan, N. Schmees, A. Pfaltz, *ibid.*, 16–22.
9 G. T. Crisp, *Chem. Soc. Rev.* **1998**, *27*, 427–436.
10 H. M. Senn, T. Ziegler, *Organometallics* **2004**, *23*, 2980–2988.
11 K. Albert, P. Gisdakis, N. Roesch, *Organometallics* **1998**, *17*, 1608–1616.
12 K. Fagnou, M. Lautens, *Angew. Chem. Int. Ed.* **2002**, *41*, 26–47.
13 C. Amatore, A. Jutand, *Acc. Chem. Res.* **2000**, *33*, 314–321.
14 S. Kozuch, S. Shaik, A. Jutand, C. Amatore, *Chem. Eur. J.* **2004**, *10*, 3072–3080.
15 L. J. Gooßen, D. Koley, H. Hermann, W. Thiel, *Chem. Comm.* **2004**, 2141–2143
16 H. von Schenck, B. Kermark, M. Svensson, *J. Am. Chem. Soc.* **2003**, *125*, 3503–3508.
17 R. Grigg, V. Sridharan, P. Stevenson, T. Worakun, *J. Chem. Soc., Chem. Commun.* **1986**, 1697–1699.
18 A. Spencer, *J. Organomet. Chem.* **1983**, *258*, 101–108.
19 T. Jeffery, *Chem. Comm.* **1984**, 1287–1289; T. Jeffery, *Tetrahedron Lett.* **1985**, *26*, 2667–2670.
20 T. Jeffery, *Tetrahedron* **1996**, *52*, 10113–10130.
21 M. T. Reetz, J. G. de Vries, *Chem. Comm.* **2004**, 1559–1563.
22 A. H. M. de Vries, J. M. C. A. Mulders, J. H. M. Mommers, H. J. W. Henderickx, J. G. de Vries, *Org. Lett.* **2003**, *5*, 3285–3288.
23 L. F. Tietze, K. Heitmann, T. Raschke, *Synlett* **1997**, 35–37.
24 Y. Zhang, G. Z. Wu, G. Agnel, E. Negishi, *J. Am. Chem. Soc.* **1990**, *112*, 8590–8592.
25 Y. Sato, M. Sodeoka, M. Shibasaki, *J. Org. Chem.* **1989**, *54*, 4738–4739.
26 O. Loiseleur, P. Meier, A. Pfaltz, *Angew. Chem. Int. Ed. Engl.* **1996**, *35*, 200–202.
27 A. Ashimori, B. Bachand, L. E. Overman, D. J. Poon, *J. Am. Chem. Soc.* **1998**, *120*, 6477–6487.
28 A. Spencer, *J. Organomet. Chem.* **1984**, *270*, 115–120.
29 M. Portnoy, Y. Ben-David, D. Milstein, *Organometallics* **1993**, *12*, 4734–4735.
30 W. A. Herrmann, M. Elison, J. Fischer, C. Köcher, G. R. J. Artus, *Angew. Chem. Int. Ed. Engl.* 1995, 34, 2371–2373; W. A. Herrmann, V. P. W. Böhm, C.-P. Reisinger, *J. Organomet. Chem.* **1999**, *576*, 23–41.
31 A. F. Littke, G. C. Fu, *Angew. Chem. Int. Ed.* **2002**, *41*, 4176–4211.
32 K. Kikukawa, K. Maemura, Y. Kiseki, F. Wada, T. Matsuda, *J. Org. Chem.* **1981**, *46*, 4885–4888.
33 M. Beller and K. Kühlein, *Synlett*, **1995**, 441–442.
34 H.-U. Blaser, A. Spencer, *J. Organomet. Chem.* **1982**, *233*, 267–274.
35 T. Sugihara, T. Satoh, M. Miura, M. Nomura, *Angew. Chem. Int. Ed.* **2003**, *42*, 4672–4674.
36 M. S. Stephan, A. J. J. M. Teunissen, G. K. M. Verzijl, J. G. De Vries, *Angew. Chem. Int. Ed.* **1998**, *37*, 662–664.
37 L. J. Gooßen, J. Paetzold, *Angew. Chem. Int. Ed.* **2002**, *41*, 1237–1241;

L. J. Gooßen, J. Paetzold, *Angew. Chem. Int. Ed.* **2004**, *43*, 1095–1098.

38 A. G. Myers, D. Tanaka, M. R. Mannion, *J. Am. Chem. Soc.* **2002**, *124*, 11250–11251.

39 A. Inoue, H. Shinokubo, K. Oshima, *J. Am. Chem. Soc.* **2003**, *125*, 1484–1485.

40 M. M. S. Andappan, P. Nilsson, H. von Schenck, M. Larhed, *J. Org. Chem.* **2004**, *69*, 5212–5218.

41 M. Tani, S. Sakaguchi, Y. Ishii, *J. Org. Chem.* **2004**, *69*, 1221–1226.

42 M. D. K. Boele, G. P. F. van Strijdonck, A. H. M. de Vries, P. C. J. Kamer, J. G. de Vries, P. W. N. M. van Leeuwen, *J. Am. Chem. Soc.* **2002**, *124*, 1586–1587.

43 J. G. de Vries, *Can. J. Chem.* **2001**, *79*, 1086–1092.

44 D. C. Caskey, WO 9010617 (**1990**).

45 http://www.albemarle.com/core_heck_right.htm; T.-C. Wu, US 5536870 (**1996**).

46 P. Baumeister, G. Seifert, H. Steiner, EP 584043 A1 (**1994**).

47 C. E. Garrett, K. Prasad, *Adv. Synth. Catal.* **2004**, *346*, 889–900.

48 M. Villa, V. Cannata, A. Rosi, P. Allegrini, WO 9851646 (**1998**).

49 A. F. Littke, G. C. Fu, *J. Am. Chem. Soc.* **2001**, *123*, 6989–7000.

References and Notes to Section 2.2

1 J. Smidt, W. Hafner, R. Jira, J. Sedlmeier, R. Sieber, R. Rüttinger, H. Kojer, *Angew. Chem.* **1959**, *71*, 176–182.

2 J. Smidt, W. Hafner, R. Jira, R. Sieber, J. Sedlmeier, A. Sabel, *Angew. Chem.* **1962**, *74*, 93–102.

3 J. Smidt, W. Hafner, R. Jira, R. Rüttinger, to *Consortium für elektrochemische Industrie*, DP 1049845, **1959**.

4 R. Jira in: *Applied Homogeneous Catalysis with Organometallic Compounds*, B. Cornils, W. A. Herrmann (Eds.), 2nd ed., Vol. 1, pages 386–405, Wiley-VCH **2002**.

5 I. I. Moiseev, M. N. Vargaftik, Y. K. Syrkin, *Dokl. Akad. Nauk SSSR* **1960**, *133*, 377–380. *Chem. Abstr.* **1960**, *54*, 24350g.

6 W. H. Clement, C. M. Selwitz, *J. Org. Chem.* **1964**, *29*, 241–243.

7 J. Tsuji, I. Shimizu, K. Yamamoto, *Tetrahedron Lett.* **1976**, *34*, 2975–2976.

8 J. Tsuji, *Synthesis* **1984**, 369–384.

9 J. Tsuji, H. Nagashima, H. Nemoto, *Organic Syntheses* **1984**, *62*, 9–13, and *Coll. Vol. VII*, 137–139.

10 J. Tsuji in *Comprehensive Organic Synthesis*, B. M. Trost, I. Fleming (Eds.), Vol. 7, pages 449–468, Pergamon Press **1991**.

11 W. G. Lloyd, B. J. Luberoff, *J. Org. Chem.* **1969**, *34*, 3949–3952.

12 T. Hosokawa, S.-I. Murahashi, *Acc. Chem. Res.* **1990**, *23*, 49–54.

13 T. Hosokawa, S.-I. Murahashi in *Handbook of Organopalladium Chemistry for Organic Synthesis*, E.-I. Negishi (Ed.), pages 2141–2192, John Wiley & Sons **2002**.

14 I. I. Moiseev, M. N. Vargaftik in *Applied Homogeneous Catalysis with Organometallic Compounds*, B Cornils, W. A. Herrmann (Eds.), 2nd ed., Vol. 1, pages 406–412, Wiley-VCH **2002**.

15 Y. Uozumi, K. Kato, T. Hayashi, *J. Am. Chem. Soc.* **1997**, *119*, 5063–5064.

16 M. A. Arai, M. Kuraishi, T. Arai, H. Sasai, *J. Am. Chem. Soc.* **2001**, *123*, 2907–2908.

17 R. M. Trend, Y. K. Ramtohul, E. M. Ferreira, B. M. Stoltz, *Angew. Chem. Int. Ed.* **2003**, *42*, 2892–2895.

18 J. M. Takacs, S. Vayalakkada in: *Science of Synthesis, Organometallics*, Vol. 1., pages 319–387, Georg Thieme **2001**.

19 P. M. Henry in *Handbook of Organopalladium Chemistry for Organic Synthesis*, E.-I. Negishi (Ed.), pages 2119–2139, John Wiley & Sons **2002**.

20 J. E. Bäckvall, B. Åkermark, S. O. Ljunggren, *J. Am. Chem. Soc.* **1979**, *101*, 2411–2416.

21 J. K. Stille, R. Divakaruni, *J. Organomet. Chem.* **1979**, *169*, 239–248.

22 I. I. Moiseev, O. G. Levanda, M. N. Vargaftik, *J. Am. Chem. Soc.* **1974**, *96*, 1003–1007.

23 O. Hamed, P. M. Henry, C. Thompson, *J. Org. Chem.* **1999**, *64*, 7745–7750.

24 T. Hayashi, K. Yamasaki, M. Mimura, Y. Uozumi, *J. Am. Chem. Soc.* **2004**, *126*, 3036–3037.

25 H. Pellissier, P.-Y. Michellys, M. Santelli, *Tetrahedron* **1997**, *53*, 7577–7586.

26 H. Pellissier, P.-Y. Michellys, M. Santelli, *Tetrahedron* **1997**, *53*, 10733–10742.
27 E. C. Alyea, S. A. Dias, G. Ferguson, A. J. McAlees, R. McCrindle, P. J. Roberts, *J. Am. Chem. Soc.* **1977**, *99*, 4985–4989.
28 M. J. Gaunt, J. Yu, J. B. Spencer, *Chem. Commun.* **2001**, 1844–1845.
29 P. E. M. Siegbahn, *J. Phys. Chem.* **1996**, *100*, 14672–14680.
30 D. D. Kragten, R. A. van Santen, J. J. Lerou, *J. Phys. Chem. A* **1999**, *103*, 80–88.
31 R. Paczkowski, C. Maichle-Mössmer, M. E. Maier, *Org. Lett.* **2000**, *2*, 3967–3969.
32 A. B. Smith, Y. S. Cho, G. K. Friestad, *Tetrahedron Lett.* **1998**, *39*, 8765–8768.
33 C. Schneider, F. Tolksdorf, M. Rehfeuter, *Synlett* **2002**, 2098–2100.
34 A. N. Ajjou, H. Alper, *Macromolecules* **1996**, *29*, 5072–5074.
35 A. N. Thadani, V. H. Rawal, *Org. Lett.* **2002**, *4*, 4321–4323.
36 D. G. Miller, D. D. M. Wayner, *J. Org. Chem.* **1990**, *55*, 2924–2927.
37 S. S. Stahl, *Angew. Chem. Int. Ed.* **2004**, *43*, 3400–3420.
38 M. D. Bercich, R. C. Cambie, P. S. Rutledge, *Aust. J. Chem.* **1999**, *52*, 241–257.
39 F. Derdar, J. Martin, C. Martin, J. M. Bregeault, J. Mercier, *J. Organomet. Chem.* **1988**, *338 (2)*, C21–C26.
40 A. Yamashita, A. Toy, T. A. Scahill, *J. Org. Chem.* **1989**, *54*, 3625–3634.
41 B. Betzemeier, F. Lhermitte, P. Knochel, *Tetrahedron Lett.* **1998**, *39*, 6667–6670.
42 M. Roussel, H. Mimoun, *J. Org. Chem.* **1980**, *45*, 5387–5390.
43 A. K. Bose, L. Krishnan, D. R. Wagle, M. S. Manhas, *Tetrahedron Lett.* **1986**, *27*, 5955–5958.
44 S.-K. Kang, K.-Y. Jung, J.-U. Chung, E.-Y. Namkoong, T.-H. Kim, *J. Org. Chem.* **1995**, *60*, 4678–4679.
45 R. Stragies, S. Blechert, *J. Am. Chem. Soc.* **2000**, *122*, 9584–9591.
46 J. Lai, X. Shi, L. Dai, *J. Org. Chem.* **1992**, *57*, 3485–3487.
47 T.-L. Ho, M. H. Chang, C. Chen, *Tetrahedron Lett.* **2003**, *44*, 6955–6957.
48 B. L. Feringa, *J. Chem. Soc., Chem. Commun.* **1986**, 909–910.
49 H. Takikawa, K. Isono, M. Sasaki, F. A. Macias, *Tetrahedron Lett.* **2003**, *44*, 7023–7025.
50 E. Keinan, K. K. Seth, R. Lamed, *J. Am. Chem. Soc.* **1986**, *108*, 3474–3480.
51 A. Fettes, E. M. Carreira, *J. Org. Chem.* **2003**, *68*, 9274–9283.
52 H. Nagashima, K. Sato, J. Tsuji, *Tetrahedron* **1985**, *41*, 5645–5651.
53 J. Tsuji, N. Nagashima, K. Hori, *Chem. Lett.* **1980**, 257–260.
54 H. Usuda, M. Kanai, M. Shibasaki, *Org. Lett.* **2002**, *4*, 859–862.
55 M. Tanaka, H. Urata, T. Fuchikami, *Tetrahedron Lett.* **1986**, *27*, 3165–3168.
56 A. I. Roshchin, S. M. Kel'chevski, N. A. Bumagin, *J. Organomet. Chem.* **1998**, *560*, 163–167.
57 K. Matsui, S. Uchimumi, A. Iwayama, T. Umeza, to *Ube-Industries*, EP 55108, **1982**.
58 T. Hosokawa, T. Yamanaka, M. Itotani, S.-I. Murahashi, *J. Org. Chem.* **1995**, *60*, 6159–6167.
59 P. Ambrosi, A. Arnone, P. Bravo, L. Bruché, A. De Cristofaro, V. Francardi, M. Frigerio, E. Gatti, G. S. Germinara, W. Panzeri, F. Pennacchio, C. Pesenti, G. Rotundo, P. F. Roversi, C. Salvadori, F. Viani, M. Zanda, *J. Org. Chem.* **2001**, *66*, 8336–8343.

3
C–H Transformation at Aldehydes and Imines

3.1
Inter- and Intramolecular Hydroacylation
Chul-Ho Jun and Young Jun Park

3.1.1
Introduction and Fundamental Examples

The hydroacylation of olefins with aldehydes is one of the most promising transformations using a transition metal-catalyzed C–H bond activation process [1–4]. It is, furthermore, a potentially environmentally-friendly reaction because the resulting ketones are made from the whole atoms of reactants (aldehydes and olefins), i.e. it is atom-economic [5]. A key intermediate in hydroacylation is a acyl metal hydride generated from the oxidative addition of a transition metal into the C–H bond of the aldehyde. This intermediate can undergo the hydrometalation of the olefin followed by reductive elimination to give a ketone or the undesired decarbonylation, driven by the stability of a metal carbonyl complex as outlined in Scheme 1.

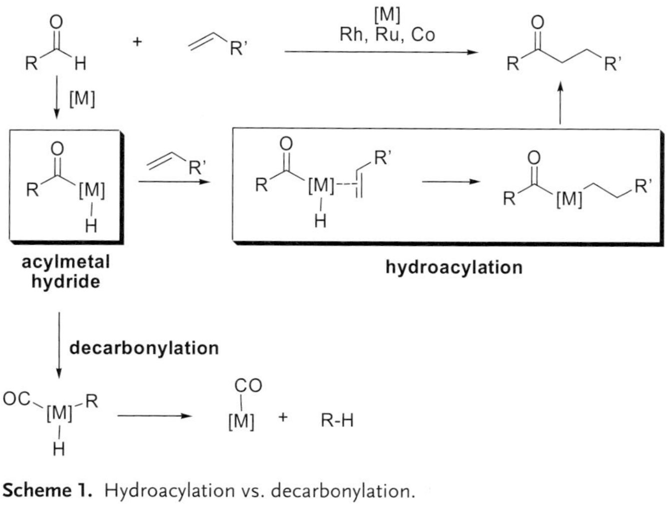

Scheme 1. Hydroacylation vs. decarbonylation.

Handbook of C–H Transformations. Gerald Dyker (Ed.)
Copyright © 2005 WILEY-VCH Verlag GmbH & Co. KGaA, Weinheim
ISBN: 3-527-31074-6

Suppression of the competitive decarbonylation reaction is essential to make the hydroacylation reaction predominant. For this reason, early intermolecular hydroacylation reactions required rather harsh conditions or specific reagents, leading to substrate limitation (Scheme 2) [6–10]. To circumvent this limitation, stabilization of the acyl metal hydride is required and two approaches, though they are conceptually identical (cyclometallation), exist [11, 12] –attachment of an additional coordinating group on the aldehyde to form a stable five-membered metalacycle[13], and placement of both aldehyde and olefin in a molecule to make a reaction proceed intramolecularly [14] (Scheme 3). The former strategy is usually used in intermolecular reactions and the latter in intramolecular reactions. In contrast with earlier examples which employ high temperatures and/or high pressures of carbon monoxide and ethylene, these approaches provide milder, more efficient, general and selective reaction environments.

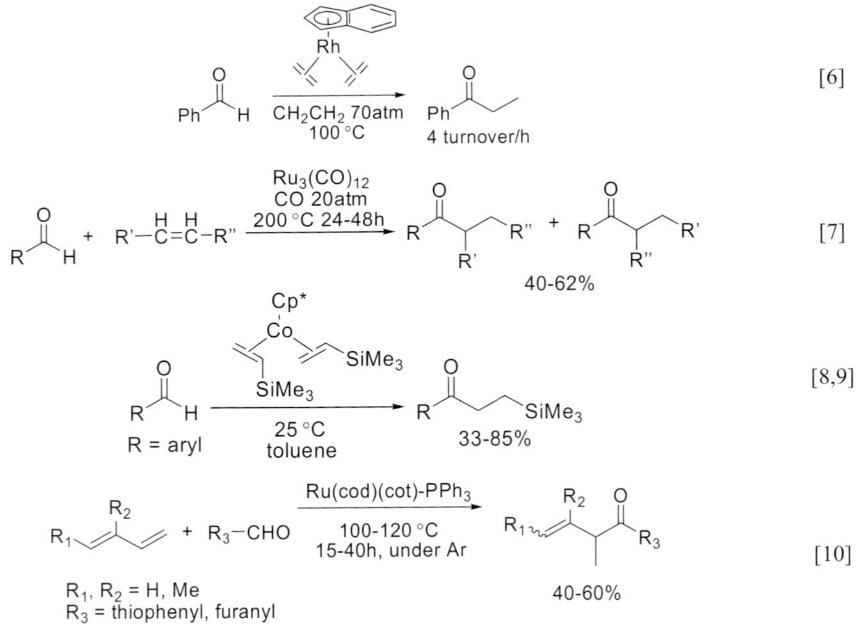

Scheme 2. Early examples.

An interesting example is the hydroiminoacylation reaction, a good alternative to hydroacylation reactions, using aldimines as a synthetic equivalent to aldehydes (Scheme 4) [4]. The rhodium-catalyzed hydroiminoacylation of an olefin with aldimines produced a ketimine which could be further acid-hydrolyzed to give the ketone. The reaction proceeded via the formation of a stable iminoacylrhodium(III) hydride (this will be discussed in the mechanism section), production of which is facilitated by initial coordination of the rhodium complex to the pyridine moiety of the aldimine. This hydroiminoacylation procedure opened up the direct use of aldehydes as substrates for hydroacylation based on the in-situ generation of aldimines from the aldehyde and 2-amino-3-picoline, a reaction which was called chelation-assisted hydroacylation (Scheme 4).

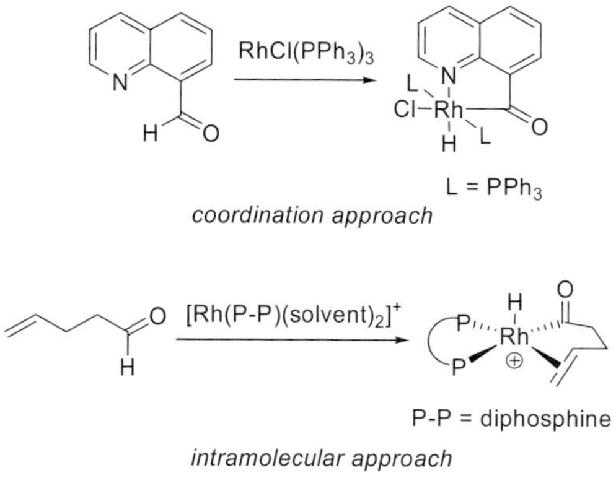

Scheme 3. Cyclometalation models.

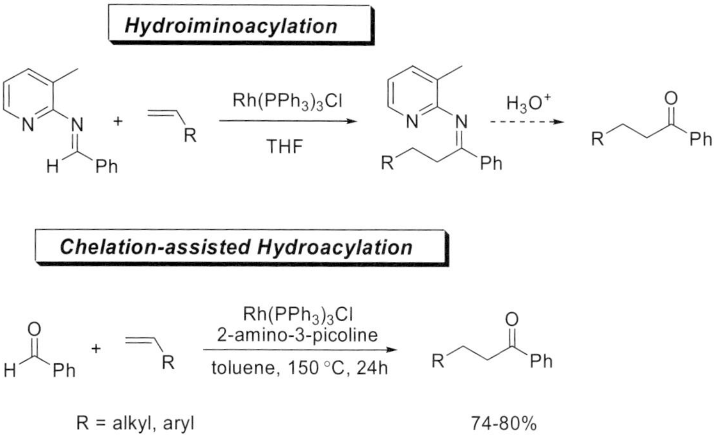

Scheme 4. Hydroiminoacylation and chelation-assisted hydroacylation.

Intramolecular hydroacylation, on the other hand, is an attractive catalytic process because it produces cyclic ketones. Furthermore, with appropriate chiral phosphine ligands, this reaction could convert prochiral 4-pentenals into chiral cyclopentanones (Scheme 5) [14].

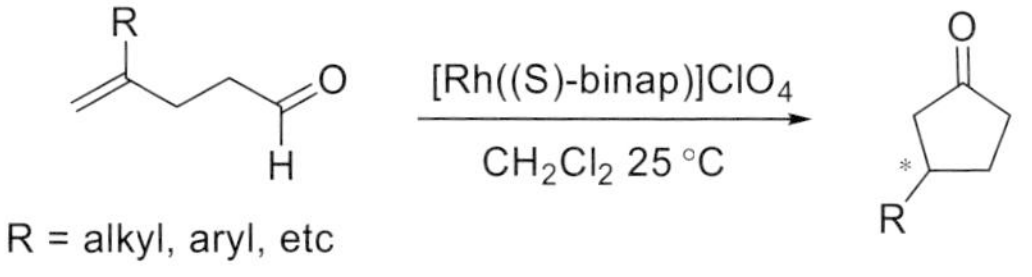

R = alkyl, aryl, etc

Scheme 5. Intramolecular hydroacylation.

3.1.2
Mechanism

Since the decarbonylation of aldehydes by Wilkinson's complex, $Rh(PPh_3)_3Cl$ (9), was discovered, the formation of acylmetal hydride species by oxidative addition of aldehydic C–H bonds to transition metals is a key intermediate in hydroacylation reactions [15, 16]. Iminohydroacylation reactions also involve similar iminoacylmetal intermediates. First, benzaldehyde condenses with 2 to form aldimine 3, which then undergoes hydroiminoacylation to afford ketimine 7, via formation of acylrhodium(III) hydride 4, then hydrometalation of the olefin to form 6 and reductive elimination. Because the ketimine is more susceptible than the aldimine toward hydrolysis, it is readily hydrolyzed to ketone 8 by water generated during the condensation step, thus regenerating 2 (Scheme 6). Therefore, amine 2 is used as a catalyst to assist chelation with the rhodium complex. In fact, in the absence of 2, no hydroacylation occurred and aldehydes underwent the rapid decarbonylation. In this mechanism, the imine condensation step is rate determining, because whole reactions could be dramatically accelerated by acid catalysis and the transimination reaction. Thus, a very efficient catalyst, which consists of 9, 2, benzoic acid (10), and aniline (11) was developed. With this catalytic system the reaction time could be reduced from 24 h to 1 h even at 130 °C [17]. The mechanism of this reaction is depicted in Scheme 7. Initially, an aldehyde condenses with a more reactive aniline to form aldimine 12. Subsequent transimination with 2 generates 3 which participates in hydroiminoacylation of an olefin. The high reactivity of this catalyst system implies that the condensation of an aldehyde with 11 followed by transimination of 12 into 3 is more facile than direct condensation of the aldehyde with 2.

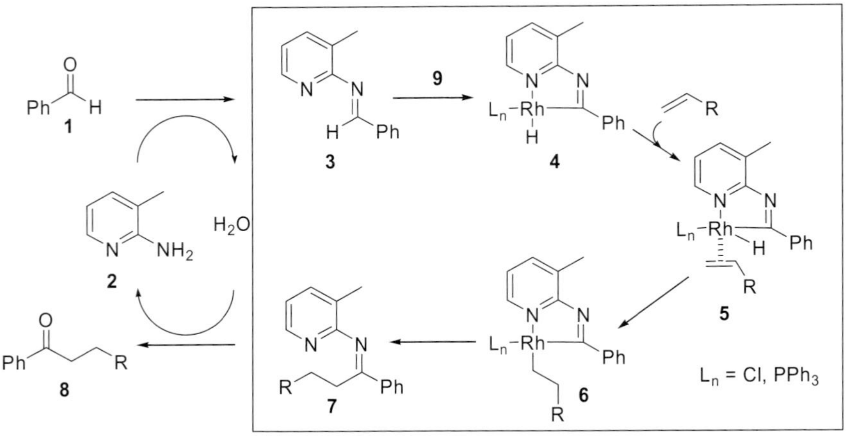

Hydroiminoacylation

Scheme 6. Mechanism of hydroiminoacylation.

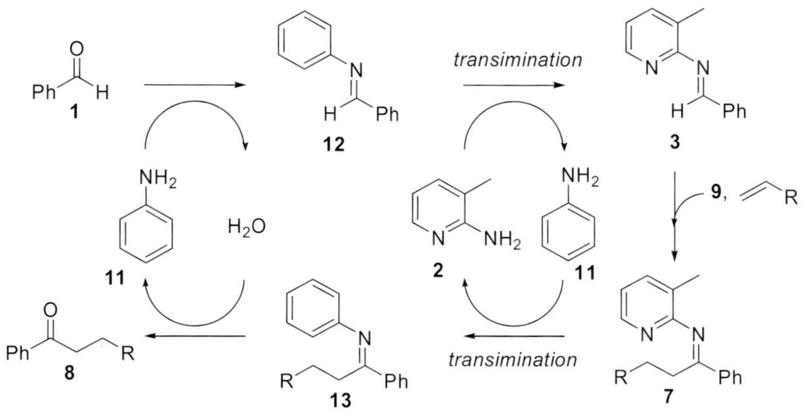

Scheme 7. Mechanism of hydroiminoacylation by transimination.

In the intramolecular hydroacylation, on the other hand, the choice of catalyst is important. As in the intermolecular hydroacylation, Wilkinson's catalyst **9** and its analog could not promote intramolecular hydroacylation efficiently because catalytically inactive rhodium carbonyl complexes form rapidly. Thus, cationic species and chelate phosphines are usually needed to prevent the decarbonylation process, as in Scheme 3 [14]. The efficiency of the catalyst is related to the rapid displacement of the solvent molecules and to the availability of at least three vacant coordination sites on the catalyst for the hydride, acyl, and olefin ligands. In the Rh(I)-catalyzed intramolecular hydroacylation of 4-alkenal derivatives, both five- and six-membered metallacycles have been generated during the reaction, although the six-membered metallacycle selectively produces the cyclopentanone (Scheme 8). The main origin of chiral induction is expected to arise from the spatial arrangements of alkyl or aryl groups at the phosphorous atom and the chiral configuration is determined by the chirality of the chelating ring.

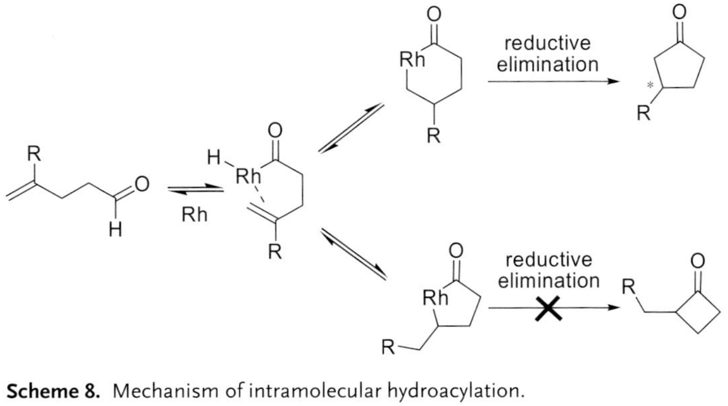

Scheme 8. Mechanism of intramolecular hydroacylation.

Table 1. Examples for inter- and intramolecular hydroacylation.

Substrate	Product	Catalytic system	Ref.
RCHO + CH₂=CHR'	R'CH₂CH₂C(O)R, 71–99%	Rh(PPh$_3$)$_3$Cl 2-amino-3-picoline Aniline, benzoic acid	[17]
HOCH$_2$R + CH₂=CHR'	R'CH₂CH₂C(O)R, 22–86%	RhCl$_3$·xH$_2$O PPh$_3$ 2-amino-4-picoline	[18]
PhCH$_2$NH$_2$ + CH₂=CHR	RCH$_2$CH$_2$C(O)Ph, 70–96%	Rh(PPh$_3$)$_3$Cl 2-amino-3-picoline	[19]
RCHO + HC≡CR'	R–C(O)–C(=CH$_2$)–R', 76–96%	Rh(PPh$_3$)$_3$Cl 2-amino-3-picoline Benzoic acid	[20]
salicylaldehyde + 1,5-hexadiene	iso / normal products, 100% (iso/normal=4/1)	Rh(PPh$_3$)$_3$Cl	[21]
MeS-CH$_2$CH$_2$CHO + CH$_2$=CHR	MeS-CH$_2$CH$_2$C(O)CH$_2$CH$_2$R, 33–84%	[Rh(dppe)]ClO$_4$	[22]
Ph-CH(CH$_2$CHO)-CH=CH-cyclopropyl	Ph-cyclooctenone, 65%	[Rh(dppe)]ClO$_4$ Ethylene	[23]
4-methylene-pentanal (R)	cyclopentanone, high ee >90%	[Rh(chiral P-P)]ClO$_4$	[14]
R-C≡C-CR'(R'')-CHO	cyclopentenone, 67–88%	[Rh(dppe)]$_2$(BF$_4$)$_2$	[24]
R-CH=CH-CH=CH-CH$_2$CH$_2$CHO	cycloheptenone 62%, cyclopentanone 13%, cyclopentanone(ene) 6%	[Rh(dppe)]ClO$_4$	[25]

3.1.3
Scope and Limitations

A primary alcohol and amines can be used as an aldehyde precursor, because it can be oxidized by transfer hydrogenation. For example, the reaction of benzyl alcohol with excess olefin afforded the corresponding ketone in good yield in the presence of Rh complex and 2-amino-4-picoline [18]. Similarly, primary amines, which were transformed into imines by dehydrogenation, were also employed as a substrate instead of aldehydes [19]. Although various terminal olefins, alkynes [20], and even dienes [21] have been commonly used as a reaction partner in hydroiminoacylation reactions, internal olefins were ineffective. Recently, methyl sulfide-substituted aldehydes were successfully applied to the intermolecular hydroacylation reaction [22]. Also in the intramolecular hydroacylation, extension of substrates such as cyclopropane-substituted 4-enal [23], 4-alkynal [24], and 4,6-dienal [25] has been developed (Table 1).

Experimental

Heptanophenone by Hydroiminoacylation
A screw-capped pressure vial (1 mL) was charged with freshly purified benzaldehyde (0.5 mmol), 2-amino-3-picoline (0.1 mmol), benzoic acid (0.03 mmol), aniline (0.3 mmol), 1-hexene (2.5 mmol), and toluene (80 mg). After stirring the mixture at room temperature for some minutes, [(PPh$_3$)$_3$RhCl] (0.01 mmol) was added and the mixture was then stirred in an oil bath preheated to 130 °C for 1 h. After cooling to room temperature, the reaction mixture was purified by column chromatography (SiO$_2$, *n*-hexane–ethyl acetate, 4:1) to yield pure heptanophenone (0.49 mmol, 98 % yield).

3.2
Cyclization of Aldehydes and Imines via Organopalladium Intermediates
Xiaoxia Zhang and Richard C. Larock

3.2.1
Introduction

Palladium reagents have been used extensively to prepare various carbo- and heterocyclic compounds by cyclic carbopalladation and annulation [1]. One of the most important factors contributing to the widespread application of palladium catalysts in organic synthesis is their tolerance of most important organic functional groups, i.e. aldehydes, ketones, imines, carboxylic acids, and their derivatives are usually readily accommodated by organopalladium intermediates under conditions in which carbon–carbon bond formation is facile [2]. Under some reaction conditions, however, organopalladium intermediates undergo several intramolecular C–H activation reactions with aldehydes and imines that they would not normally undergo if the reaction were intermolecular. For example, the palladium-catalyzed reaction of o-iodo- or o-bromobenzaldehyde with internal alkynes has proven to be an effective method for synthesis of indenones [3]. The key ring-closure step involves the C–H transformation of an aldehyde (Scheme 1).

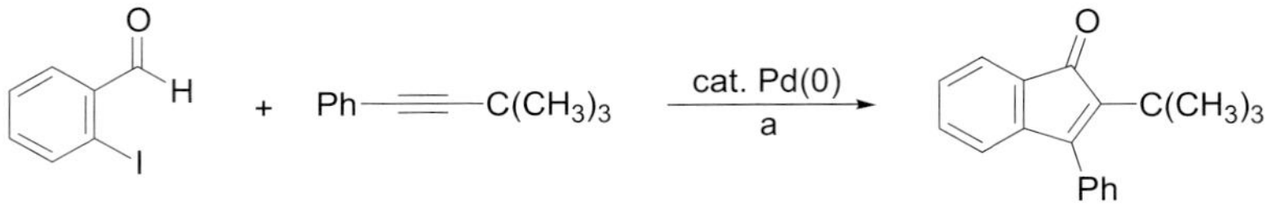

Scheme 1. C–H transformation of aldehydes: (a) 5 mol% Pd(OAc)$_2$, 4 equiv. Na$_2$CO$_3$, 1 equiv. n-Bu$_4$NCl, DMA, 100 °C, 24 h.

The C–H transformation of imines via palladacycles has recently provided a novel synthesis of fluoren-9-ones [4]. This high yielding synthesis is achieved by migration of palladium from an aryl position to an imidoyl position via C–H activation of the imine moiety, followed by intramolecular arylation (Scheme 2).

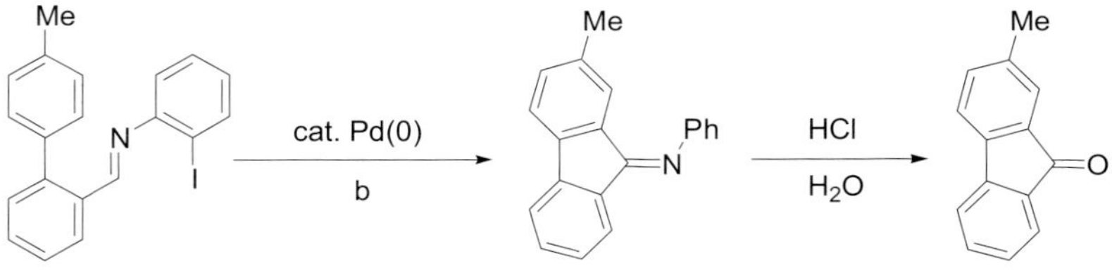

Scheme 2. C–H transformation of imines: (b) 5 mol% Pd(OAc)$_2$, 5 mol% (Ph$_3$P)$_2$CH$_2$, 2 equiv. CsO$_2$CCMe$_3$, DMF, 100 °C, 4 h.

3.2.2
Mechanism

The mechanism of the indenone synthesis (Scheme 3) seems to involve: (1) oxidative addition of the aryl iodide to Pd(0); (2) arylpalladium coordination to the alkyne and subsequent insertion of the alkyne to form a vinylpalladium intermediate (8), (3) then either the vinylic palladium intermediate adds to the carbonyl group and subsequently undergoes a β-hydride elimination (path A) or the aldehydic C–H bond may oxidatively add to the palladium to produce an organopalladium(IV) intermediate (six-membered ring palladacycle) which subsequently undergoes rapid reductive elimination of the indenone and palladium (path B). The actual mode of ring closure of the vinylic palladium intermediate to the indenone is unclear. Overall, the C–H bond of the aldehyde is transformed into a new C–C bond.

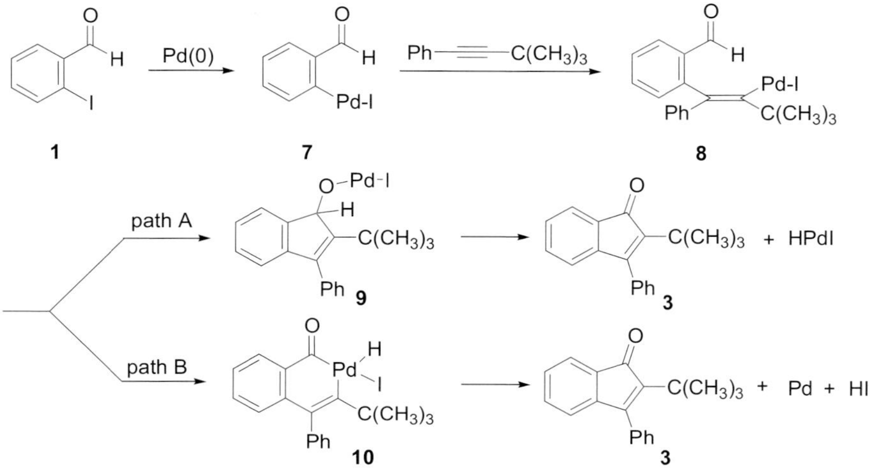

Scheme 3. Mechanism of the synthesis of indenones **3**.

Yamamoto et al. have reported that reaction of internal alkynes with *o*-bromobenzaldehydes [5] or analogous ketones [6] under different reaction conditions affords indenols (Scheme 4). It is believed the mechanism is similar to that of Scheme 3, but using path A, except that the protonolysis of intermediate **9** in Scheme 3 occurs, rather than β-hydride elimination, to form indenols. Therefore, no C–H bond transformation is involved in this indenol process.

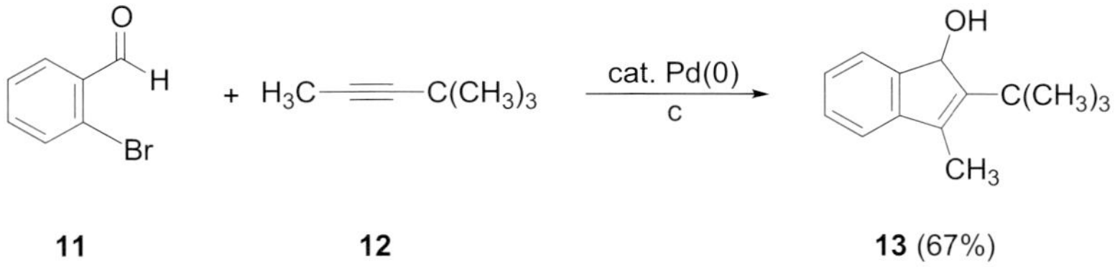

Scheme 4. Synthesis of indenols by nucleophilic attack of a vinylic palladium species on a carbonyl: (c) 5 mol% Pd(OAc)$_2$, 2 equiv. KOAc, 10 equiv. EtOH, DMF, 60 °C, 24 h.

It appears that the biarylcarboxaldehyde imine **4** affords *N*-fluoren-9-ylideneaniline **5** via the following mechanistic steps (Scheme 5). After oxidative addition of the C–I bond to Pd(0), the palladium undergoes a 1,4-palladium migration from the *ortho* position of the aniline to the imidoyl position (through path a), followed by arylation at the 2′ position of the biaryl to generate the desired migration/arylation product **5**. Alternatively, the palladium intermediate may undergo what seems to be a rather unfavorable process (path b) to generate a highly strained complex **A**, which after two reductive eliminations affords the final migration product. The high yield suggests that the palladium intermediate did, in fact, migrate from the aryl position to the imidoyl position by a complete through-space 1,4-shift to generate the imidoylpalladium intermediate **B**. The imidoyl palladium intermediate **B** apparently then undergoes either an insertion into the C–H bond of the neighboring arene (path c) or electrophilic aromatic substitution (path d). After elimination of HI, both pathways give a six-membered ring palladacycle **C**. Subsequent reductive elimination and hydrolysis then afford the observed ketone product **6**.

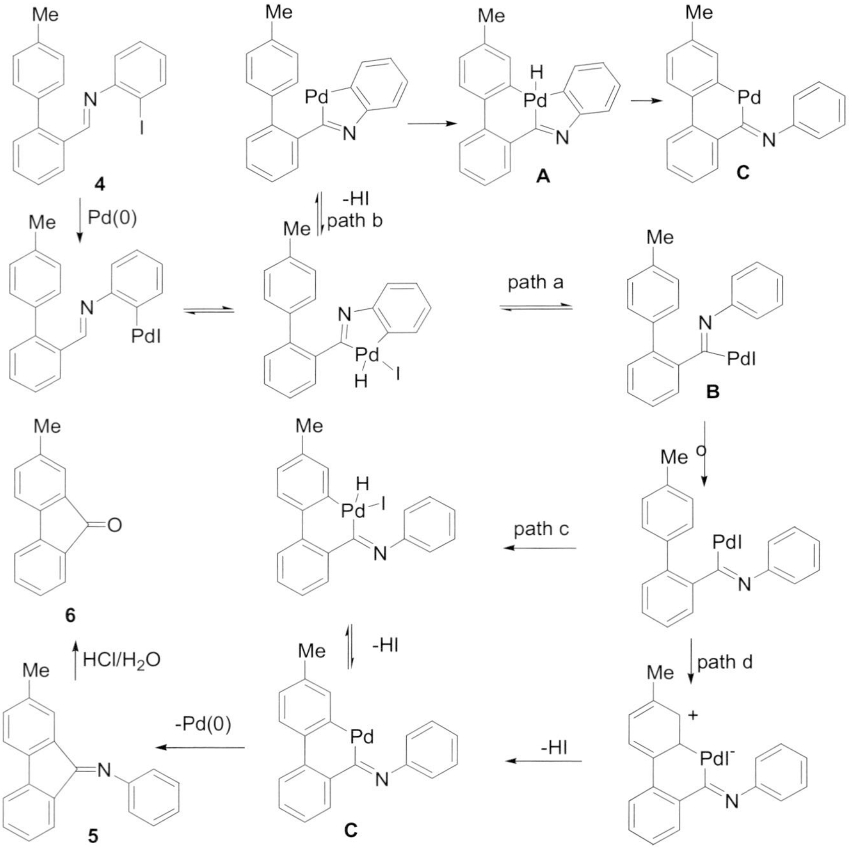

Scheme 5. Mechanistic pathway for C–H activation of imines: synthesis of fluoren-9-ones.

3.2.3
Scope and Limitations

o-Iodobenzaldehydes usually give modest to good yields of indenones by the process shown in Scheme 1. *o*-Bromobenzaldehydes can also be employed successfully in this annulation process, although in a slightly lower yield, and longer reaction times are needed (Table 1, entry 1). This process is reasonably general for diaryl, dialkyl, and aryl–alkyl internal alkynes providing indenones with a variety of substituents in the 2 and 3 positions (Table 1, entries 2–4). Hydroxy and silyl groups are accommodated (Table 1, entries 2 and 3). Isomerization of some of the products to β,γ-enones is observed with indenones bearing a primary alkyl group in the 3-position (Table 1, entry 4). The ease of isomerization is attributed to the antiaromaticity of the indenone ring system.

Table 1. Examples of C–H transformations of aldehydes: synthesis of indenones.

Entry	Benzaldehyde	Alkyne	Product(s)	Yield (%)
1	2-Br-benzaldehyde (11)	Ph—≡—Ph (14)	2,3-diphenylindenone (15)	82
2	2-I-benzaldehyde (1)	Ph—≡—C(CH₃)₂OH (16)	2-Ph-3-C(CH₃)₂OH-indenone (17)	58
3	2-I-benzaldehyde (1)	Ph—≡—Si(CH₃)₃ (18)	2-Ph-3-Si(CH₃)₃-indenone (19)	42
4	2-I-benzaldehyde (1)	H₃C—≡—C(CH₃)₃ (12)	2-CH₃-3-C(CH₃)₃-indenone (20) + β,γ-isomer (21)	26 + 26

The C–H transformation reactions of biarylcarboxaldehyde imines bearing different functional groups on the aromatic rings has been investigated. Electron-donating groups, for example methoxy groups, on the biaryl facilitate the arylation step and reduce the reaction time to 2 h (Table 2, entry 2). Although the presence of an electron-withdrawing group, for example a nitro group, on the ring undergoing substitution makes the aromatic ring more electron-deficient, and therefore makes electrophilic aromatic substitution more difficult, electron-withdrawing groups do not affect the overall yield, although substantially longer reaction times are required (Table 2, entry 3). After 1,4-migration of the palladium from the aryl to the imidoyl position, subsequent substitution of a furan also proceeds smoothly (Table 2, entry 4). The substrate with the electron-rich furan moiety requires a shorter reaction time. Migration of the palladium in the imidoylpalladium intermediate to another aryl position before trapping is also possible, although a higher temperature and much longer reaction time are required and the yield is lower (Table 2, entry 5). The reluctance of the palladium in this imidoylpalladium species to migrate to the sterically hindered *ortho* position of the phenoxy group might account for these observations.

Table 2. Synthesis of fluorenones via 1,4-palladium migration and subsequent arylation of imines.

Entry	Imines			Time (h)	Product		Yield (%)
1		X = Me	4	4		6	97
2		X = OMe	22	2		23	100
3		X = NO$_2$	24	12		25	100
4			26	2		27	82
5			28	168		29	35

Experimental

2-*tert*-Butyl-3-phenyl-1-indenone (3)

Pd(OAc)$_2$ (6 mg, 0.027 mmol), Na$_2$CO$_3$ (212 mg, 2.0 mmol), *n*-Bu$_4$NCl (150 mg, 0.54 mmol), 2-iodobenzaldehyde (116 mg, 0.5 mmol), *tert*-butylphenylacetylene (158 mg, 1.0 mmol), and 10 mL DMA were placed in a 4-dram vial which was heated in an oil bath at 100 °C for 24 h. The reaction was monitored by TLC to establish completion. The reaction mixture was cooled, diluted with 30 mL ether, washed with saturated NH_4Cl solution (2 × 45 mL), dried over anhydrous Na_2SO_4, and filtered. The solvent was evaporated under reduced pressure. The reaction mixture was chromatographed using 15:1 hexane–EtOAc to yield a yellow solid (mp 114–116 °C, from *n*-hexane): ^1H NMR ($CDCl_3$) δ 1.16 (s, 9 H, CH_3), 6.47 (d, J = 7.2 Hz, 1 H, aryl), 7.0–7.6 (m, 8 H, aryl); ^{13}C NMR ($CDCl_3$) δ 30.6, 33.6, 120.3, 121.7, 127.8, 128.03, 128.08, 128.1, 129.8, 133.3, 135.3, 141.4, 147.6, 153.9, 198.4; IR ($CHCl_3$) 1699 (C=O) cm^{-1}; mass spectrum m/z 262.13617 (calcd for $C_{19}H_{18}O$, 262.13577).

2-Methylfluoren-9-one (6)

The *N*-(4′-methylbiphenyl-2-ylmethylene)-2-iodoaniline (0.25 mmol), $Pd(OAc)_2$ (2.8 mg, 0.0125 mmol), *bis*(diphenylphosphino)methane (dppm; 4.8 mg, 0.0125 mmol), and CsO_2CCMe_3 (0.117 g, 0.5 mmol) in DMF (4 mL) under Ar were heated at 100 °C with stirring for 4 h. The reaction mixture was then cooled to room temperature, diluted with diethyl ether (35 mL), and washed with brine (30 mL). The aqueous layer was re-extracted with diethyl ether (15 mL). The combined organic layers were dried over $MgSO_4$, filtered, and the solvent was removed to afford the crude imine product. To an acetone (5 mL) solution of the crude imine product, 1.0 M HCl (2 mL) was added. The resulting reaction mixture was stirred until disappearance of the imine was indicated by thin layer chromatography. The mixture was then diluted with H_2O and extracted with diethyl ether (2 × 15 mL). The organic extract was dried over $MgSO_4$, filtered, and the solvent was removed under reduced pressure to afford the crude fluoren-9-one, which was purified by flash chromatography using 7:1 hexanes–ethyl acetate to afford a yellow solid: m.p. 90–91 °C (lit. [7] m.p. 92 °C); ^1H NMR ($CDCl_3$) δ 2.33 (s, 3H), 7.19–7.24 (m, 2H), 7.33 (d, J = 7.6 Hz, 1H), 7.04–7.41 (m, 3H), 7.58 (d, J = 7.2 Hz, 1H); ^{13}C NMR ($CDCl_3$) δ 21.4, 120.0, 120.2, 124.2, 125.0, 128.6, 134.3, 134.4, 134.7, 135.1, 139.3, 141.8, 144.7, 194.2. The spectral properties were identical with those previously reported [7].

References to Chapter 3 – C–H Transformation at Aldehydes and Imines

References and Notes to Section 3.1

1 G. Dyker, *Angew. Chem. Int. Ed.* **1999**, 38, 1699–1712.
2 J. A. Labinger, J. E. Bercaw, *Nature* **2002**, 417, 507–514.
3 V. Ritleng, C. Sirlin, M. Pfeffer, *Chem. Rev.* **2002**, 102, 1731–1769.
4 C.-H. Jun, C. W. Moon, D.-Y. Lee, *Chem. Eur. J.* **2002**, 8, 2423–2428.
5 B. M. Trost, *Acc. Chem. Res.* **2002**, 35, 695–705.
6 T. B. Marder, D. C. Roe, D. Milstein, *Organomet.* **1988**, 7, 1451–1453.
7 T. Kondo, M. Akazome, Y. Tsuji. Y. Watanabe, *J. Org. Chem.* **1990**, 55, 1286–1291.
8 C. P. Lenges, M. Brookhart, *J. Am. Chem. Soc.* **1997**, 119, 3165–3166.
9 C. P. Lenges, P. S. White, M. Brookhart, *J. Am. Chem. Soc.* **1998**, 120, 6965–6979.
10 T. Kondo, N. Hiraishi, Y. Morisaki., K. Wada, Y. Watanabe, T. Mitsudo, *Organomet.* **1998**, 17, 2131–2134.
11 B. Breit, *Chem. Eur. J.* **2000**, 6, 1519–1524
12 M. I. Bruce, *Angew. Chem. Int. Ed.* **1977**, 16, 73.
13 J. W. Suggs, *J. Am. Chem. Soc.* **1978**, 100, 640–641.
14 B. Bosnich, *Acc. Chem. Res.* **1998**, 31, 667–674.
15 C. F. Lochow, R. G. Miller, *J. Am. Chem. Soc.* **1976**, 98, 1281–1283.
16 J. Tsuji, K. Ohno, *J. Am. Chem. Soc.* **1968**, 90, 94.
17 C.-H. Jun, D.-Y. Lee, H. Lee, J.-B. Hong, *Angew. Chem. Int. Ed.* **2000**, 39, 3070.
18 C.-H. Jun, C.-W. Huh, S.-J. Na, *Angew. Chem. Int. Ed.* **1998**, 37, 145–147.
19 C.-H. Jun, K.-Y. Chung, J.-B. Hong, *Org. Lett.* **2001**, 3, 785–787.
20 C.-H. Jun, H. Lee, J.-B. Hong, B.-I. Kwon, *Angew. Chem. Int. Ed.* **2002**, 41, 2146–2147.
21 M. Tanaka, M. Imai, Y. Yamamoto, K. Tanaka, M. Shimowatari, S. Nagumo, N. Kawahara, H. Suemune, *Org. Lett.* **2003**, 5, 1365–1367.
22 M. C. Willis, S. J. McNally, P. J. Beswick, *Angew. Chem. Int. Ed.* **2004**, 43, 340–343.
23 A. D. Aloise, M. E. Layton, M. D. Shair, *J. Am. Chem. Soc.* **2000**, 122, 12610–12611.
24 K. Tanaka, G. C. Fu, *J. Am. Chem. Soc.* **2001**, 123, 11492–11493.
25 Y. Sato, Y. Oonishi, M. Mori, *Angew. Chem. Int. Ed.* **2002**, 41, 1218–1221.

References to Section 3.2

1 (a) Negishi, E.; Copéret, C.; Ma, S.; Liou, S.-Y.; Liu, F. *Chem. Rev.* **1996**, 96, 365; (b) Li, J. J.; Gribble, G. W. Palladium in Heterocyclic Chemistry, Pergamon: Oxford, 2000; (c) Bäckvall, J.–E. *Pure Appl. Chem.* **1992**, 64, 429; (d) Larock, R. C. *J. Organomet. Chem.* **1999**, 576, 111; (e) Collins, I. *J. Chem. Soc., Perkin Trans. 1* **2000**, 2845.
2 (a) Tsuji, J. *Palladium Reagents and Catalysts: Innovations in Organic Synthesis*; John Wiley: Chichester, U.K., 1995; (b) Soderberg, B. C. In *Comprehensive Organometallic Chemistry II*; Abel, E. W., Stone, F. G. A., Wilkinson, G., Eds.; Pergamon: Oxford, 1995; Vol. 12, p 241.
3 Larock, R. C.; Doty, M. J.; Cacchi, S. *J. Org. Chem.* **1993**, 58, 4579.
4 Yue, D.; Larock, R. C., work in progress.
5 Gevorgyan, V.; Quan, L. G.; Yamamoto, Y. *Tetrahedron Lett.* **1999**, 40, 4089.
6 Quan, L. G.; Gevorgyan, V.; Yamamoto, Y. *J. Am. Chem. Soc.* **1999**, 121, 3545.
7 Campo, M. A.; Larock, R. C. *J. Org. Chem.* **2002**, 67, 5616.